Discrete
Mathematics

Discrete Mathematics

Sherwood Washburn

Seton Hall University

Thomas Marlowe

Seton Hall University

Charles T. Ryan

Formerly of Seton Hall University

 **ADDISON-WESLEY**

An imprint of Addison Wesley Longman, Inc.

Reading, Massachusetts • Menlo Park, California • New York • Harlow, England
Don Mills, Ontario • Sydney • Mexico City • Madrid • Amsterdam

Sponsoring Editor: Carolyn Lee-Davis
Senior Production Supervisor: Peggy McMahon
Design Supervisor: Barbara T. Atkinson
Cover Design: Barbara T. Atkinson
Production Services: Jennifer Bagdigian
Composition and Technical Art Illustration: Techsetters, Inc.
Cover Photographs: Copyright © 1999 by Photo Disc, Inc. and International Stock/Gregory Edwards

CREDIT: endpapers: Normal Table, provided by Neil Weiss, author of *Introductory Statistics*, Fifth Edition (Addison Wesley Longman) and *Elementary Statistics*, Fourth Edition (Addison Wesley Longman)

Library of Congress Cataloging-in-Publication Data

Washburn, Sherwood.
    Discrete mathematics / Sherwood Washburn, Thomas Marlowe, Charles
T. Ryan
      p.   cm.
    Includes bibliographical references and index.
    ISBN 0-201-88336-8
    1. Mathematics.   2. Computer science–Mathematics.   I. Marlowe,
Thomas.   II. Ryan, Charles T.   III. Title.
QA39.2.W368   1999
511–dc21   97-48417
                                                        CIP

1 2 3 4 5 6 7 8 9 10 QUD 02010099

To Our Students and Our Families

Preface

GOALS OF A DISCRETE MATHEMATICS COURSE

There are two types of courses in discrete mathematics, which have been adopted almost universally. Almost every university has introduced a one-semester course at the freshman-sophomore level. In many universities, an upper-division course has also been introduced.

The elementary course serves the mathematical needs of computer science majors and mathematics majors, many of whom take computer science courses. The combined requirements of mathematics and computer science have created curricular difficulties, which are particularly severe in sophomore year. The discrete mathematics course is looked to more and more as a resource for dealing with them. One such case: Computer science students often do not have room in their curriculum for a logic course, so the learn it in discrete math.

In the discrete mathematics course, the student learns essential topics: induction and recursion, combinatorics, graph theory, and proofs and logic. Some of these topics are treated in more depth in the upper-division course.

This text is fully adequate for all discrete mathematics courses. It serves as a text for the freshman-sophomore course and also as a text for the yearlong upper-division course.

OUR TEXT

This text developed from the "Blue Notes" (so named due to their blue cover) that were used for years as the text for an elementary discrete mathematics course at Seton Hall.

Starting with these notes, we began writing because we saw a need for a new discrete mathematics textbook that would give students a sense of the historical development and conceptual unity of the subject. Years of classroom experience shape our treatment of induction, recursion, proof techniques, counting, and graph theory.

This is the result of our efforts: a book that serves the mathematical needs of students taking computer science courses, with a historical perspective, and a logical, yet flexible, organization of topics.

A TEXT INFORMED BY COMPUTER SCIENCE

While this is not a computer science text, we believe that a discrete mathematics text should be informed by computer science. Two of us are computer scientists. In writing this book, we have kept in mind the needs of computer science students, and mathematics students who will take computer science courses.

THE FUTURE OF DISCRETE MATH

Discrete mathematics is a developing subject, and students should be given a view of its future. With this in mind, we discuss new ideas in our book. For example, throughout the book we discuss several new ideas about genetic code. In Chapter 2.3, in the application in 5.2, and throughout Chapter 10, students have the opportunity to read about and apply discrete mathematics to this current issue.

A FLEXIBLE ORGANIZATION

The first six chapters are designed as a highly flexible text for a one-semester course in discrete mathematics at the freshman-sophomore level. The topics in these six chapters are covered in a way that allows the instructor to choose the emphasis of the course.

At the beginning of Chapter 7 the level of the text rises slightly. The last four chapters can be used as a text for a one-semester upper-division course or, for students with less preparation, for a yearlong upper-division course that begins with material from the first six chapters.

Additionally, by omitting some proofs, material from the last four chapters can be used in a course at the freshman-sophomore level. For example, we have successfully taught the material on Polya's theorem in Section 9.3 to Seton Hall freshmen.

FEATURES

Comprehensive Coverage of Topics   We cover an unusually wide set of topics. Our treatment of combinatorics, such a basic topic, is more thorough than the treatment in comparable texts. With an eye towards the future of discrete mathematics, we have also chosen to cover topics (e.g. lattices) that are expected to be very important in the years ahead.

Our treatment of proof techniques is more thorough than many texts in this market. We initially concentrate on induction and recursion, but choose to cover proof techniques and logic later in the book, when we feel that students are prepared better.

Integrated Historical Perspective   The historical material is fully integrated in our text, not separated in boxes or sidelined. Reading other texts, students often have the impression that the subject is artificial or unnatural. Our book corrects this impression. Many interesting examples, not found in other texts, are included and these examples have great value in themselves, and will appeal to instructors and students alike.

Carefully Graduated Exercises   The exercise sets, found at the end of each section, include Exercises, Advanced Exercises, and Computer Exercises. With a graduated level of difficulty, these high-quality exercises enable students to practice their skills as they work up to more difficult exercises.

Basic and Unique Computer Exercises   The computer exercises are excellent for both math and computer science majors, adding to the foundation of computer science knowledge. We have been careful to cover standard topics, but we have also included computer exercises that are not found in other texts. Our exercises on the 3X+1 problem in Section 1.5 are an example, as are our exercises on Godel numbers in Section 10.3.

Frequent, Necessary Applications   Each section ends with the description of a significant real-world application of the ideas and techniques of that section. Many topics in computer science are discussed. In the first chapter, the student is introduced to the Tower of Hanoi, the Knapsack Problem, error-correcting codes, and shift registers. In later chapters, we discuss structural induction, register allocation, network reliability, and much more.

Chapter Review   At the end of each chapter there is a chapter summary with supplementary exercises, which help students both review and practice what they have learned in the preceding sections.

Guide to Literature   A great resource for students, this appendix is organized by topic and lists additional resources that students can call upon to gain further understanding and insight in a particular area.

ANCILLARIES

Students' Solutions Manual—This manual offers students further help in learning and practicing their skills by including full solutions to the odd Exercises, Advanced Exercises, and Computer Exercises. (ISBN: 0-201-61924-5)

Instructor's Solutions Manual—This manual includes full solutions to the odd and even Exercises, Advanced Exercises, and Computer Exercises. (ISBN: 0-201-61924-5)

ACKNOWLEDGMENTS

This book grew out of a long and successful relationship with our students at Seton Hall, and we have dedicated it to them. To thank them would hardly be enough: without them, our book would not exist.

Among our students we particularly want to thank Joe DeVito for his hard work and enthusiasm. Joe's program gave us new insight into the 3X+1 problem. We would like to express our thanks and appreciation to the reviewers who gave us their time and advice:

Curtis Bennett
Bowling Green State University

Dean Hoffman
Auburn University

Douglas Campbell
Brigham Young University

Kenneth Johnson
North Dakota State University

Eddie Cheng
Oakland University

Gopal Lakhani
Texas Tech University

Allan C. Cochran
University of Arkansas

Joe Malkevitch
York College (CUNY)

Vladimir Drobot
San Jose State University

Ho Kuen Ng
San Jose State University

Herbert Enderton
University of California at Los Angeles

Mary K. Patton
University of Illinois at Springfield

Las C. Goonetilleke
Middlesex County College

Ralph Selfridge
University of Florida

Harvey Greenberg
University of Colorado at Denver

Douglas Shier
Clemson University

Laxmi Gupta
Rochester Institute of Technology

Denise Troxell
Babson College

Russell Gusack
Suffolk County Community College

Over the years our project received strong and consistent support from our colleagues at Seton Hall. We particularly want to thank John Saccoman for his encouragement. Saccoman remarked to one of us that the text for the discrete mathematics course we were teaching might develop into a book: He was right. Dan Gross, Esther Guerin, John Masterson, Laura Schoppmann, and others in our department gave us advice and help.

Our deans, Jerry Hirsch and Jim VanOosting, looked with a benevolent eye on a project that sometimes seemed as if it would never be finished.

Eberhard Grosse, the university printer at Seton Hall, printed the Blue Notes and many versions of the text for our book.

Bob Hallissey, the grants director at Seton Hall, helped one of us obtain a Summer Research Grant which furthered work on our book. Wolmer Vasconcelos of Rutgers was helpful to one of us during a sabbatical year, which allowed much work on this book to be completed. One of us is grateful to Marvin Paull of Rutgers for his help and guidance, and also to Barbara Ryder.

Neil Sloane generously shared ideas about codes with us. We learned from Joseph Brennan's incisive views about combinatorics.

We thank Peter Guidon for his explanations of the structure of the genetic code.

We thank Matthew Arnold and Atanas Rountev for their work on our Computer Exercises. We thank Frank Purcell for his work on the Exercises and Advanced Exercises.

The editorial staff at Addison Wesley Longman supported us with imagination and unvarying professionalism.

Contents

7

Graphs and Relations   225

8

Algorithms   253

9

Combinatorics   283

10

Models of Computation   327

A

B

Sets, Subsets, Induction, and Recursion

1.1 THE PASCAL TRIANGLE

"Just the place for a Snark!" the Bellman cried

As he landed his crew with care ...

Lewis Carroll, The Hunting of the Snark

On May 22, 1649, Blaise Pascal (1623–1662) was granted a royal *Privilege* (a monopoly) for the manufacture and sale of a calculating machine:

> *Louis, by the grace of God, king of France and of Navarre, to our ... Bailiffs, Senechals, Provosts ... greetings. Our ... Pascal ... has made a number of inventions, in particular a machine, by means of which one can carry out ... Additions, Subtractions, Multiplications, Divisions, and all the other Laws of Arithmetic ... we permit by these presents ... the said Pascal ... to construct or fabricate ... the machine which he has invented ... and we expressly forbid all other persons ... to make such a machine ... For such is our pleasure ... Given at Compiègne, the twenty-second day of May, in the year of grace one thousand six hundred and forty-nine, and the seventh of our reign.*

Pascal's father, a scientist and administrator, at one time had the duty of examining all the tax records of the Province of Normandy. While helping his father with this enormous task, Pascal had the idea of building a machine which would add and subtract by moving a system of gears. Under Pascal's supervision, a crew of workers completed the machine after building more than fifty preliminary models. It was the first computing machine ever manufactured and offered for sale. Perhaps it is not an exaggeration to say that Pascal invented the computer.

In 1654 Pascal completed the *Traité du triangle arithmétique* (*Treatise on the Arithmetic Triangle*). Figure 1.1 shows the Arithmetic Triangle as it appears in Pascal's *Treatise* (the triangle can be continued indefinitely).

```
1   1   1   1   1   1   1   1   1   1
1   2   3   4   5   6   7   8   9
1   3   6   10  15  21  28  36
1   4   10  20  35  56  84
1   5   15  35  70  126
1   6   21  56  126
1   7   28  84
1   8   36
1   9
1
```

Figure 1.1   The Arithmetic Triangle

In this section we shall discuss some basic ideas in the theory of counting. We shall define *graphs*, and an important type of graph called a *tree*. All of these concepts are of basic importance in computer science. At the end of the section we shall state and solve a counting problem which involves trees.

Today the Pascal Triangle is usually drawn as it appears in Figure 1.2. Notice that each row begins and ends with a 1, and that each of the other entries is the sum of the two entries diagonally above it.

DEFINITION 1.1.1   Let $n = 0, 1, 2, \ldots$ be a non-negative integer, and let k be an integer such that $0 \leq k \leq n$.

i. The top row of the Pascal Triangle is counted as the $0 - th$ row, and the 1 at the beginning of each row is counted as the $0 - th$ entry of that row.

ii. The $k - th$ entry of the nth row of the Pascal Triangle is the *binomial coefficient*

$$\binom{n}{k}.$$

EXAMPLE 1   What is the fourth row of the Pascal Triangle?

$$
\begin{array}{ccccccccccccccccccc}
 & & & & & & & & & 1 & & & & & & & & & \\
 & & & & & & & & 1 & & 1 & & & & & & & & \\
 & & & & & & & 1 & & 2 & & 1 & & & & & & & \\
 & & & & & & 1 & & 3 & & 3 & & 1 & & & & & & \\
 & & & & & 1 & & 4 & & 6 & & 4 & & 1 & & & & & \\
 & & & & 1 & & 5 & & 10 & & 10 & & 5 & & 1 & & & & \\
 & & & 1 & & 6 & & 15 & & 20 & & 15 & & 6 & & 1 & & & \\
 & & 1 & & 7 & & 21 & & 35 & & 35 & & 21 & & 7 & & 1 & & \\
 & 1 & & 8 & & 28 & & 56 & & 70 & & 56 & & 28 & & 8 & & 1 & \\
1 & & 9 & & 36 & & 84 & & 126 & & 126 & & 84 & & 36 & & 9 & & 1 \\
\end{array}
$$

Figure 1.2   The Pascal Triangle

Solution   The entries of the fourth row are

$$
\binom{4}{0} = 1 \quad \binom{4}{1} = 4 \quad \binom{4}{2} = 6 \quad \binom{4}{3} = 4 \quad \binom{4}{4} = 1
$$

●

EXAMPLE 2   Find the binomial coefficient

$$
\binom{8}{4}
$$

Solution   It is the fourth entry of the eighth row of the Pascal Triangle, where the 1 at the beginning of the row is counted as the 0th entry, so

$$
\binom{8}{4} = 70
$$

●

A *recursion* is a rule which generates numbers or other objects in terms of numbers or objects which were given previously. The rule which generates the entries of the Pascal Triangle is an important example of a recursion.

The Recursion for the Binomial Coefficients   Let $n = 0, 1, 2, \ldots$ be a non-negative integer, and let k be an integer such that $0 \le k < n$. Then the binomial coefficients satisfy the equations

$$
\binom{n}{0} = 1 = \binom{n}{n}
$$

and

$$
\binom{n}{k} + \binom{n}{k+1} = \binom{n+1}{k+1}
$$

Recursions are important in computer science for many reasons. They are used to construct computer codes, and computers can be easily programmed to run recursions.

DEFINITION 1.1.2 If n is a positive integer then $n!$, n *factorial*, is the product of the integers from 1 to n:

$$n! = 1 \cdot 2 \cdot \ldots \cdot n$$

We make the convention that $0! = 1$.

Figure 1.3 shows the values of the factorial function in the range $0 \leq n \leq 20$.

$$
\begin{array}{rcl}
0! & = & 1 \\
1! & = & 1 \\
2! & = & 2 \\
3! & = & 6 \\
4! & = & 24 \\
5! & = & 120 \\
6! & = & 720 \\
7! & = & 5{,}040 \\
8! & = & 40{,}320 \\
9! & = & 362{,}880 \\
10! & = & 3{,}628{,}880 \\
11! & = & 39{,}916{,}800 \\
12! & = & 479{,}001{,}600 \\
13! & = & 6{,}227{,}020{,}800 \\
14! & = & 87{,}178{,}291{,}200 \\
15! & = & 1{,}307{,}674{,}368{,}000 \\
16! & = & 20{,}922{,}789{,}888{,}000 \\
17! & = & 355{,}687{,}428{,}096{,}000 \\
18! & = & 6{,}402{,}373{,}705{,}728{,}000 \\
19! & = & 121{,}645{,}100{,}408{,}832{,}000 \\
20! & = & 2{,}432{,}902{,}008{,}176{,}640{,}000 \\
\end{array}
$$

Figure 1.3   The Factorial Function

The following formula, *Pascal's Formula*, was suggested to Pascal by a friend named de Gagnières.

Pascal's Formula   Let $n = 0, 1, 2, \ldots$ be a non-negative integer, and let k be an integer such that $0 \leq k \leq n$. Then

$$\binom{n}{k} = \frac{n!}{k!(n-k)!} = \frac{n \cdot (n-1) \cdot \ldots \cdot (n-k+1)}{k!}$$

EXAMPLE 3   Compute the binomial coefficient using Pascal's Formula.

$$\binom{4}{2}$$

Solution

$$\binom{4}{2} = \frac{4!}{2!(4-2)!} = \frac{24}{2 \cdot 2} = \frac{24}{4} = 6$$ ●

EXAMPLE 4   Compute the binomial coefficient using Pascal's Formula.

$$\binom{8}{4}$$

Solution

$$\binom{8}{4} = \frac{8!}{4!(8-4)!} = \frac{8!}{4!4!} = \frac{8 \cdot 7 \cdot 6 \cdot 5}{4!} = \frac{8 \cdot 7 \cdot 6 \cdot 5}{24} = 2 \cdot 7 \cdot 5 = 70$$ ●

A *set* is a collection of objects. We shall discuss sets and their properties in Section 1.3 and Section 1.4.

A *graph* G is a set of points called *nodes*, with a set of lines called *edges* connecting some pairs of the nodes. An example of a graph is shown in Figure 1.4.

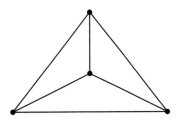

Figure 1.4   A Graph

A special type of graph called a *tree* is very useful in studying other graphs, and also in applications to computer science. A tree is a *connected* graph (each pair of nodes can be connected by a path of edges), which has no *cycles* (a cycle is a path which begins at one node v and returns to v without repeating any edges or nodes except for v). Trees will always be denoted by the letter T.

The most important type of tree is called a *rooted binary tree*. The *root* of the tree is a selected node which is denoted by an open circle, and is always drawn at the top of the tree. As we move down from the root, each edge that we draw moves to the right or to the left, and at each node the number of edges moving down from that node is zero, one, or two. A node which has no edges emerging from it as we move down is called a *terminal node*.

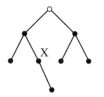

Figure 1.5 shows a rooted binary tree. The appearance of the tree is highly dependent on the choice of the root. To illustrate this, the same tree is shown in Figure 1.6, with a different root (the node marked X in Figure 1.5). In the new tree we have made new assignments of right and left to the edges.

Figure 1.5
A Rooted Binary Tree

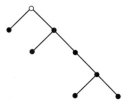

Figure 1.6   Rooted Binary Tree with a Different Choice of Root

The *level* of a node in a tree T is the number of edges from that node to the root of T. The node is at the 0-th level. The *full binary tree with n levels* is the rooted binary tree in which each node has two edges emerging from it as we move downward, except for the nodes at level *n*, which have no edges emerging from them.

The full binary tree is important for many reasons. Figure 1.7 shows the full binary tree with four levels, with a path from the root to a terminal vertex. Such paths can be used to store information in the tree. A *binary string* is a sequence of 0s and 1s, such as 0101. If we symbolize 1 by an edge moving to the right, and 0 by an edge moving to the left, we can store the binary string 0101 in the full binary tree with four levels by representing it as the path in Figure 1.7. When used in this way, a binary tree is called a *binary search tree.*

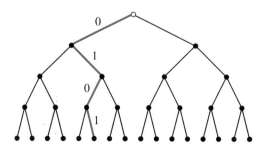

Figure 1.7   The Full Binary Tree with Four Levels

Binary strings are of basic importance in computer science. The machine code of a computer is written in binary strings. Information which is to be transmitted over a long distance is usually encoded first as a binary string. Binary strings have a wide variety of other interpretations. For example, consider a coin-tossing game where a head is symbolized by H and a tail is symbolized by T. Suppose that the coin is tossed four times, and heads comes up first, then tails, then heads, then tails. We can record the result of this game by the string HTHT. By replacing 0 by H and 1 by T, we can record the result of the game by the path in Figure 1.8.

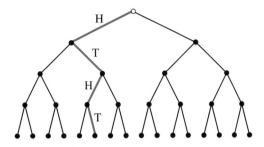

Figure 1.8   The Full Binary Tree Used to Store a Binary String

APPLICATION   *A Counting Problem*

Combinatorics is the science of counting. A form of this word was used first by Gottfried Wilhelm Leibniz (1646–1716) in the title of his *Dissertatio de arte combinatoria* (Dissertation on the art of combinations), which was published in 1666.

In Pascal's *Treatise*, Leibniz's *Dissertatio*, Jakob Bernoulli's *Ars Conjectandi*, and Montmort's *On Games of Chance*, we can see the origins of combinatorics as a science. Counting problems are very important in computer science.

A Counting Problem   In the full binary tree with n levels

(i)   How many paths are there from the root to the terminal nodes?

(ii)   If $0 \le k \le n$, how many of these paths have exactly k edges moving to the left?

THEOREM 1.1.1   In the full binary tree T with n levels

(i)   There are 2^n paths from the root to the terminal nodes.

(ii)   The number of such paths which have exactly k edges moving to the left is

$$\binom{n}{k}$$

Proof   Let T be the full binary tree with n levels. If $n = 1$ then T consists of the root and two terminal nodes, so there are $2^1 = 2$ paths from the root to these terminal nodes. Each time we move down one level we must multiply the number of nodes at that level by 2 to obtain the number of nodes at the next level, so at the nth level there will be 2^n nodes.

Let $N(n, k)$ be the number of paths from the root of T to the nth level which have exactly k edges moving to the left. Since a path which always moves to the left or always moves to the right is unique, we have

$$N(n, 0) = N(n, n) = 1$$

If T is the full binary tree with $n + 1$ levels, how can we obtain a path from the root to the $(n + 1)$-th level with exactly $k + 1$ edges to the left? We can choose a path from the root of T to the nth level with k edges to the left and add another edge to the left, and there are $N(n, k)$ of these; or we can choose a path from the root of T to the nth level with $k + 1$ edges to the left, and add an edge to the right, and there are $N(n, k + 1)$ of these. In all,

$$N(n + 1, k + 1) = N(n, k) + N(n, k + 1).$$

Thus the numbers $N(n, k)$ satisfy exactly the same recursion as the binomial coefficients, and it follows that

$$N(n, k) = \binom{n}{k}$$

and this proves the theorem.

COROLLARY 1.1.1   The full binary tree with n levels has 2^n terminal nodes.

There is a unique path from the root to each terminal node, and each such path has a certain number of edges moving to the left. This gives the next corollary.

COROLLARY 1.1.2   The sum of the entries of the nth row of the Pascal Triangle is 2^n:

$$\binom{n}{0} + \binom{n}{1} + \ldots + \binom{n}{n} = 2^n$$

Proofs of the kind we have just given can be better organized as *proofs by induction*. In the next section we introduce proofs by induction, or *inductive proofs*.

EXERCISES

Draw the Pascal Triangle as far as the fourteenth row, and use your triangle to evaluate the following binomial coefficients.

1. $\binom{0}{0}$ **2.** $\binom{5}{2}$ **3.** $\binom{5}{3}$

4. $\binom{7}{4}$ **5.** $\binom{10}{5}$ **6.** $\binom{11}{8}$

7. $\binom{12}{6}$

8. What do you notice about the binomial coefficients

$$\binom{14}{4}, \binom{14}{5}, \binom{14}{6} ?$$

9–16. Use the Pascal Formula to evaluate the binomial coefficients in Exercises 1–8.

17. Draw the full binary tree with three levels three times, and draw the

$$\binom{3}{1} = 3$$

paths in this tree with exactly one edge moving to the left.

18. Draw the full binary tree with three levels three times, and draw the

$$\binom{3}{2} = 3$$

paths in this tree with exactly two edges moving to the left.

19. Draw the full binary tree with four levels four times, and draw the

$$\binom{4}{1} = 4$$

paths in this tree with exactly one edge moving to the left.

20. Draw the full binary tree with four levels six times, and draw the

$$\binom{4}{2} = 6$$

paths in this tree with exactly two edges moving to the left.

21. Verify that the sum of the entries of the nth row of the Pasal Triangle is 2^n for $n = 0, 1, 2, 3, 4, 5, 6, 7$.

ADVANCED EXERCISES

1. Draw the Pascal Triangle as far as the fourteenth row and underline all the even entries. What pattern do they form?

2. Draw the Pascal Triangle as far as the sixteenth row and underline all the multiples of 3. What pattern do they form?

3. Draw the Pascal Triangle as far as the fourteenth row and underline all the multiples of 4. What pattern do they form?

4. Draw the Pascal Triangle as far as the thirteenth row and underline all the multiples of 5. What pattern do they form?

5. Draw the Pascal Triangle as far as the fourteenth row and underline all the multiples of 6. What pattern do they form?

6. Draw the Pascal Triangle as far as the twelfth row and underline all the multiples of 7. What pattern do they form?

7. Formulate a conjecture on the basis of your work in problems 1–6.

COMPUTER EXERCISES

You can do these Exercises using *Derive*, *Maple*, or *Mathematica*.

1. Make a table of factorials $n!$ for $0 \leq n \leq 40$.

2. In Montmort's book *On Games of Chance*, there is a discussion of the game of *Pharaoh*. In this game a deck of 52 cards is shuffled, and a *Banker* then turns the cards up one at a time while bets are made. There are as many games of Pharaoh as there are orderings of 52 cards, and it turns out that there are 52! of these. Compute 52!.

3. Make a table of the middle binomial coefficients

$$\binom{2n}{n}$$

for $0 \leq n \leq 25$.

4. The nth middle binomial coefficient has the property that it is always divisible by $n + 1$, and the

numbers

$$C_n = \frac{1}{n+1}\binom{2n}{n}$$

are called *Catalan numbers*. These numbers are named for the Belgian mathematician Eugène-Charles Catalan (1814–1894). Using your results in the previous exercise, make a table of Catalan numbers for $0 \leq n \leq 25$.

Write a program in *Pascal* or another language whose output is the nth row of the Pascal Triangle. Using your program, print the following rows.

5. The 25th row.

6. The 30th row.

7. The 35th row.

8. The 40th row.

9. The 45th row.

10. The 50th row.

1.2   THE PRINCIPLE OF INDUCTION

The crew was complete: it included a Boots—
A maker of Bonnets and Hoods—
A Barrister, brought to arrange their disputes—
And a Broker, to value their goods.

In the *Treatise on the Arithmetic Triangle*, Pascal had to prove a certain Proposition which has infinitely many cases. To paraphrase Pascal's argument:

A short proof can be given of this Proposition, although there are infinitely many cases, by using the following two Lemmas.
The first, which is evident, is that the Proposition is true in the first case.
The second, that if the Proposition is true in a given case, then it is true in the next case.

On the basis of this passage, Pascal is often credited with the discovery of the *Principle of Induction*. Augustus DeMorgan (1806–1871) stated the Principle of Induction in its modern form, calling it *Mathematical Induction*.

As an application, we shall use induction to describe the *Tower of Hanoi* puzzle, which is a useful model in computer science. This puzzle was invented by the French mathematician François-Édouard-Anatole Lucas (1842–1891). Lucas' ideas about recursions and prime numbers have had a lasting influence. In the Computer Exercises we shall describe another of Lucas' problems, the *Problem of the Square Pyramid*.

The Principle of Induction   Suppose that $P(n)$ is a statement for each positive integer $n = 1, 2, 3, \ldots$.
Suppose that

(i) $P(1)$ is true, and

(ii)   If $P(n)$ is true, then $P(n+1)$ is true for each positive integer $n = 1, 2, 3, \ldots$.
Then $P(n)$ is true for each $n = 1, 2, 3, \ldots$.

Proving that $P(1)$ is true is called *starting the induction*. Proving that $P(n)$ implies $P(n+1)$ is called the *inductive step*. It is possible to start the induction with any integer, for example with 0. In that case, a proof by induction proves the statement $P(n)$ for each $n = 0, 1, 2, \ldots$.

EXAMPLE 1   Prove by induction that

$$1 + 2 + 3 + \ldots + n = \frac{n(n+1)}{2}$$

Solution   ***Starting the Induction:***   If $n = 1$ then $1 = \frac{1(1+1)}{2}$. This starts the induction.

The Inductive Step: We assume that the statement $P(n)$ is true, and this is the statement that

$$1 + 2 + 3 + \ldots + n = \frac{n(n+1)}{2}$$

If we add $n + 1$ to both sides of this equation, we obtain

$$1+2+3+\ldots+(n+1) = \frac{n(n+1)}{2}+(n+1) = \frac{n(n+1)}{2}+\frac{2n+2}{2} = \frac{(n+1)(n+2)}{2}$$

This is the statement $P(n+1)$, (that is, the statement $P(n)$ with $n+1$ substituted for each occurence of n). We have proved that if $P(n)$ is true then $P(n+1)$ is true, and this completes the proof. $P(n)$ is true for all n by the Principle of Induction. ●

EXAMPLE 2   Prove by induction that

$$1 + 3 + 5 + \ldots + (2n - 1) = n^2$$

Solution   ***Starting the Induction:***   If $n = 1$ then $1 = 1^2$. This starts the induction.

The Inductive Step: We assume that the statement $P(n)$ is true. This is the statement that

$$1 + 3 + 5 + \ldots + (2n - 1) = n^2$$

If we add $2(n + 1) - 1 = 2n + 1$ to both sides of this equation, we obtain

$$1 + 3 + 5 + \ldots + (2n - 1) + (2n + 1) = n^2 + (2n + 1) = (n + 1)^2$$

This is the statement $P(n + 1)$. We have proved that if $P(n)$ is true then $P(n + 1)$ is true, and this completes the proof by induction. ●

EXAMPLE 3   Prove by induction that

$$1 + 4 + 7 + \ldots + (3n - 2) = \frac{n(3n - 1)}{2}$$

Solution   ***Starting the Induction:***   If $n = 1$ then $1 = \frac{1(3 \cdot 1 - 1)}{2} = \frac{2}{2} = 1$. This starts the induction.

The Inductive Step: We assume that $P(n)$ is true. This is the statement that

$$1 + 4 + 7 + \ldots + (3n - 2) = \frac{n(3n - 1)}{2}$$

If we add $3(n + 1) - 2 = 3n + 1$ to both sides of this equation, we obtain

$$1 + 4 + 7 + \ldots + (3n - 2) + (3n + 1) = \frac{n(3n - 1)}{2} + (3n + 1)$$

$$= \frac{n(3n - 1) + 6n + 2}{2} = \frac{3n^2 + 5n + 2}{2}$$

$$= \frac{(n + 1)(3(n + 1) - 1)}{2}$$

This is the statement $P(n + 1)$. We have proved that if $P(n)$ is true then $P(n + 1)$ is true, and this completes the proof by induction.   ●

The numbers $\frac{n(n+1)}{2}$ are called *triangular numbers*; the numbers n^2 are called *square numbers*; and the numbers $\frac{n(3n-1)}{2}$ are called *pentagonal numbers*.

Since we often have to consider sums, there is a special notation for sums called *sigma notation*. For example,

$$\sum_{k=1}^{n} k = 1 + 2 + \ldots + n$$

expresses the sum of the first n integers.

The lower index, $k = 1$ in this case, indicates where the sum begins, and the upper index, n in this case, indicates where the sum ends.

Another example is:

$$\sum_{k=1}^{n} (2k - 1) = 1 + 3 + \ldots + (2n - 1)$$

which expresses the sum of the first n odd integers.

More generally, if we have a variable expression a_k we can write the sum

$$\sum_{k=1}^{n} a_k = a_1 + a_2 + \ldots + a_n$$

We can express equations using sigma notation. The equation in Example 1 can be written:

$$\sum_{k=1}^{n} k = \frac{n(n + 1)}{2}$$

and the equation in Example 2 can be written:

$$\sum_{k=1}^{n} (2k - 1) = n^2$$

If we sum the triangular, square, and pentagonal numbers we get the equations:

$$\sum_{k=1}^{n} \frac{k(k + 1)}{2} = \frac{n(n + 1)(n + 2)}{6}$$

$$\sum_{k=1}^{n} k^2 = \frac{n(n + 1)(2n + 1)}{6}$$

$$\sum_{k=1}^{n} \frac{k(3k - 1)}{2} = \frac{n^2(n + 1)}{2}$$

The numbers on the right-hand side in these expressions are called *pyramidal numbers*. They count the number of cells in a pyramid with n levels whose horizontal section is a triangle, square, or pentagon.

We will prove the second of these formulas by induction.

EXAMPLE 4   Prove by induction that

$$\sum_{k=1}^{n} k^2 = \frac{n(n+1)(2n+1)}{6}$$

Solution   ***Starting the Induction:***   If $n = 1$ then $1^2 = \frac{1(1+1)(2+1)}{6} = 1$. This starts the induction.

The Inductive Step: Assuming that $P(n)$, the statement above, is true, we must prove that $P(n+1)$ is true. If we add $(n+1)^2$ to both sides of the equation, we obtain

$$\sum_{k=1}^{n} k^2 + (n+1)^2 = \frac{n(n+1)(2n+1)}{6} + (n+1)^2$$

$$= \frac{n(n+1)(2n+1) + 6(n+1)^2}{6} = \frac{(n+1)(n(2n+1) + 6(n+1))}{6}$$

$$= \frac{(n+1)(2n^2 + 7n + 6)}{6} = \frac{(n+1)(n+2)(2(n+1)+1)}{6}$$

If we equate the first term and the last we get the statement $P(n+1)$, and this completes the proof by induction.   ●

The statement $P(n)$ which we assume in proving the inductive step is called the *inductive assumption*.

The identity

$$\sum_{k=0}^{n} x^k = 1 + x + x^2 + \ldots + x^n = \frac{x^{n+1} - 1}{x - 1}$$

is very useful. Many other identities can be derived from it by making an appropriate substitution for x.

EXAMPLE 5   Prove the above identity by induction.

Solution   ***Starting the Induction:***   We will start the induction with $n = 0$:

$$1 = \frac{x - 1}{x - 1}$$

This starts the induction.

The Inductive Step:

$$\sum_{k=0}^{n+1} x^k = \sum_{k=0}^{n} x^k + x^{n+1} = \frac{x^{n+1} - 1}{x - 1} + x^{n+1}$$

$$= \frac{x^{n+1} - 1}{x - 1} + \frac{x^{n+2} - x^{n+1}}{x - 1} = \frac{x^{n+2} - 1}{x - 1}$$

This completes the proof by induction.   ●

EXAMPLE 6   Prove the identity

$$\sum_{k=0}^{n} 2^k = 2^{n+1} - 1$$

Solution   In the above identity, let $x = 2$.   ●

The Binomial Theorem   We have

$$(X + Y)^n = \sum_{k=0}^{n} \binom{n}{k} X^{n-k} Y^k$$

for each $n = 0, 1, 2, \ldots$.

Proof by induction   ***Starting the Induction:***   Start the induction with $n = 0$. Then the Binomial Theorem says that

$$(X + Y)^0 = \binom{0}{0} X^0 Y^0$$

Since both these terms are equal to 1, this starts the induction.

The Inductive Step: We must prove that if the Binomial Theorem holds for a given integer n then it holds for $n + 1$.

$$(X + Y)^{n+1} = (X + Y)(X + Y)^n$$

Expanding $(X + Y)^n$ by the inductive assumption, we see that the term $X^{n+1-k} Y^k$ arises by multiplying $X^{n+1-k} Y^{k-1}$ by Y, or $X^{n-k} Y^k$ by X. Thus it occurs with the coefficient

$$\binom{n}{k-1} + \binom{n}{k} = \binom{n+1}{k}$$

by the recursion for the binomial coefficients. This completes the proof by induction.   ●

There are many relations between the binomial coefficients, some of which follow from the Binomial Theorem.

COROLLARY   For any $n = 0, 1, 2, \ldots$ we have
1.2.1

$$2^n = \sum_{k=0}^{n} \binom{n}{k}$$

Proof   In the Binomial Theorem, set $X = Y = 1$.   ●

EXAMPLE 7   Prove Pascal's Formula by induction.

Solution   ***Starting the Induction:***   Start the induction with $n = 0$: remember that $0! = 1$.

$$\binom{0}{0} = \frac{0!}{0!0!} = 1$$

This starts the induction.

The Inductive Step: Our inductive assumption is that

$$\binom{n}{k} = \frac{n!}{k!(n-k)!}$$

Using this assumption,

$$\binom{n}{k-1} + \binom{n}{k} = \frac{n!}{(k-1)!(n-k+1)!} + \frac{n!}{k!(n-k)!}$$

$$= \frac{n!}{(k-1)!(n-k+1)} \cdot \frac{k}{k} + \frac{n!}{k!(n-k)!} \cdot \frac{(n-k+1)}{n-k+1}$$

$$= \frac{n!k + n!(n-k+1)}{k!(n+1-k)!} = \frac{n!(n+1)}{k!(n+1-k)!} = \frac{(n+1)!}{k!(n+1-k)!}$$

$$= \binom{n+1}{k}$$

This completes the proof by induction.   ●

APPLICATION   ***The Tower of Hanoi***

In the Tower of Hanoi puzzle there are n disks of unequal size stacked on one of k pegs. The disks are in order of decreasing size, with the smallest disk on top. The puzzle is shown in Figure 1.9. The problem is to move the disks from one peg to another, stacked in the same order, by a sequence of moves. Each single disk is moved from one peg to another, never placing a larger disk on a smaller one.

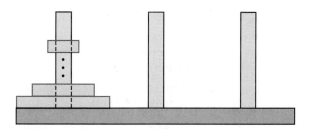

Figure 1.9   The Tower of Hanoi Puzzle

THEOREM 1.2.1   The minimal number of moves needed to solve the Tower of Hanoi puzzle is

   (i) $2^n - 1$, if there are n disks and 3 pegs.

   (ii) $2n - 1$, if there are n disks and k pegs, where $k \geq n + 1 \geq 3$.

Proof of (i)   Let $N(n)$ be the minimal number of moves required to move n disks from one peg to another if the number of pegs is 3.

Starting the Induction: If $n = 1$ it is clear that one move is required to move the one disk. Since $N(1) = 2^1 - 1 = 1$, this starts the induction.

The Inductive Step: Our inductive assumption is that $N(n) = 2^n - 1$. To move $n + 1$ disks from one peg to another, we must first move n disks off the largest disk; then move the largest disk; then replace the n disks on the largest disk. The minimal number of moves required to do this is

$$N(n + 1) = N(n) + 1 + N(n) = (2^n - 1) + 1 + (2^n - 1) = 2^{n+1} - 1$$

This completes the proof by induction.

Proof of (ii)   We will prove this statement directly. To move n disks from one peg to another, move the first $n-1$ disks to $n-1$ separate pegs. Then move the largest disk to the desired peg. Then replace the $n-1$ disks on top of the largest disk. This requires $(n-1)+1+(n-1) = 2n-1$ moves, and this number of moves is clearly minimal.   ●

The two cases of the Tower of Hanoi puzzle which we have described are very different. In the first case, the number of moves needed is an *exponential* function of the input n. In the second case, the number of moves is a *linear* function of the input.   ●

EXAMPLE 8   A computer has been programmed to solve the Tower of Hanoi puzzle. If the computer can carry out one billion moves per second, how long will it take to move 64 disks if there are 3 pegs? How long will it take to move 64 disks if there are k pegs, where $k \geq 65$?

Solution   In the case of 64 disks and 3 pegs, the number of moves is

$$2^{64} - 1 = 18{,}446{,}744{,}073{,}709{,}551{,}615$$

We assume that the computer can make 10^9 moves every second. Assuming that there are 365 days in each year, the computer will require $584.9424\ldots$ years to make all the moves!

In the case of 64 disks and 65 or more pegs, the number of moves is

$$2 \cdot 64 - 1 = 127$$

which the computer can carry out in 0.000000127 seconds!

EXERCISES

Prove the statements 1–9 by induction:

1. $2 + 4 + 6 + \ldots + 2n = n(n + 1)$

2. $2 + 5 + 8 + \ldots + (3n - 1) = \dfrac{n(3n + 1)}{2}$

3. $1 + 5 + 9 + \ldots + (4k - 3) = n(2n - 1)$

4. $\dfrac{1}{1 \cdot 2} + \dfrac{1}{2 \cdot 3} + \ldots + \dfrac{1}{n(n + 1)} = \dfrac{n}{n + 1}$

5. $\dfrac{1}{1 \cdot 3} + \dfrac{1}{3 \cdot 5} + \ldots + \dfrac{1}{(2n - 1)(2n + 1)} = \dfrac{n}{2n + 1}$

6. $\displaystyle\sum_{k=0}^{n} 2^k = 2^{n+1} - 1$

7. $\displaystyle\sum_{k=0}^{n} 3^k = \dfrac{3^{n+1} - 1}{2}$

8. $2^n > n$ for all $n = 0, 1, 2, \ldots$

9. $n! \geq 2^n$ for $n \geq 4$

The formulas in this section for the triangular, square, and pentagonal numbers are special cases of the formula

$$\sum_{k=1}^{n}(kt - (t-1)) = \frac{n(nt - (t-2))}{2}$$

when $t = 1, 2, 3$. The expression on the right is called the nth t-figurate number. It represents the number of points on an expanding regular polygon with $t+2$ sides. Thus for $t = 4$ we have hexagonal numbers, for $t = 5$ heptagonal numbers, and so on. This formula is due to Hypsicles of Alexandria (about 250 A.D.).

10. Using Hypsicles' formula with $t = 4$, write the first twelve hexagonal numbers.

11. Using Hypsicles' formula with $t = 5$, write the first twelve heptagonal numbers.

12. Using Hypsicles' formula with $t = 6$, write the first twelve octagonal numbers.

13. Prove Hypsicles' formula by induction.

14. Prove by induction that

$$\sum_{k=1}^{n}\frac{k(k+1)}{2} = \frac{n(n+1)(n+2)}{6}$$

15. Prove by induction that

$$\sum_{k=0}^{n}\frac{k(3k-1)}{2} = \frac{n^2(n+1)}{2}$$

16. Prove by induction that

$$\sum_{k=0}^{n}k^3 = \left(\frac{n(n+1)}{2}\right)^2$$

(Notice what this says: the sum of the cubes of the first n integers is equal to the square of the sum of the first n integers.)

ADVANCED EXERCISES

In this section we considered the Tower of Hanoi puzzle, in two extreme cases. The puzzle cannot be solved unless there are at least three pegs, and we computed the smallest number of moves which are needed to solve the puzzle in this case. The puzzle is easy to solve if the number of pegs is large compared with the number of disks. We considered this case, where the number of disks is n and the number of pegs is at least $n + 1$.

The other cases of the puzzle, where the number of pegs is k, the number of disks is n, and $3 < k \leq n$, are more complicated, and very interesting. There is a general formula for the smallest number of moves. In these exercises, we ask you to find the smallest number of moves in the simplest case which is different from the two extreme cases. Give an explanation of your answers in each case.

1. Find the smallest number of moves needed to solve the Tower of Hanoi puzzle with four disks and four pegs.

2. Find the smallest number of moves needed to solve the Tower of Hanoi puzzle with five disks and five pegs.

3. Find the smallest number of moves needed to solve the Tower of Hanoi puzzle with six disks and six pegs.

4. Give a formula for the smallest number of moves needed to solve the Tower of Hanoi puzzle with n disks and n pegs, and prove by induction that your formula is correct.

COMPUTER EXERCISES

In this section we gave a formula for the sum of the squares of the first n integers:

$$1^2 + 2^2 + 3^2 + \ldots + n^2 = \frac{n(n+1)(2n+1)}{6}$$

1. Check that if $n = 24$ this sum is a perfect square. That is, verify the identity

$$1^2 + 2^2 + \ldots + 24^2 = 70^2$$

Lucas discovered that the sum of the first n squares is a perfect square only if $n = 1$ or $n = 24$. He called this the Problem of the Square Pyramid, since a pyramid with a square section would contain a number of blocks which is a square only if it had one level or twenty-four levels. He gave an incomplete proof of this. It is an excellent computer exercise to check that Lucas' result is true in a large range.

2. Write a program to check for solutions of the equation

$$6y^2 = x(x + 1)(2x + 1)$$

in positive integers. Use your program to verify that the only solutions of this equation in the range $1 \le x \le 1,000$ are $(x, y) = (1, 1)$ and $(x, y) = (24, 70)$.

3. Prove that $(x, y) = (1, 1)$ and $(x, y) = (24, 70)$ are the only solutions in positive integers of the equation in the previous problem for a range for x which is as large as possible. $1 \le x \le 10,000$ is good; $1 \le x \le 100,000$ is better. If you write your program carefully, you should be able to prove this for all x in an even larger range.

Lucas also stated that the sum of the first n triangular numbers is a square only if $n = 1$, $n = 2$, or $n = 48$. The latter identity is

$$\sum_{k=1}^{48} \frac{k(k + 1)}{2} = 140^2$$

4. Verify that this identity holds.

5. Verify, in as large a range as you can, that the only solutions in positive integers to the equation

$$6y^2 = x(x + 1)(x + 2)$$

are $(x, y) = (1, 1), (2, 2), (48, 140)$.

1.3   SETS, SUBSETS, AND BINARY STRINGS

A Billiard-marker, whose skill was immense,

Might perhaps have won more than his share—

But a Banker, engaged at enormous expense,

Had the whole of their cash in his care.

The English mathematician George Boole (1815–1864) published *An Investigation of the Laws of Thought* in 1854 whose purpose was

> *... to investigate the fundamental laws of those operations of the mind by which reasoning is performed; to give expression to them in the symbolical language of a Calculus, and upon this foundation to establish the science of Logic and construct its method; to make that method itself the basis of a general method for the application of the mathematical doctrine of Probabilities ....*

In *The Laws of Thought*, Boole described a theory whose aim was to unify logic, combinatorics, and the theory of probabilities, including implications for psychology. The concept of a *set* is fundamental in Boole's theory.

We shall start to study sets in this section, and the closely related concept of *binary vectors*. At the end of this section we shall apply our results about sets and binary vectors to a basic problem in computer science, the *Knapsack Problem*. We shall give a solution to the Knapsack Problem which uses the *Standard Gray Code*.

DEFINITION 1.3.1

i. A *set* is a collection of objects. We will denote sets by letters

$$A, B, C, \ldots$$

or by letters with subscripts

$$A_1, A_2, A_3, \ldots$$

ii. The objects that a set contains are its *elements*. The notation

$$a \in A$$

means that a is an element of the set A, and the notation

$$a \notin A$$

means that a is not an element of the set A.

iii. A set can be described by putting its elements in curly brackets. For example, the set whose elements are the integers 1, 2, 3 is denoted by $\{1, 2, 3\}$.

iv. A set can be described by giving a logical condition which its elements must satisfy. The set of elements x which satisfy the condition $P(x)$ is denoted

$$A = \{x : P(x)\}$$

This is called *set-builder notation*. For example, the set $A = \{1, 2, 3\}$ can be obtained using set-builder notation by taking $P(x)$ to be the condition "x is an integer which is greater than or equal to 1 and less than or equal to 3".

v. A set is determined by its elements. Two sets A and B are equal if and only if they have the same elements.

vi. Since sets are determined by their elements, there is a unique set $\emptyset$ which has *no* elements. $\emptyset$ is called the *empty set* or the *null set*.

vii. A set is *finite* if its elements can be listed by a list that stops. A set which is not finite is *infinite*. The set

$$A = \{1, 2, 3, \ldots, n\}$$

is a finite set with n elements. The set of natural numbers

$$N = \{1, 2, 3, \ldots\}$$

is an infinite set.

The phrase "if and only if" between two statements means that they are *logically equivalent*.

EXAMPLE 1 Describe the set of even natural numbers using set-builder notation.

Solution

$$\{2, 4, 6, 8, \ldots\} = \{a : a \in N \quad \text{and} \quad a = 2b \quad \text{for some } b \in N\}.$$ ●

EXAMPLE 2 Describe the empty set $\emptyset$ using set-builder notation.

Solution   The empty set ∅, which has no elements, can be described by any logically contradictory condition (that is, by any condition which no object can satisfy). For example,

$$\emptyset = \{a : a \in N \quad \text{and} \quad a \le 1 \quad \text{and} \quad a \ge 2\}$$ ●

DEFINITION 1.3.2

i. Let A and B be sets. B is a *subset* of A if each element of B is an element of A. The notation

$$B \subset A$$

means that B is a subset of A.

ii. If A is a set, the set of all subsets of A is denoted $P(A)$ and is called the *power set* of A.

iii. If A and B are sets then the *product set* $A \times B$ is the set of ordered pairs (a, b), where $a \in A$ and $b \in B$. In set-builder notation,

$$A \times B = \{(a, b) : a \in A \text{ and } b \in B\}$$

If $A_1, A_2, \ldots, A_n$ are sets, then the *n-fold product* is

$$A_1 \times A_2 \times \ldots \times A_n = \{(a_1, a_2, \ldots, a_n) : a_1 \in A_1, a_2 \in A_2, \ldots, a_n \in A_n\}$$

If A is any set, it is clear that

$$A \subset A$$

For any set A, it is also true that

$$\emptyset \subset A$$

Since the empty set has no elements, it cannot fail to be true that each element of the empty set is an element of A.

EXAMPLE 3   If $A = \{1, 2\}$, describe $P(A)$.

Solution

$$P(A) = \{\emptyset, \{1\}, \{2\}, \{1, 2\}\}$$ ●

We shall often assume that the sets $A, B, C, \ldots$ are subsets of a single set X. For such subsets, we will define four operations.

DEFINITION 1.3.3

i. The *union* of two subsets $A \subset X$ and $B \subset X$:

$$A \cup B = \{x \in X : x \in A \quad \text{or} \quad x \in B\}$$

ii. The *intersection* of two subsets $A \subset X$ and $B \subset X$:

$$A \cap B = \{x \in X : x \in A \quad \text{and} \quad x \in B\}$$

iii. The *difference* of two subsets $A \subset X$ and $B \subset X$:

$$A - B = \{x \in X : x \in A \quad \text{and} \quad x \notin B\}$$

iv. The *complement* of a subset $A \subset X$:

$$A^c = \{x \in X : x \notin A\}$$

Figure 1.10 shows diagrams which illustrate the basic operations on sets. Diagrams of this kind are called *Venn diagrams*.

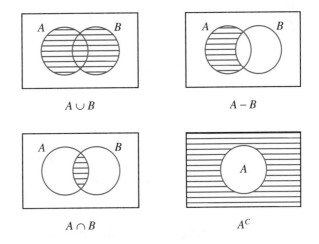

$A \cup B$ 　　　　　 $A - B$

$A \cap B$ 　　　　　 A^C

Figure 1.10　Operations on Sets

EXAMPLE 4　If $X = \{1, 2, 3, 4, 5, 6, 7, 8, 9, 10\}$, $A = \{1, 5, 9\}$, and $B = \{2, 3, 6, 10\}$, find $A \times B$, $A \cup B$, $A \cap B$, $A - B$, A^c, and B^c.

Solution

$$
\begin{aligned}
A \times B = \{&(1, 2), \ (1, 3), \ (1, 6), \ (1, 10), \\
&(5, 2), \ (5, 3), \ (5, 6), \ (5, 10), \\
&(9, 2), \ (9, 3), \ (9, 6), \ (9, 10)\} \\
A \cup B = \{&1, 2, 3, 5, 6, 9, 10\} \\
A \cap B = \ &\emptyset \\
A - B = \{&1, 5, 9\} \\
A^c = \{&2, 3, 4, 6, 7, 8, 10\} \\
B^c = \{&1, 4, 5, 7, 8, 9\}
\end{aligned}
$$

●

Remember that the expression "$a \in A$" means that a is an element of the set A. Are expressions such as "$A \in A$" meaningful?

Suppose that A is the set of all logical concepts. The set of logical concepts is itself a logical concept, so we might say that the expression $A \in A$ should be meaningful. We must not allow expressions such as $A \in A$ and $A \notin A$ to be meaningful, however, as the following argument shows. This argument was discovered by Bertrand Arthur William Russell (1872–1970).

Russell's Paradox   Consider the set

$$X = \{A : A \notin A\}$$

then $X \in X$ and $X \notin X$.

Proof   Either $X \in X$ or $X \notin X$ must be true. If $X \in X$ then X satisfies the condition for membership in X, which is that $X \notin X$. If $X \notin X$ then X satisfies the condition for membership in X, so $X \in X$.   ●

Notice how the paradox arises through a combination of self-reference and negation. It is because paradoxes like this will occur that we must insist expressions like $A \in A$ and $A \notin A$ are not meaningful. A paradox like Russell's Paradox is called a *paradox of self-reference*. One of the themes of this book is to explain why paradoxes of this kind are important in computer science.

In computer science, codes consisting of sequences of 0s and 1s are used constantly.

> **DEFINITION 1.3.4**
>
> **i.** A *binary vector of length n* is an expression $(a_1, a_2, \ldots, a_n)$, where $a_i = 0$ or $a_i = 1$ for each $1 \le i \le n$.
>
> **ii.** A *binary string* is a sequence of 0s and 1s whose length may vary.

For example,

$$(1, 1, 1, 0, 1, 0, 0, 0)$$

is a binary vector of length 8. Often the parentheses and commas are not used when describing binary vectors or binary strings. For example, the binary vector above might be written as

$$11101000$$

The *machine code* of a computer is written in binary strings. Before a computer program can run, it is necessary to *compile* it, that is, to turn the instructions in the program into a binary string such as

$$1100100111010011001111000101010011101011111\ldots$$

There is a close relationship between subsets of a set and binary vectors.

Let

$$A = \{a_1, a_2, \ldots, a_n\}$$

be a finite set and $B \subset A$ a subset of A. The *characteristic function* of B is defined as follows. If $a \in A$,

$$\chi(B) = (\chi_B(a_1), \chi_B(a_2), \ldots, \chi_B(a_n))$$
$$\chi_B(a) = 1 \text{ if } a \in B, \text{ and } \chi_B(a) = 0 \text{ if } a \notin B$$

EXAMPLE 5   Let $A = \{1, 2, 3, 4, 5, 6, 7, 8\}$ and $B = \{1, 2, 3, 5\}$. Find $\chi(B)$.

Solution

$$\chi(B) = (\chi_B(1), \chi_B(2), \chi_B(3), \chi_B(4), \chi_B(5), \chi_B(6), \chi_B(7), \chi_B(8))$$
$$= (1, 1, 1, 0, 1, 0, 0, 0)$$ ●

Binary vectors and strings can be added using *binary addition*. We will give the definition now, and use it in the next section to give a better explanation of the relation between subsets and binary vectors. Binary addition is the addition which is used in a computer.

Binary Addition   In binary addition, 0 and 1 are added by the rules:

$$0 + 0 = 0, \ 1 + 0 = 1, \ 0 + 1 = 1, \ 1 + 1 = 0$$

In other words, binary addition is the same as ordinary addition except for the rule that $1 + 1 = 0$.

The subsets of $\{1, 2, 3, \ldots, n\}$ can be represented as paths in the full binary tree with n levels. An edge to the left means that the corresponding element is not in the subset, and an edge to the right means that the corresponding element is in the subset. Figure 1.11 shows how the subset $\{1, 3\} \subset \{1, 2, 3, 4\}$ can be represented as a path in the full binary tree with four levels. Figure 1.12 shows how the binary vector $(1, 0, 1, 0)$ can be represented as a path in the same tree.

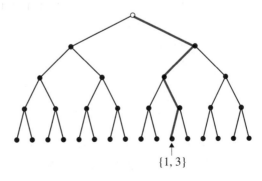

$\{1, 3\}$

Figure 1.11   Subsets as Paths in the Full Binary Tree

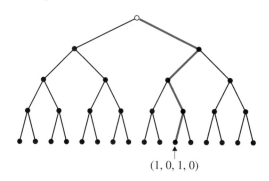

$(1, 0, 1, 0)$

Figure 1.12   Binary Vectors as Paths in the Full Binary Tree

THEOREM 1.3.1

    (i)   If A is a set with n elements and $0 \leq k \leq n$, then the number of k-element subsets of A is $\binom{n}{k}$.

    (ii)   If A is a set with n elements, then A has 2^n subsets.

    (iii)   The number of binary vectors of length n with exactly k 1s is $\binom{n}{k}$.

    (iv)   There are 2^n binary vectors of length n.

Proof   We have seen that subsets of a set with n elements can be represented as paths in the full binary tree with n levels, where an edge to the left means that the corresponding element is not in the subset, and an edge to the right means that the element is in the subset. In Section 1.1 we proved that the number of paths in the full binary tree with exactly k edges to the left is $\binom{n}{k}$: this proves (i).

    In Section 1.1 we proved that the total number of paths in the full binary tree from the root to the nth level is 2^n, and this proves (ii).

    Just as in the proof of (i), binary vectors of length n can be represented as paths in the full binary tree from the root to the nth level. Binary vectors with exactly k 1s correspond to paths of this kind with exactly k edges to the right. By Theorem 1.1.1, there are $\binom{n}{k}$ such paths, and this proves (iii).

    To prove (iv), we observe again that the total number of paths in the full binary tree from the root to the nth level is 2^n.   ●

DEFINITION 1.3.5   An *algorithm* is a set of rules for constructing or listing objects of a given type, or for determining whether an object has a given property.

DEFINITION 1.3.6   A *data structure* is a construct used for organizing and accessing information in a meaningful and useful way.

Data structures are used in computer science to enter and store data in the computer. Algorithms are then used to process the data and obtain useful results.

We have seen how the full binary tree can be used to store subsets and binary strings. When used in this way, the full binary tree is a basic example of a data structure. Now we shall give an example of an algorithm, which gives the *Standard Gray Code*. This code was invented by Frank Gray, a scientist at the Bell Telephone Laboratories. Its original purpose was to avoid errors in signals transmitted by pulse code modulation.

The Standard Gray Code   This algorithm gives a list of the binary vectors of length n in such a way that successive vectors in the list differ by only one element.

    (i)   Begin with the list

$$(0)$$
$$(1)$$

    (ii) If a list of binary vectors of length n has been constructed, make a list of binary vectors of length $n + 1$ by preceding each vector on the list by a 0, and then listing the vectors in reverse order, each preceded by a 1.

It is clear from the construction that the algorithm produces lists of binary vectors such that successive vectors in the lists differ in exactly one element. If we list the subsets of $\{1, 2, \ldots, n\}$ corresponding to these binary vectors by the characteristic function, we also get a list of the subsets such that successive subsets on the list differ by only one element.

EXAMPLE 6   List the binary vectors of length 3 using the Standard Gray Code. Use the same list to list the subsets of $\{1, 2, 3\}$.

Solution   Begin with a list of the binary vectors of length 1:

$$(0)$$
$$(1)$$

Next make a list of the binary vectors of length 2:

$$(0, 0)$$
$$(0, 1)$$
$$(1, 1)$$
$$(1, 0)$$

Finally, make a list of the binary vectors of length 3, and the subsets of $\{1, 2, 3\}$ corresponding to these by the characteristic function.

$(0, 0, 0)$	$\emptyset$
$(0, 0, 1)$	$\{3\}$
$(0, 1, 1)$	$\{2, 3\}$
$(0, 1, 0)$	$\{2\}$
$(1, 1, 0)$	$\{1, 2\}$
$(1, 1, 1)$	$\{1, 2, 3\}$

$$(1, 0, 1) \quad \{1, 3\}$$
$$(1, 0, 0) \quad \{1\}$$

⬤

Since there are 2^n binary vectors of length n, it would take a long time to list them using the Standard Gray Code. There are 2^n subsets of an n-element set, and these would take an equally long time to list. For example, by the same calculation as at the end of the previous section, a computer which can list a billion binary vectors per second will take more than 584 years to list all the binary vectors of length 64.

Similarly, it would take a long time to carry out the solution to the Knapsack Problem.

APPLICATION   *The Knapsack Problem*

Many problems in computer science involve examining a large set of objects to find a single object with desirable properties. In order to do this, it is usually necessary to *sort* the objects in some way, in order to *list* them. One then *searches* through the list to find the desired object. There are a number of *sorting algorithms* and *searching algorithms* which accomplish these two goals.

The Knapsack Problem   A hiker has a knapsack which will hold a total of N pounds. The hiker has a number of objects weighing n_1 pounds, n_2 pounds, ..., n_r pounds which he wishes to take with him. He wants to fill the knapsack as full as possible, that is, to pack it with objects having the maximum total weight.

This is equivalent to asking for a subset of the r objects whose total weight is as large as possible, subject to the condition that the total weight is no greater than N.

We can solve the Knapsack Problem by using the Standard Gray Code. If we have r objects which are to go into the Knapsack, we use the Standard Gray Code to list the binary vectors of length r, and the corresponding sums of weights ("1" means the weight is included in the sum; "0" means that it is not). Then we search through the list for the greatest sum not greater than N.

EXAMPLE 7   Solve the Knapsack problem if $N = 28$, and we have four objects with weights

$$n_1 = 4, \ n_2 = 8, \ n_3 = 14, \ \text{and} \ n_4 = 15.$$

We list the binary vectors of length 4 and the corresponding sums of weights:

$$(0, 0, 0, 0) \quad 0$$
$$(0, 0, 0, 1) \quad 15$$
$$(0, 0, 1, 1) \quad 14 + 15 = 29$$
$$(0, 0, 1, 0) \quad 14$$
$$(0, 1, 1, 0) \quad 8 + 14 = 22$$
$$(0, 1, 1, 1) \quad 8 + 14 + 15 = 37$$
$$(0, 1, 0, 1) \quad 8 + 15 = 23$$
$$(0, 1, 0, 0) \quad 8$$
$$(1, 1, 0, 0) \quad 4 + 8 = 12$$
$$(1, 1, 0, 1) \quad 4 + 8 + 15 = 27$$

$$(1, 1, 1, 1) \quad 4 + 8 + 14 + 15 = 41$$
$$(1, 1, 1, 0) \quad 4 + 8 + 14 = 26$$
$$(1, 0, 1, 0) \quad 4 + 14 = 18$$
$$(1, 0, 1, 1) \quad 4 + 14 + 15 = 33$$
$$(1, 0, 0, 1) \quad 4 + 15 = 19$$
$$(1, 0, 0, 0) \quad 4$$

Searching through the list, we see that 27 is the largest sum which is not greater than 28. The first, second, and fourth objects should be put into the knapsack.

EXERCISES

Which of the following statements is true and which is false? Which has no meaning at all?

1. $1 \in \{1, 2, 3\}$.

2. $1 \notin \{1, 2, 3\}$.

3. $\{1\} \in \{1, 2, 3\}$.

4. $\{1\} \notin \{1, 2, 3\}$.

5. $\{1\} \subset \{1, 2, 3\}$.

6. $\{1\} \not\subset \{1, 2, 3\}$.

7. $A \in A$.

8. $A \notin A$.

9. $A \in B$.

10. $A \notin B$.

Use set-builder notation to describe the following sets:

11. $A = \{2, 4, 6, 8, 10, 12\}$

12. $A = \{3, 6, 9, 12, 15, 18\}$

13. $A = \{4, 8, 12, 16, 20, 24\}$

14. $A = \{2, 4, 8, 16, 32, 64\}$

15. $A = \{1, 3, 7, 15, 31, 63\}$

16. $A = \{3, 5, 9, 17, 33, 65\}$

17. The empty set $\emptyset$.

18. The positive odd integers.

19. The positive integers which give a remainder of 1 when divided by 4.

20. The positive integers which give a remainder of 3 when divided by 4.

Given a set X and subsets A and B of X, find $A \times B$, $A \cup B$, $A \cap B$, $A - B$, A^c, and B^c:

21. $X = \{1, 2, 3, 4, 5, 6, 7, 8, 9, 10\}$, $A = \{1, 3, 5, 7, 9\}$, $B = \{1, 2, 8, 9\}$.

22. $X = \{1, 2, 3, 4, 5, 6, 7, 8, 9, 10\}$, $A = \{3, 6, 9\}$, $B = \{2, 4, 6, 8, 10\}$.

23. $X = \{1, 2, 3, 4, 5, 6, 7, 8, 9, 10, 11, 12, 13, 14, 15, 16\}$, $A = \{1, 4, 7, 10, 13, 16\}$, $B = \{4, 8, 12, 16\}$.

24. $X = \{1, 2, 3, 4, 5, 6, 7, 8, 9, 10, 11, 12, 13, 14, 15, 16\}$, $A = \{5, 9, 13\}$, $B = \{3, 7, 11, 15\}$.

In the next three problems, find the characteristic function $\chi(B)$ for the given subset $B \subset A$.

25. $B = \{1, 3, 5\} \subset A = \{1, 2, 3, 4, 5, 6\}$.

26. $B = \{1, 2, 6, 7\} \subset A = \{1, 2, 3, 4, 5, 6, 7, 8\}$.

27. $B = \{3, 4, 5\} \subset A = \{1, 2, 3, 4, 5, 6, 7\}$.

Represent the following subsets as paths in the full binary tree:

28. $\{2, 3\} \subset \{1, 2, 3\}$

29. $\{1, 3, 4\} \subset \{1, 2, 3, 4\}$

30. $\{1, 3, 5\} \subset \{1, 2, 3, 4, 5\}$

Represent the following binary vectors as paths in the full binary tree:

31. $(1, 1, 0, 1)$

32. $(0, 1, 0, 1, 0)$

33. $(1, 1, 1, 0, 0, 0)$

List the subsets of $\{1, 2, \ldots, n\}$ in columns, where each column consists of subsets with the same number of elements, if

34. $n = 3$.

35. $n = 4$.

36. $n = 5$.

List the binary vectors of length n in columns, where each column consists of binary vectors with the same number of 1s, if

37. $n = 3$.

38. $n = 4$.

39. $n = 5$.

40. Using the Standard Gray Code, list the binary vectors of length five.

41. Using the Standard Gray Code, list the binary vectors of length six.

ADVANCED EXERCISES

Solve the Knapsack Problem by using the Standard Gray Code, in the following cases:

1. $n_1 = 6, n_2 = 7, n_3 = 16, n_4 = 18, N = 32$.

2. $n_1 = 10, n_2 = 11, n_3 = 14, n_4 = 26, N = 46$.

3. $n_1 = 12, n_2 = 18, n_3 = 21, n_4 = 28, N = 66$.

4. $n_1 = 4, n_2 = 6, n_3 = 9, n_4 = 25, n_5 = 30, N = 48$.

5. $n_1 = 7, n_2 = 14, n_3 = 16, n_4 = 19, n_5 = 25, N = 54$.

COMPUTER EXERCISES

Write a program to generate the Standard Gray Code, and use it to print a list of binary vectors of length n in the following cases:

1. $n = 10$.

2. $n = 11$.

3. $n = 12$.

In the next three problems, solve the Knapsack Problem in the case where the weights are the triangular numbers $\frac{n(n+1)}{2}$ and

4. $1 \leq n \leq 10$ and $N = 140$.

5. $1 \leq n \leq 11$ and $N = 200$.

6. $1 \leq n \leq 12$ and $N = 230$.

1.4   SET OPERATIONS

There was also a Beaver, that paced on the deck,

Or would sit making lace in the bow:

And had often (the Bellman said) saved them from wreck,

Though none of the sailors knew how.

The English logician John Venn (1834–1923) wrote *Symbolic Logic* which was published in 1881. Venn was strongly influenced by Boole's work, which he clarified through the use of logic diagrams. These diagrams are now called *Venn diagrams*. The convention of enclosing the diagram in a rectangle was first used systematically by Lewis Carroll (the pen name of Charles Lutwidge Dodgson, (1832–1898), the author of *Alice in Wonderland* and *The Hunting of the Snark*).

There are many identities between expressions involving sets. For example, the identities

$$(A \cup B)^c = A^c \cap B^c$$

and

$$(A \cap B)^c = A^c \cup B^c$$

are called *DeMorgan's Laws*.

In this section we shall describe Venn diagrams and how they can be used to prove set identities. In the Advanced Exercises, a procedure will be given for drawing a Venn diagram for any finite number of sets. We shall also describe *Boolean algebras*, which are important in computer science. At the end of the section we shall describe an *error-correcting code*, the *Hamming Code H_8*, using binary vectors, and also using subsets of an eight-element set.

In the previous section we gave the Venn diagrams for the operations of union, intersection, difference, and complement. Venn diagrams can be used to represent these and other set operations. Figure 1.13 shows the Venn diagrams for one, two, and three sets, where the sets are represented by circles. Venn had noticed that there is a diagram for four sets, as in Figure 1.14, where the sets are represented by ellipses.

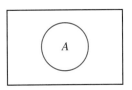 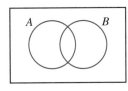 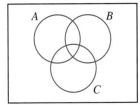

Figure 1.13   Venn Diagrams for One, Two, and Three Sets

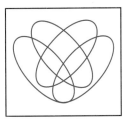

Figure 1.14   A Venn Diagram for Four Sets

The problem in constructing these diagrams is to draw circles, ellipses, or other regions corresponding to the sets in such a way that there is a region in the diagram corresponding to every possible intersection of the sets involved.

We shall use Venn diagrams for small numbers of sets. If the number of sets is large, Venn diagrams become awkward to use. *Boolean functions*, which we shall describe in Chapter 6, can be used instead of Venn diagrams, regardless of the number of sets involved.

Venn diagrams can be used to give informal proofs of set identities by drawing a diagram for each of the two sides of the identity, and comparing them to see if they are equal.

Proving Set Identities with Venn Diagrams   To prove that a set identity is correct using Venn diagrams, draw two Venn diagrams, sketch the regions corresponding to the expressions on each side of the identity, and verify that these two regions are the same.

In the previous section we stated that two sets are equal if and only if they have the same elements. In a more formal theory of sets, this is the *Axiom of Extension*.

The Axiom of Extension   Two sets A and B are equal if and only if $A \subset B$ and $B \subset A$.

The importance of the Axiom of Extension is not only theoretical but practical, since the best way to prove that two sets are equal is often to show that each is a subset of the other.

Arguments with Elements   To prove that two sets A and B are equal, prove that $A \subset B$ and $B \subset A$.

EXAMPLE 1   Prove that the identity

$$A \cap (B \cup C) = (A \cap B) \cup (A \cap C)$$

holds by using Venn diagrams, and also by using an argument with elements.

Proof   In Figure 1.15 are two Venn diagrams, in which are sketched the regions corresponding to the left-hand and right-hand sides of the identity. Since they are the same, the identity holds.

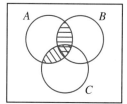 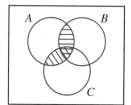

Figure 1.15   A Set Identity Proved with Venn Diagrams

Now we will give an argument with elements. If $a \in A \cap (B \cup C)$ then $a \in A$ and $a \in B$ or $a \in C$. This implies that either $a \in A$ and $a \in B$ or $a \in A$ and $a \in C$, which means that $a \in (A \cap B) \cup (A \cap C)$. We have proved that $A \cap (B \cup C) \subset (A \cap B) \cup (A \cap C)$. Now assume that $a \in (A \cap B) \cup (A \cap C)$. This means that either $a \in A$ and $a \in B$, or $a \in A$ and $a \in C$. This implies that $a \in A$, and $a \in B$ or $a \in C$, and this means that $a \in A \cap (B \cup C)$. We have proved that $(A \cap B) \cup (A \cap C) \subset A \cap (B \cup C)$.   ●

DEFINITION 1.4.1   If A and B are sets, the *symmetric difference* is

$$A \triangle B = (A - B) \cup (B - A) = (A \cup B) - (A \cap B)$$

Notice that the elements of $A \triangle B$ are the elements which belong to exactly one of the sets A and B.

EXAMPLE 2   If $A = \{1, 5, 9, 10\}$ and $B = \{2, 3, 5, 6, 10\}$, then

$$A \triangle B = (A - B) \cup (B - A) = \{1, 9\} \cup \{2, 3, 6\} = \{1, 2, 3, 6, 9\}$$
$$= (A \cup B) - (A \cap B) = \{1, 2, 3, 5, 6, 9, 10\} - \{5, 10\}$$ ●

There are many identities involving sets. Here is a list of a few identities which arise frequently.

Set Identities

The following identities involve sets A, B, and C, which are subsets of a fixed set X.

   I.   Associativity

   (i) $A \cup (B \cup C) = (A \cup B) \cup C$
   (ii) $A \triangle (B \triangle C) = (A \triangle B) \triangle C$
   (iii) $A \cap (B \cap C) = (A \cap B) \cap C$

   II.   Identity

   (i) $\emptyset \cup A = A \cup \emptyset = A$
   (ii) $\emptyset \triangle A = A \triangle \emptyset = A$
   (iii) $X \cap A = A \cap X = A$

   III. Inverse

   (i) $A \triangle A = \emptyset$

   IV.   Commutativity

   (i) $A \cup B = B \cup A$
   (ii) $A \triangle B = B \triangle A$
   (iii) $A \cap B = B \cap A$

   V.   Idempotence

   (i) $A \cup A = A$
   (ii) $A \cap A = A$

   VI.   Distributivity

   (i) $A \cap (B \cup C) = (A \cap B) \cup (A \cap C)$
   (ii) $A \cap (B \triangle C) = (A \cap B) \triangle (A \cap C)$

   VII. Complement

   (i) $A \cup A^c = X$
   (ii) $A \triangle A^c = X$
   (iii) $A \cap A^c = \emptyset$

VIII. Involution

(i) $(A^c)^c = A$

IX. DeMorgan's Laws

(i) $(A \cup B)^c = A^c \cap B^c$

(ii) $(A \cap B)^c = A^c \cup B^c$

X. Duality

If, in any valid set identity, the symbols $\cup$, $\cap$, X, and $\emptyset$ are replaced by the symbols $\cap$, $\cup$, $\emptyset$, and X, respectively, the result is another valid set identity.

Boole had the idea that the union of two sets could be viewed as a kind of sum, and that the intersection of two sets could be viewed as the product of those sets. His ideas were considered very shocking at the time, and gave rise to a long controversy. Gradually his ideas were accepted, and now we think of the identities in the list above as describing the properties of this addition and multiplication, with the added idea that we might substitute the symmetric difference Δ for the union $\cup$ as the notion of addition in some cases.

DEFINITION 1.4.2   A *Boolean algebra* is a set X together with the operations $\cup$, $\cap$, and c on the subsets of X.

Boolean algebras are very useful in computer science. The reason that we sometimes want to choose the symmetric difference Δ as the sum in a Boolean algebra instead of the union $\cup$ is explained by the following theorem.

THEOREM 1.4.1 $\chi(A \Delta B) = \chi(A) + \chi(B)$, where the addition is binary addition.

Proof   If $a \in A \Delta B$ then $a \in A$ or $a \in B$ but not both, and $\chi_{A \Delta B}(a) = 1$. Then $\chi_A(a) + \chi_B(a) = 1 + 0 = 1$ or $= 0 + 1 = 1$. If $a \notin A \Delta B$ then $\chi_{A \Delta B}(a) = 0$, and either $a \in A$ and $a \in B$, or $a \notin A$ and $a \notin B$. Then $\chi_A(a) + \chi_B(a) = 1 + 1 = 0$ or $= 0 + 0 = 0$.   ●

If A is a finite set we shall use the notation $|A|$ to denote the number of elements of A. Suppose that we have two finite sets A and B, and we want to know how many elements are in A or in B, that is, we want to find a formula for $|A \cup B|$. The formula is

$$|A \cup B| = |A| + |B| - |A \cap B|$$

To understand the reason for this, look at the Venn diagram for two sets in Figure 1.13. The elements of $A \cup B$ consist of the elements of A and B, but in the sum $|A| + |B|$ the elements of $A \cap B$ are counted twice.

Suppose now that we have three sets A, B, and C, and we want a formula for $|A \cup B \cup C|$. Referring to the Venn diagram in Figure 1.13 we see that in $|A| + |B| + |C|$ the elements of $A \cap B$, $A \cap C$, and $B \cap C$ are counted twice, so we must subtract the number of elements in these pairwise intersections. Then we must add the number of elements in $A \cap B \cap C$, since these elements have so far not been counted. In this way we get the formula

$$|A \cup B \cup C| = |A| + |B| + |C| - |A \cap B| - |A \cap C| - |B \cap C| + |A \cap B \cap C|$$

Perhaps you notice a pattern in these two formulas. The number of elements in the union of a finite number of sets is the sum of the numbers of elements in the sets, minus the numbers of elements in the pairwise intersections, plus the numbers of elements in the triple intersections, and so on. These two formulas are examples of the *Principle of Inclusion-Exclusion*, which counts the number of elements in the union of any finite number of sets. We shall state the general Principle of Inclusion-Exclusion in Chapter 3.

A *directed graph*, or *digraph*, is a graph with arrows on each edge to indicate direction. Digraph models are used often in Discrete Mathematics. We shall give a digraph model for the problem in the next two examples, which will complement our use of Inclusion-Exclusion.

EXAMPLE 3   Twelve people in a town have decided to form three clubs. Some are more social than others, and the clubs are chosen so that half of the people belong to each club, one-third of the people to each pair of clubs, and one-fourth of the people to all three clubs. How many people belong to at least one of the clubs? How many people belong to none?

Solution   Let A_1 be the set of members of the first club, A_2 the set of members of the second club, and A_3 the set of members of the third club. Since half of the twelve people belong to each club, $|A_1| = |A_2| = |A_3| = 6$. Since one-third of the people belong to each pair of clubs, $|A_1 \cap A_2| = |A_1 \cap A_3| = |A_2 \cap A_3| = 4$. Since one-fourth of the people belong to all three clubs, $|A_1 \cap A_2 \cap A_3| = 3$.

The people who belong to at least one club are the members of the union $A_1 \cup A_2 \cup A_3$. By the principle of Inclusion-Exclusion for three sets,

$$|A_1 \cup A_2 \cup A_3| = |A_1| + |A_2| + |A_3| - |A_1 \cap A_2| - |A_1 \cap A_3| - |A_2 \cap A_3|$$
$$+ |A_1 \cap A_2 \cap A_3|$$
$$= 6 + 6 + 6 - 4 - 4 - 4 + 3 = 9$$

Thus 9 people belong to at least one club, and $12 - 9 = 3$ people belong to none of the clubs.   ●

EXAMPLE 4   In the previous example, how many people belong to two or more clubs?

Solution   Let sets of members of the clubs be A_1, A_2, and A_3 as before. The people who belong to the first and second clubs are the elements of the set $B_1 = A_1 \cap A_2$, and similarly $B_2 = A_1 \cap A_3$ for the members of the first and third clubs, and $B_3 = A_2 \cap A_3$ for the members of the second and third clubs. The people who belong to two or more clubs belong to the set $B_1 \cup B_2 \cup B_3$, so our problem is to compute the number of elements of this set. As before, $|B_1| = |B_2| = |B_3| = 4$. The pairwise and triple intersections of the sets B_i are all equal to $A_1 \cap A_2 \cap A_3$. Now using the Principle of Inclusion-Exclusion,

$$|B_1 \cup B_2 \cup B_3| = |B_1| + |B_2| + |B_3| - |B_1 \cap B_2| - |B_1 \cap B_3| - |B_2 \cap B_3|$$
$$+ |B_1 \cap B_2 \cap B_3|$$
$$= 4 + 4 + 4 - 3 - 3 - 3 + 3 = 6$$

According to the previous two examples, the Principle of Inclusion-Exclusion predicts that three people each belong to exactly 0 clubs, 1 club, 2 clubs, or 3 clubs.   ●

The calculations that we have done do not make it clear that it is possible to find three clubs satisfying the conditions of the problem. To see that it is possible, refer to the digraph in Figure 1.16. Let the twelve vertices correspond to the twelve people. Represent the members of the first club by the vertices that can be reached by a directed path (a path following the arrows) from vertex 1, and similarly for the second and third clubs. You will see that this definition of the three clubs satisfies all the conditions of the problem.

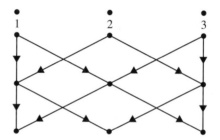

Figure 1.16   A Town With Three Clubs

The people who belong to 0, 1, 2, or 3 clubs correspond to the four rows of three vertices, exactly as predicted by the Principle of Inclusion-Exclusion.

APPLICATION   *An Error-Correcting Code*

In the theory of communication, it is a basic principle that any information can be encoded as a sequence of 0s and 1s. This fact was recognized long ago, and is used in Morse Code, where the letters of the English alphabet are encoded as sequences of dots and dashes.

There is a problem which arises when we transmit information in this way. Errors may occur in the transmission, which cause 0s and 1s to be interchanged. Such errors are typically caused by noise in the channel.

A method which is often used to deal with this difficulty is to use an error-correcting code. Instead of sending a single bit 0 or 1, or a short binary vector, one sends a longer binary vector.

Figure 1.17 lists the vectors of a useful error-correcting code, the *Hamming Code H_8*, together with the subsets of the set $\{1, 2, 3, 4, 5, 6, 7, 8\}$ corresponding to these binary vectors by the characteristic function.

$(1, 1, 1, 1, 1, 1, 1, 1)$	$\{1, 2, 3, 4, 5, 6, 7, 8\}$
$(1, 1, 1, 0, 1, 0, 0, 0)$	$\{1, 2, 3, 5\}$
$(1, 1, 0, 1, 0, 0, 0, 1)$	$\{1, 2, 4, 8\}$
$(1, 0, 1, 0, 0, 0, 1, 1)$	$\{1, 3, 7, 8\}$
$(1, 1, 0, 0, 0, 1, 1, 0)$	$\{1, 2, 6, 7\}$
$(1, 0, 0, 0, 1, 1, 0, 1)$	$\{1, 5, 6, 8\}$
$(1, 0, 0, 1, 1, 0, 1, 0)$	$\{1, 4, 5, 7\}$
$(1, 0, 1, 1, 0, 1, 0, 0)$	$\{1, 3, 4, 6\}$
$(0, 0, 0, 0, 0, 0, 0, 0)$	$\emptyset$
$(0, 0, 0, 1, 0, 1, 1, 1)$	$\{4, 6, 7, 8\}$
$(0, 0, 1, 0, 1, 1, 1, 0)$	$\{3, 5, 6, 7\}$
$(0, 1, 0, 1, 1, 1, 0, 0)$	$\{2, 4, 5, 6\}$
$(0, 0, 1, 1, 1, 0, 0, 1)$	$\{3, 4, 5, 8\}$
$(0, 1, 1, 1, 0, 0, 1, 0)$	$\{2, 3, 4, 7\}$
$(0, 1, 1, 0, 0, 1, 0, 1)$	$\{2, 3, 6, 8\}$
$(0, 1, 0, 0, 1, 0, 1, 1)$	$\{2, 5, 7, 8\}$

Figure 1.17   The Hamming Code H_8

The Hamming codes (there is an infinite set of such codes) were discovered by Richard Wesley Hamming, a scientist at Bell Laboratories. Notice how the code H_8 is constructed. Each of the first eight vectors consists of a 1 followed by the vector $(1, 1, 0, 1, 0, 0, 0)$ or a vector obtained from this by successively placing first elements last. The next eight vectors are the first eight vectors with 0s and 1s interchanged. Each of the last eight vectors is the binary sum of one of the first eight vectors and the vector $(1, 1, 1, 1, 1, 1, 1, 1)$. The vectors in a code are called *codewords*.

The subsets of $\{1, 2, 3, 4, 5, 6, 7, 8\}$ which correspond to the codewords of H_8 form what is called a *design*. Designs are useful, as the name implies, in the design of experiments.

A set of binary vectors which contains the zero vector $(0, 0, \ldots, 0)$ and also contains the binary sum of any pair of vectors in it is called a *linear code*. The code H_8 has the remarkable property that it is a linear code. For example,

$$(1, 1, 1, 0, 1, 0, 0, 0) + (1, 0, 0, 1, 1, 0, 1, 0)$$

$$= (1 + 1, 1 + 0, 1 + 0, 0 + 1, 1 + 1, 0 + 0, 0 + 1, 0 + 0)$$

$$= (0, 1, 1, 1, 0, 0, 1, 0)$$

which is again a vector in the code.

The corresponding property for the subsets in Figure 1.17 is that the symmetric difference of any two of them is again a subset on the list. For example,

$$\{1, 2, 3, 5\} \triangle \{1, 4, 5, 7\} = \{2, 3, 4, 7\}$$

You will find that each vector differs from each of the other vectors in the code in at least four entries. The minimum number of entries by which two different vectors in a linear code differ is called the *Hamming distance* of the code. For a binary code it is always equal to the minimal number of 1s in a non-zero vector of the code.

Now suppose that we wish to send a message using the code H_8. First we encode the message as a binary string (a sequence of 0s and 1s). Then we break the string up into blocks, for example, blocks of length four. Then we replace each block with a vector in the code H_8. Since there are sixteen binary vectors of length four and sixteen vectors in H_8 it is possible to do this.

Then we send the message. If a codeword is received intact, we assume that no error has occurred. If a codeword is received which differs from a codeword v in H_8 by a single entry (notice that v is unique), we decode the codeword as v. This procedure is called *maximum likelihood decoding*. It is effective if the number of errors in the transmission of the message is small. In particular, when used as an error-correcting code, the code H_8 will correct one error.

There is an extensive theory of codes and designs. Sometimes it is best to think of the codewords of a binary code as binary vectors; sometimes it is best to think of them as subsets of a finite set.

EXERCISES

If A, B, and C are subsets of X, describe the following subsets with Venn diagrams.

1. $(A \cup B) - C$

2. $(A \cap B) - C$

3. $(A - B) - C$

4. $(A \triangle B) - C$

5. $A \cap (B \cup C)$

6. $(A \cap B) \cup (A \cap C)$

7. $A \cap (B \triangle C)$

8. $(A \cap B) \triangle (A \cap C)$

If A, B, and C are subsets of X, prove each of the following identities with Venn diagrams, and then with an argument with elements.

9. $A \cup (B \cup C) = (A \cup B) \cup C$

10. $A \triangle (B \triangle C) = (A \triangle B) \triangle C$

11. $A \cap (B \triangle C) = (A \cap B) \triangle (A \cap C)$

12. $(A \cup B)^c = A^c \cap B^c$

13. $A \cup A^c = X$

14. $(A \cup B \cup C)^c = A^c \cap B^c \cap C^c$

15. $(A \cap B \cap C)^c = A^c \cup B^c \cup C^c$

16. $(A - B)^c = A^c \cup (A \cap B)$

17. $(A \triangle B)^c = (A \cup B)^c \cup (A \cap B)$

18. $(A \cup B) - (B \cup C) = A - (B \cup C)$

19. $(A \cap B) - (B \cap C) = (A \cap B) - C$

20. $A \cap (A \cup B) = A$

If A_1, A_2, and A_3 are sets, use the Principle of Inclusion-Exclusion to find the number of elements in $A_1 \cup A_2 \cup A_3$.

21. $|A_1| = |A_2| = |A_3| = 12$, $|A_1 \cap A_2| = |A_1 \cap A_3| = |A_2 \cap A_3| = 8$, $|A_1 \cap A_2 \cap A_3| = 6$

22. $|A_1| = 7$, $|A_2| = 3$, $|A_3| = 6$, $|A_1 \cap A_2| = 2$, $|A_1 \cap A_3| = 5$, $|A_2 \cap A_3| = 2$, $|A_1 \cap A_2 \cap A_3| = 2$

23. $|A_1| = |A_2| = |A_3| = 5$, $|A_1 \cap A_2| = 1 = |A_2 \cap A_3|$, $|A_1 \cap A_3| = 3$, $|A_1 \cap A_2 \cap A_3| = 1$

24. Twenty-four people in a town have decided to form three clubs. If one-half the people belong to each club, one-third of the people to each pair of clubs, and one-fourth of the people to all three clubs, how many people belong to at least one club? How many people belong to none?

25. In the previous problem, how many people belong to two or more clubs?

Take the binary sum of each of the following vectors in the Hamming code H_8, and verify that the sum is again a vector in H_8.

26. $(1, 1, 0, 1, 0, 0, 0, 1)$ and $(0, 1, 1, 0, 0, 1, 0, 1)$

27. $(1, 0, 0, 0, 1, 1, 0, 1)$ and $(1, 0, 1, 1, 0, 1, 0, 0)$

28. $(0, 0, 1, 0, 1, 1, 1, 0)$ and $(0, 0, 0, 1, 0, 1, 1, 1)$

ADVANCED EXERCISES

Here is a procedure for finding a Venn diagram for any finite number of sets. For one, two, or three sets we use the usual diagrams in Figure 1.13 of this section. In Figure 1.18 a Venn diagram for four sets has been constructed, consisting of three rectangles and a circle. In Figure 1.19 a Venn diagram for five sets has been constructed, by adding a square to the four sets in Figure 1.18. This construction can be continued by adding a regular octagon to the diagram in Figure 1.19, then a regular 16-sided figure, then a regular 32-sided figure, and so on.

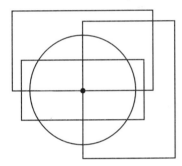

Figure 1.18
A Venn Diagram for Four Sets

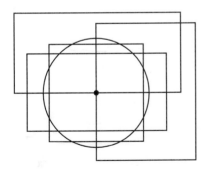

Figure 1.19
A Venn Diagram for Five Sets

1. Draw a Venn diagram for six sets by adding a regular octagon to the diagram in Figure 1.19.

A Venn diagram is always divided into *primitive regions*: regions in the diagram which are not further subdivided. In a Venn diagram for n sets there are always 2^n primitive regions (counting the region outside all the sets).

2. For the Venn diagrams for one, two, and three sets, verify that there are two, four, and eight primitive regions respectively.

3. For the Venn diagram for four sets in Figure 1.18, verify that there are sixteen primitive regions.

4. For the Venn diagram for five sets in Figure 1.19, verify that there are thirty-two primitive regions.

In the previous section we described the Standard Gray Code. In general, a *Gray Code* is a list of all the binary vectors of length n such that each vector on the list differs from the previous vector in exactly one element. The Standard Gray Code is an example of a Gray Code, but there are many others. A Gray Code could start with any binary vector, but we shall assume that the first vector is always the zero vector (0), $(0, 0)$, $(0, 0, 0)$, etc.

5. Prove that there are exactly two Gray Codes for binary vectors of length two: one starts $(0, 0)$, $(0, 1)$ and the other starts $(0, 0)$, $(1, 0)$.

6. Find all Gray codes for binary vectors of length three which begin with the vector $(0, 0, 0)$. How many such Gray Codes are there?
   Hint: You can represent these codes as paths in a labelled tree. Label the root with the vector $(0, 0, 0)$, then join the root to three vertices labelled $(1, 0, 0)$, $(0, 1, 0)$, $(0, 0, 1)$. Continue, in such a way that the vector which labels each new vertex differs from the vector which labels an adjacent vertex in exactly one entry. As you go down branches of the tree (paths from the root), no vector should be repeated.
   This tree has more than thirty terminal vertices, so you may find it easier to draw if you break it up into three subtrees with roots labelled $(1, 0, 0)$, $(0, 1, 0)$, $(0, 0, 1)$.

COMPUTER EXERCISE

1. Write a program to list the Gray Codes for binary vectors of length four which begin with the vector $(0, 0, 0, 0)$. How many such codes are there?

1.5   RECURSIONS

There was one who was famed for the number of things

He forgot when he entered the ship:

His umbrella, his watch, all his jewels and his rings,

And the clothes he had bought for the trip.

The loss of his clothes hardly mattered, because

He had seven suits on when he came,

With three pairs of boots—but the worst of it was,

He had wholly forgotten his name.

Leonardo Fibonacci, or Leonardo of Pisa, was born about 1170 A.D. In 1202 A.D. Fibonacci published the *Liber Abaci*. This book contains the first description of the *Fibonacci sequence*, in the solution to the *Rabbit Problem*, which we shall state in the exercises. There is a journal devoted to the Fibonacci sequence, the *Fibonacci Quarterly*, and several books have been devoted to the applications of the Fibonacci sequence in computer science and elsewhere.

The Fibonacci numbers, and the closely related sequence of *Lucas numbers*, are examples of *recursive sequences*, and more particularly of *linear recursive sequences*. In this section we shall begin to study recursive sequences.

At the end of this section we shall illustrate the importance of recursive sequences by describing a basic item of computer hardware, a *shift register*. Many computer codes are constructed by taking shift register output and modifying it. A shift register is a machine whose output is a binary recursion.

DEFINITION 1.5.1   A *recursive sequence*

$$a_1, a_2, a_3, \ldots, a_n, \ldots$$

is a sequence which is defined as follows:

i. A number of terms of the sequence $a_1, a_2, \ldots, a_r$ are given. These are the *initial values*.

ii. A rule called the *recursion* is given, which explains how a_n is to be computed in terms of previous terms in the sequence, if $n > r$.

The terms of a recursive sequence can be numbers, graphs, or other objects.

The Fibonacci Numbers   The *Fibonacci numbers* F_n have the initial values

$$F_0 = 0, \, F_1 = 1$$

and the recursion

$$F_n = F_{n-1} + F_{n-2}$$

if $n \geq 2$.

EXAMPLE 1   Compute the Fibonacci numbers F_n for $0 \leq n \leq 10$.

Solution $F_0 = 0, F_1 = 1, F_2 = 0 + 1 = 1, F_3 = 1 + 1 = 2, F_4 = 1 + 2 = 3, F_5 = 2 + 3 = 5,$
$F_6 = 3 + 5 = 8, F_7 = 5 + 8 = 13, F_8 = 8 + 13 = 21, F_9 = 13 + 21 = 34,$
$F_{10} = 21 + 34 = 55.$ ●

The Lucas Numbers   The *Lucas numbers* L_n have the initial values

$$L_0 = 2, L_1 = 1$$

and the recursion

$$L_n = L_{n-1} + L_{n-2}$$

if $n \geq 2$.

EXAMPLE 2   Compute the Lucas numbers L_n for $0 \leq n \leq 10$.

Solution $L_0 = 2, L_1 = 1, L_2 = 2 + 1 = 3, L_3 = 1 + 3 = 4, L_4 = 3 + 4 = 7, L_5 = 4 + 7 = 11,$
$L_6 = 7 + 11 = 18, L_7 = 11 + 18 = 29, L_8 = 18 + 29 = 47, L_9 = 29 + 47 = 76,$
$L_{10} = 47 + 76 = 123.$ ●

Notice that the recursion that defines the Lucas numbers is the same as the recursion that defines the Fibonacci numbers. Only the initial values are different.

DEFINITION 1.5.2   The *Tribonacci numbers* T_n have the initial values

$$T_0 = 0, T_1 = 0, T_2 = 1$$

and the recursion

$$T_n = T_{n-1} + T_{n-2} + T_{n-3}$$

if $n \geq 3$.

EXAMPLE 3   Compute the Tribonacci numbers T_n for $0 \leq n \leq 10$.

Solution $T_0 = 0, T_1 = 0, T_2 = 1, T_3 = 0 + 0 + 1 = 1, T_4 = 0 + 1 + 1 = 2, T_5 = 1 + 1 + 2 = 4,$
$T_6 = 1 + 2 + 4 = 7, T_7 = 2 + 4 + 7 = 13, T_8 = 4 + 7 + 13 = 24, T_9 = 7 + 13 + 24 = 44,$
$T_{10} = 13 + 24 + 44 = 81.$ ●

The *Bernoulli numbers* were introduced by Jakob Bernoulli in the *Ars Conjectandi*.

> **DEFINITION 1.5.3**   The *Bernoulli numbers* B_n are defined by the initial value $B_0 = 1$ and the recursion
>
> $$\binom{n+1}{0} B_0 + \binom{n+1}{1} B_1 + \ldots + \binom{n+1}{n} B_n = 0$$

This recursion has a simple expression in terms of the Pascal Triangle (Figure 1.2).

$$
\begin{aligned}
B_0 &= 1 \\
B_0 + 2B_1 &= 0 \\
B_0 + 3B_1 + 3B_2 &= 0 \\
B_0 + 4B_1 + 6B_2 + 4B_3 &= 0 \\
B_0 + 5B_1 + 10B_2 + 10B_3 + 5B_4 &= 0
\end{aligned}
$$

(handwritten) $B_0 + 6B_1 + 15B_2 + 20B_3 + 15B_4 + 6B_5 = 0$

Figure 1.20   The Recursion for the Bernoulli Numbers

(handwritten) $1 - 3 + \frac{15}{6} + 0 \quad - \quad \frac{1}{2} + 6B_5 = 0$

EXAMPLE 4   Find the Bernoulli numbers B_1, B_2, B_3, and B_4.

Solution   We will refer to the recursion in Figure 1.20. First,

$$B_0 + 2B_1 = 1 + 2B_1 = 0,$$

so $B_1 = -\frac{1}{2}$. Next,

$$B_0 + 3B_1 + 3B_2 = 1 + 3\left(-\frac{1}{2}\right) + 3B_2 = 0,$$

so $B_2 = \frac{1}{6}$. Next,

$$B_0 + 4B_1 + 6B_2 + 4B_3 = 1 + 4\left(-\frac{1}{2}\right) + 6\left(\frac{1}{6}\right) + 4B_3 = 0,$$

so $B_3 = 0$. Finally,

$$B_0 + 5B_1 + 10B_2 + 10B_3 + 5B_4 = 1 + 5\left(-\frac{1}{2}\right) + 10\left(\frac{1}{6}\right) + 10(0) + 5B_4 = 0,$$

so $B_4 = -\frac{1}{30}$.   ●

The *Collatz sequence* is named for the German mathematician Lothar Collatz (1910–1990). It is also called the $3X + 1$ *sequence*.

The Collatz sequence is defined recursively as follows. First, choose a positive integer $C[0]$ and let this be the initial value of the sequence. Then, if $C[n]$ has been determined, define $C[n + 1] = 3C[n] + 1$ if $C[n]$ is odd, and $C[n + 1] = C[n]/2$ if $C[n]$ is even. Of course, there will be a different sequence for each initial value $C[0]$. We will stop writing terms for the sequence if it ever reaches 1, for then the sequence will endlessly go through the cycle 1, 4, 2, 1, 4, 2, 1, 4, 2, 1, ...

EXAMPLE 5   Find the terms of the Collatz sequence if $C[0] = 13$.

Solution $C[0] = 13$, so $C[1] = 3 \cdot 13 + 1 = 40$, $C[2] = 40/2 = 20$, $C[3] = 20/2 = 10$, $C[4] = 10/2 = 5$, $C[5] = 3 \cdot 5 + 1 = 16$, $C[6] = 16/2 = 8$, $C[7] = 8/2 = 4$, $C[8] = 4/2 = 2$, $C[9] = 2/2 = 1$.   ●

We see that the Collatz sequence reaches 1 after a finite number of steps if $C[0] = 13$. Does the Collatz sequence always reach 1 after a finite number of steps? This is a famous unsolved problem.

The Collatz Problem   Let $C[0]$ be any positive integer. Does the Collatz sequence with initial value $C[0]$ reach 1 after a finite number of steps?

It is known that the Collatz sequence does reach 1 after a finite number of steps in the range $1 \leq C[0] \leq 10^9$, and for many other values of $C[0]$. It is thought that more computer time has been used to study the Collatz Problem than any other mathematical problem.

We can define sequences like the Collatz sequence using any odd integer other than 3 in the definition of the recursion. We might think that these sequences would behave just like the Collatz sequence. They do not.

EXAMPLE 6   Define the $5X + 1$ sequence, and find an initial value $C[0]$ for which this sequence never reaches 1.

Solution   Define the $5X + 1$ sequence by choosing a positive integer $C[0]$ as the initial value, with the recursion: $C[n + 1] = 5C[n] + 1$ if $C[n]$ is odd, and $C[n + 1] = C[n]/2$ if $C[n]$ is even.

Choose $C[0] = 13$, and compute the terms of the sequence. $C[1] = 5 \cdot 13 + 1 = 66$, $C[2] = 66/2 = 33$, $C[3] = 5 \cdot 33 + 1 = 166$, $C[4] = 166/2 = 83$, $C[5] = 5 \cdot 83 + 1 = 416$, $C[6] = 416/2 = 208$, $C[7] = 208/2 = 104$, $C[8] = 104/2 = 52$, $C[9] = 52/2 = 26$, $C[10] = 26/2 = 13$.

After ten terms the sequence has returned to the initial value. This cycle will recur over and over, and the sequence will never reach 1.   ●

It is possible to give recursive definitions of graphs as well as numbers. For example, we shall define the *Fibonacci tree*.

DEFINITION 1.5.4   The *Fibonacci tree with n levels T* is defined as follows:
  i. The 0-th level of T consists of a single node, the root of T.
  ii. Join two nodes to the root of T to get the first level of T (one edge to the left and one edge to the right).
  iii. Thereafter, join two nodes to each node at the end of an edge to the left, and alternately join one or two nodes to each node at the end of an edge to the right.

We give the first four levels of the Fibonacci tree in Figure 1.21.

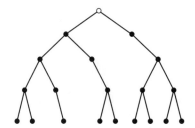

Figure 1.21   The Fibonacci Tree

DEFINITION 1.5.5   If T is a tree, the *level sequence* of T is the sequence

$$a_0, a_1, \ldots, a_n, \ldots$$

where a_n is the number of nodes v of T such that there is a path of length exactly n from v to the root of T.

For the Fibonacci tree $a_0 = 1$, and from the definition above we see that

$$a_n = a_{n-1} + a_{n-2}$$

if $n \geq 2$. It follows that

$$a_n = F_{n+1}$$

if $n \geq 0$.

In other words, *the level sequence of the Fibonacci tree is the Fibonacci sequence.*

So far we have given examples of recursive sequences which are generated by a certain rule. Suppose that we are given a sequence which may or may not be recursive. How can we decide whether the sequence is recursive, and if so, which rule generates the sequence?

The first thing to do when one encounters an unknown sequence is to construct the *difference table*. The first row of the difference table is the given sequence. The second row is the sequence of differences of successive terms of this sequence. The third row is the differences of the differences, and so on. More formally, the *difference operator* is defined by the expression

$$\Delta a_n = a_{n+1} - a_n$$

and the *k-th difference operator* is

$$\Delta^k a_n = \Delta^{k-1} a_{n+1} - \Delta^{k-1} a_n$$

If any row in the difference table becomes constant (consists of the same number repeated over and over) we can reconstruct the sequence. Figure 1.22 gives the sequence of squares and its first and second differences. We could find the next square (if we did not know it) by starting in the third row, which consists of 2s. Add 2 to the last element in the second row, then add this element to the last element in the first row.

$$
\begin{array}{ccccccc}
 & 1 & 4 & 9 & 16 & 25 & 36 & \boxed{49} \\
\Delta & & 3 & 5 & 7 & 9 & 11 & \boxed{13} \\
\Delta^2 & & & 2 & 2 & 2 & 2 & \boxed{2}
\end{array}
$$

Figure 1.22   Using the Difference Table

Figure 1.23 shows the sequences of triangular, square, and pentagonal numbers. Notice that the constant entries in the second rows of these three difference tables are the integers 1, 2, and 3, respectively.

$$
\begin{array}{ccccccc}
 & 1 & 3 & 6 & 10 & 15 & 21 \\
\Delta & & 2 & 3 & 4 & 5 & 6 \\
\Delta^2 & & & 1 & 1 & 1 & 1
\end{array}
$$

$$
\begin{array}{ccccccc}
 & 1 & 4 & 9 & 16 & 25 & 36 \\
\Delta & & 3 & 5 & 7 & 9 & 11 \\
\Delta^2 & & & 2 & 2 & 2 & 2
\end{array}
$$

$$
\begin{array}{ccccccc}
 & 1 & 5 & 12 & 22 & 35 & 51 \\
\Delta & & 4 & 7 & 10 & 13 & 16 \\
\Delta^2 & & & 3 & 3 & 3 & 3
\end{array}
$$

Figure 1.23   The Triangular, Square, and Pentagonal Numbers

DEFINITION 1.5.6   A sequence of numbers

$$a_0, a_1, a_2, \ldots, a_n, \ldots$$

is a *linear recursive* sequence if there is an integer r, called the *degree* of the sequence, and constants $c_1, c_2, \ldots, c_r$ such that

$$a_n = c_1 a_{n-1} + c_2 a_{n-2} + \ldots + c_r a_{n-r}$$

if $n \geq r$.

For example, the Fibonacci numbers and the Lucas numbers are linear recursive sequences of degree 2, and the Tribonacci numbers are a linear recursive sequence of degree 3.

Recall that we defined binary addition in the previous section. Binary addition is just like ordinary addition, except that $1 + 1 = 0$. Suppose that we consider the Fibonacci sequence as a binary sequence. The initial values are $F_0 = 0$ and $F_1 = 1$, and the next values are

$$F_2 = 0 + 1 = 1, \ F_3 = 1 + 1 = 0, \ F_4 = 1 + 0 = 1, \ F_5 = 0 + 1 = 1, \ldots.$$

Notice what has happened: the binary sequence is $0, 1, 1, 0, 1, 1, \ldots$. That is, the binary sequence is periodic of period 3. It turns out that binary linear recursive sequences are always periodic.

EXAMPLE 7   Find the period of the binary recursive sequence with initial values $a_0 = 0, a_1 = 0, a_2 = 1$ and recursion

$$a_n = a_{n-2} + a_{n-3}$$

if $n \geq 3$.

Solution

$$a_3 = 0 + 0 = 0, \ a_4 = 1 + 0 = 1, \ a_5 = 0 + 1 = 1, \ a_6 = 1 + 0 = 1,$$
$$a_7 = 1 + 1 = 0, \ a_8 = 1 + 1 = 0, \ a_9 = 0 + 1 = 1.$$

Notice that the pattern of initial values $0, 0, 1$ has repeated. The period of the sequence is 7: the pattern $0, 0, 1, 0, 1, 1, 1$ repeats over and over.   ●

APPLICATION   <u>*Shift Registers*</u>

The binary recursive sequences which we have just described are very useful in computer science. In designing a computer or a communications device, one often wants to generate a binary string which will have a certain effect. First, however, it is often necessary to generate a "random" binary string. One then modifies this string until it has the desired properties.

A computer circuit element called a *shift register* will generate any given binary linear recursive sequence. The shift register pictured in Figure 1.24 will generate the binary recursion in the previous example. Before explaining this, we will mention a result about the period of a binary recursion.

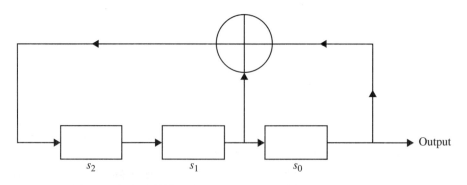

Figure 1.24   A Shift Register

THEOREM 1.5.1 The period of a binary linear recursive sequence of degree d is a divisor of $2^d - 1$.

In particular, the period cannot be larger than $2^d - 1$. In our two examples, the periods were $3 = 2^2 - 1$, and $7 = 2^3 - 1$.

There are numerous recursions whose period has length $2^d - 1$. For such recursions, the length of the period grows rapidly with d. Within a period, such a sequence satisfies many tests for randomness. Thus a shift register with a degree of moderate size, say $d = 10$, will generate an essentially random binary string of considerable length.

A shift register is composed of two types of circuit elements. One is a storage element ⊔ in which binary bits 0 and 1 can be stored. The other is a binary adder ⊕ with two inputs (for the two bits to be added) and one output (for the binary sum). By combining these two circuit elements in various ways, a shift register can be constructed whose output is any desired binary recursion.

The shift register in Figure 1.24 works as follows. The three storage elements s_0, s_1, s_2 are filled with the first three elements of the sequence, the binary bits $a_0 = 0$, $a_1 = 0$, $a_2 = 1$. In the first shift, $a_0 = 0$ becomes the output, $a_1 = 0$ is shifted to s_0, $a_2 = 1$ is shifted to s_1, and the element generated by the recursion $a_3 = a_1 + a_0 = 0 + 0 = 0$ becomes the new entry of s_2. Later shifts occur in a similar way. The result is as follows:

	Input			*Output*
Initial State	$a_2 = 1$	$a_1 = 0$	$a_0 = 0$	
First Shift	$a_3 = 0$	$a_2 = 1$	$a_1 = 0$	0
Second Shift	$a_4 = 1$	$a_3 = 0$	$a_2 = 1$	0
Third Shift	$a_5 = 1$	$a_4 = 1$	$a_3 = 0$	1
Fourth Shift	$a_6 = 1$	$a_5 = 1$	$a_4 = 1$	0
Fifth Shift	$a_7 = 0$	$a_6 = 1$	$a_5 = 1$	1
Sixth Shift	$a_8 = 0$	$a_7 = 0$	$a_6 = 1$	1
Seventh Shift	$a_9 = 1$	$a_8 = 0$	$a_7 = 0$	1

You can see that the output, given in the right-hand column, is the recursive sequence of the previous example.

The sequence described here has interesting properties. Since the degree of the recursion is small, the period is short. If we chose a recursion of a much larger degree, the output in the right-hand column would appear to be essentially random for a long time.

EXERCISES

1. Compute the Fibonacci numbers F_n for $0 \le n \le 20$.

2. Compute the Lucas numbers L_n for $0 \le n \le 20$.

3. Compute the Tribonacci numbers $T(n)$ for $0 \le n \le 20$.

Find the following Bernoulli numbers:

4. B_5.

5. B_6.

6. B_7.

7. B_8.

Find all the terms of the Collatz sequence for the following values of $C[0]$:

8. $C[0] = 7$. *Hint:* you should compute at least 16 terms.

9. $C[0] = 19$. *Hint:* you should compute at least 20 terms.

10. $C[0] = 39$. *Hint:* you should compute at least 34 terms.

11. $C[0] = 123$. *Hint:* you should compute at least 46 terms.

12. $C[0] = 159$. *Hint:* you should compute at least 54 terms.

13. $C[0] = 251$. *Hint:* you should compute at least 65 terms.

14. $C[0] = 263$. *Hint:* you should compute at least 78 terms.

15. $C[0] = 103$. *Hint:* you should compute at least 87 terms.

16. $C[0] = 91$. *Hint:* you should compute at least 92 terms.

17. $C[0] = 27$. *Hint:* you should compute at least 111 terms.

18. Compute the first ten terms of the $5X + 1$ sequence with initial value $C[0] = 17$. Does this sequence ever reach 1?

19. Sketch the Fibonacci tree with seven levels.

20. In how many ways can a man climb a flight of n stairs if he climbs one or two stairs at a time?

21. In how many ways can a man climb a flight of n stairs if he climbs one, two, or three stairs at a time?

22. How many pairs of rabbits can be bred from one pair of rabbits in a year if each pair of rabbits breeds another pair each month, and begins breeding in the second month after birth?

Use the difference table to find the next term of each of the following sequences.

23. $2, 5, 10, 17, 26, 37, 50$

24. $3, 7, 13, 21, 31, 43, 57$

25. $1, 4, 11, 22, 37, 56, 79$

26. $1, 8, 27, 64, 125, 216, 343$

27. $1, 7, 25, 61, 121, 211, 337$

For each of the following binary recursions, find as many terms as the period.

28. $a_0 = 1, a_1 = 1, a_n = a_{n-1} + a_{n-2}$, if $n \geq 2$.

29. $a_0 = 0, a_1 = 0, a_2 = 1, a_n = a_{n-1} + a_{n-3}$, if $n \geq 3$.

Compare the terms in a period of this sequence with the same terms in the last example in this section.

30. $a_0 = 0, a_1 = 0, a_2 = 0, a_3 = 1, a_n = a_{n-3} + a_{n-4}$, if $n \geq 4$.

ADVANCED EXERCISE

There is a tree associated with the Collatz Problem, the *Collatz tree*, pictured in Figure 1.25. The Collatz tree is a binary tree, defined recursively as follows.

  (i) Begin with a single node labelled with the integer 1.

  (ii) Suppose that a node of the tree is given, labelled with the integer n. Join another node to this one and label the new node with the integer $2n$. If there is an odd integer k such that $n = 3k + 1$, join another node to the node labelled n, and let the label of this new node be k.

  The resulting graph has a trivial cycle $1 \rightarrow 4 \rightarrow 2 \rightarrow 1$, sketched with a dotted line. Except for this, the graph is a tree. Another way to state the Collatz Problem is this:

  Does every positive integer occur as the label of some node in the Collatz tree?

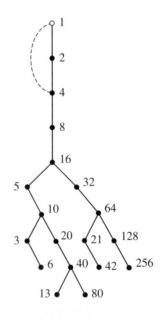

Figure 1.25
The Collatz Tree

Recall that we defined the level sequence of a tree in this section. Level sequences of trees often have interesting recursive properties.

1. Draw sixteen levels of the Collatz tree, and give the first sixteen terms of the level sequence of this tree.

Unlike the level sequence of the Fibonacci tree, which is essentially the Fibonacci sequence, very little is known about the level sequence of the Collatz tree. The following problem is unsolved:

Is the level sequence of the Collatz tree a linear recursive sequence?

COMPUTER EXERCISES

1. Write a program to compute the Fibonacci numbers (some computer algebra systems have programs to do this). Print the Fibonacci numbers F_n for $0 \le n \le 119$ in two columns. In the first column, list the first 60 Fibonacci numbers F_n for $0 \le n \le 59$. In the second column, list the next 60 Fibonacci numbers F_n for $60 \le n \le 119$. Compare the 1s digits in the first and second columns. What do you notice about them?

2. Write a program to compute the Lucas numbers. Print the Lucas numbers L_n for $0 \le n \le 119$ in two columns. In the first column, list the Lucas numbers L_n for $0 \le n \le 59$. In the second column, list the Lucas numbers L_n for $60 \le n \le 119$. Compare the 1s digits in the first and second columns. What do you notice about them?

3. In the previous two problems, you noticed that the 1s digits of the Fibonacci numbers and the Lucas numbers are *periodic*. The period of the 1s digits of the Fibonacci numbers is 60, and the period of the 1s digits of the Lucas numbers is 12.

The 1s digits of the Tribonacci numbers are also periodic. What is the period? The problem is to list these numbers and to find the first time that the 1s digits repeat their initial pattern 0, 0, 1.

4. The *length* of the Collatz sequence is the smallest integer N such that $C[N] = 1$ (if such an integer exists). Write a program whose output is the length of the Collatz sequence for a given initial value $C[0]$, and print a list of initial values and lengths in the range $1 \le C[0] \le 1000$.

A *Collatz k-tuple* is a sequence of k consecutive initial values $C[0]$ for which the Collatz sequence has the same length.

Use the table that you made in Problem 4 to check that:

5. The Collatz sequence has length 27 if $C[0] = 65,66,67$.

6. The Collatz sequence has length 25 if $C[0] = 98,99,100,101,102$.

7. The Collatz sequence has length 117 if $C[0] = 290,291,292,293,294$.

8. The Collatz sequence has length 120 if $C[0] = 386,387,388,389,390,391$.

9. The Collatz sequence has length 41 if $C[0] = 840,841,842,843,844,845$.

10. Use the program you wrote for Problem 4 to check that the Collatz sequence has length 224 for all initial values in the interval $4,585,418 \le C[0] \le 4,585,471$, so that these integers form a Collatz 54-tuple.

11. Use the program you wrote for Problem 4 to check that the Collatz sequence has length 247 for all initial values in the interval $136,696,632 \le C[0] \le 136,696,751$, so that these integers form a Collatz 120-tuple.

12. Use the program you wrote for Problem 4 to check that the Collatz sequence has length 214 for all initial values in the interval $1,202,088,120 \le C[0] \le 1,202,088,320$, so that these integers form a Collatz 201-tuple.

1.6   CHAPTER SUMMARY AND SUPPLEMENTARY EXERCISES

He would answer to "Hi!" or to any loud cry,

Such as "Fry me!" or "Fritter my wig!"

To "What-you-may-call-um!" or "What-was-his-name!"

But especially "Thing-um-a-jig!"

"His form is ungainly—his intellect small—"

(So the Bellman would often remark)

"But his courage is perfect! And that, after all,

Is the thing that one needs with a Snark."

Chapter Summary

Section 1.1

We began with some remarks about Pascal's Calculating Machine, the first "computer." We defined the Pascal Triangle, and showed how to compute its elements, the binomial coefficients, either directly, using recursion, or using the Pascal Formula. We discussed the full binary tree.

Section 1.2

We described the Principle of Induction and gave several examples of proofs by induction. We proved Pascal's Formula by induction. We discussed Lucas' Tower of Hanoi puzzle.

Section 1.3

In this section we discussed sets and their subsets, and the relation between subsets of a set and binary vectors and strings. We described the full binary tree, gave an example of an algorithm, the Standard Gray Code, and used it to solve the Knapsack Problem.

Section 1.4

In this section we discussed set operations and identities, and proof methods for identities. We gave an example of an error-correcting code.

Section 1.5

In this section we discussed recursive sequences of numbers, and also gave examples of recursive definitions of trees. We gave several examples of recursive sequences, including the Fibonacci sequence. We showed how to use the difference table to analyze a recursive sequence, and how shift registers can be used to generate binary strings.

SUPPLEMENTARY EXERCISES

Draw the Pascal Triangle through the fourteenth row, and use it to evaluate the following binomial coefficients. Then evaluate each binomial coefficient using Pascal's Formula.

1. $\binom{6}{3}$ **2.** $\binom{8}{3}$ **3.** $\binom{8}{4}$

4. $\binom{12}{5}$ **5.** $\binom{12}{7}$ **6.** $\binom{14}{7}$

Prove by induction that

7. $\sum_{k=1}^{n}(5k-4) = \dfrac{n(5n-3)}{2}$

8. $\dfrac{\sum_{k=1}^{n}k^2}{\sum_{k=1}^{n}k} = \dfrac{2n+1}{3}$

9. $\sum_{k=1}^{n}\dfrac{1}{4k^2-1} = \dfrac{n+1}{2n+3}$

Use set-builder notation to describe the following sets.

10. The set of positive integers which give a remainder of 1 when divided by 8.

11. The set of positive integers which give a remainder of 7 when divided by 8.

Given a set X and subsets A and B of X, find $A \times B$, $A \cup B$, $A \cap B$, $A - B$, A^c and B^c.

12. $X = \{1, 2, 3, 4, 5, 6, 7, 8\}$, $A = \{1, 3, 5, 7\}$, $B = \{2, 3, 5, 8\}$

13. $X = \{1, 2, 3, 4, 5, 6, 7, 8, 9\}$, $A = \{2, 6, 7, 8, 9\}$, $B = \{1, 4, 7, 9\}$

14. $X = \{1, 2, 3, 4, 5, 6, 7, 8, 9, 10\}$, $A = \{1, 4, 5, 6, 10\}$, $B = \{1, 2, 4, 6, 8, 9\}$

Represent the following subsets as paths in the full binary tree.

15. $B = \{1, 3\} \subset A = \{1, 2, 3\}$

16. $B = \{2, 3, 4\} \subset A = \{1, 2, 3, 4\}$

17. $B = \{1, 4, 5\} \subset A = \{1, 2, 3, 4, 5\}$

18. Using the Standard Gray Code, list the subsets of $\{1, 2, 3, 4, 5\}$.

19. Using the Standard Gray Code, list the subsets of $\{1, 2, 3, 4, 5, 6\}$.

Prove each of the following set identities by using Venn diagrams, and then by an argument with elements.

20. $(A \cup B) \cap (A \cup B^c) = A$

21. $(A \cap B) \cup (A \cap B^c) = A$

22. $(B \cap C) - (A \cap B) = (B \cap C) - A$

If A_1, A_2, and A_3 are sets, use the Principle of Inclusion-Exclusion to find the number of elements in $A_1 \cup A_2 \cup A_3$.

23. $|A_1| = |A_2| = |A_3| = 4$, $|A_1 \cap A_2| = |A_2 \cap A_3| = 0$, $|A_1 \cap A_3| = 3$, $|A_1 \cap A_2 \cap A_3| = 0$

24. $|A_1| = |A_2| = |A_3| = 4$, $|A_1 \cap A_2| = |A_1 \cap A_3| = 1$, $|A_2 \cap A_3| = 2$, $|A_1 \cap A_2 \cap A_3| = 1$

25. $|A_1| = |A_3| = 6$, $|A_2| = 5$, $|A_1 \cap A_2| = |A_1 \cap A_3| = |A_2 \cap A_3| = |A_1 \cap A_2 \cap A_3| = 4$

In Figure 1.17 of Section 1.4, we listed the elements of the Hamming Code H_8, viewed as a design. For each of the following pairs of sets in this design, verify that the symmetric difference of the two sets is again in the design.

26. $\{1, 2, 3, 5\}$ and $\{1, 4, 5, 7\}$

27. $\{1, 3, 7, 8\}$ and $\{1, 2, 6, 7\}$

28. $\{1, 5, 6, 8\}$ and $\{3, 4, 5, 8\}$

The following sequences have difference tables with rows which eventually become constant. Use the difference table to find the next term in each sequence.

29. $0, 3, 8, 15, 24, 35, 48$

30. $-1, -2, -1, 2, 7, 14, 23$

31. $-1, -2, 3, 20, 55, 114, 203$

32. $1, 2, 17, 82, 257, 626, 1297$

ADVANCED EXERCISES

Lucas noticed that the Fibonacci numbers and the binomial coefficients are related by the formula

$$F_{n+1} = \binom{n}{0} + \binom{n-1}{1} + \binom{n-2}{2} + \cdots$$

(the sum continues as long as the lower index is less than or equal to the upper index). The formula is known as *Lucas' Formula.*

1. Use Lucas' Formula to evaluate the Fibonacci numbers F_n for $1 \le n \le 15$.

2. Draw the Pascal Triangle as it appears in Figure 1.2 of Section 1.1, and show how Lucas' Formula can be obtained by drawing parallel bands in the triangle, and summing the binomial coefficients in these bands.

The Lucas numbers can be obtained as the row sums of the triangle in Figure 1.26.

3. Give a precise description of the rows of this triangle, and prove that the row sums are the Lucas numbers $L_1 = 1$, $L_2 = 3$, $L_3 = 4$, $L_4 = 7$, $L_5 = 11$, $L_6 = 18, \ldots$.

$$
\begin{array}{ccccccccccc}
 & & & & & 1 & & & & & \\
 & & & & 1 & & 2 & & & & \\
 & & & 1 & & 1 & & 2 & & & \\
 & & 1 & & 1 & & 3 & & 2 & & \\
 & 1 & & 1 & & 4 & & 3 & & 2 & \\
1 & & 1 & & 5 & & 4 & & 5 & & 2
\end{array}
$$

Figure 1.26
A Triangle for the Lucas Numbers

COMPUTER EXERCISES

The following exercises demonstrate the growth rate of recursive sequences.

1. Write a program to generate the Fibonacci numbers F_n, and print F_{100}, F_{200}, F_{300}, F_{400}, F_{500}, F_{600}, F_{700}, F_{800}, F_{900}, and F_{1000}.

2. Write a program to generate the Lucas numbers L_n, and print L_{100}, L_{200}, L_{300}, L_{400}, L_{500}, L_{600}, L_{700}, L_{800}, L_{900}, and L_{1000}.

3. Write a program to generate the Tribonacci numbers T_n, and print T_{100}, T_{200}, T_{300}, T_{400}, T_{500}, T_{600}, T_{700}, T_{800}, T_{900}, and T_{1000}.

4. Write a program to generate the Bernoulli numbers B_n using the recursion, and print these numbers for $1 \le n \le 20$.

5. In the computer exercises in Section 1.1, we defined the Catalan numbers C_n by the formula

$$
C_n = \frac{1}{n+1} \binom{2n}{n}
$$

The Catalan numbers can also be defined by the initial value $C_0 = 1$ and the recursion

$$
C_{n+1} = C_0 C_n + C_1 C_{n-1} + \ldots + C_n C_0
$$

Write a program to generate the Catalan numbers by this recursion, and print them for $0 \le n \le 25$.

2 Integers, Remainders, and the Golden Ratio

2.1 THE INTEGERS

The last of the crew needs especial remark,

Though he looked an incredible dunce:

He had just one idea—but, that one being "Snark,"

The good Bellman engaged him at once.

About 300 B.C. Euclid's *Elements* was published in Alexandria, Egypt. The *Elements* consists of thirteen books, of which Books VII, VIII, and IX are concerned with the properties of the integers. Euclid did not consider 0, or negative numbers. At the beginning of Book VII Euclid made a number of definitions, including the definition of *prime numbers*.

DEFINITION 2.1.1 A positive integer p is *prime* if $p > 1$ and p is only divisible by 1 and p.

Every integer can be factored into primes. We shall prove this by induction, but it is convenient to use the *Strong Form* of the Principle of Induction. This form of the Principle of Induction is also called *Complete Induction*.

The Strong Form of the Principle of Induction Suppose that $P(n)$ is a statement for each positive integer $n = 1, 2, 3, \ldots$.

Suppose that

(i) $P(1)$ is true, and

(ii) For each $n = 1, 2, 3, \ldots$, if $P(k)$ is true for all $1 \le k < n$, then $P(n)$ is true.

Then $P(n)$ is true for each $n = 1, 2, 3, \ldots$. ●

THEOREM 2.1.1 Every positive integer $n > 1$ factors into primes.

Proof If n is prime, the theorem is true. If $n > 1$ is not prime, then $n = rs$, where $1 < r < n$ and $1 < s < n$. By the Strong Form of the Principle of Induction, both r and s factor into primes. Therefore n also factors into primes. ●

The Strong Form of the Principle of Induction is closely related to the *well-ordering* property of the positive integers, which we shall use to give another proof of Theorem 1.

The Well-Ordering of the Positive Integers Every non-empty set of positive integers has a least element.

Another Proof of Theorem 1 Let S be the set of positive integers $n > 1$ which do not factor into primes. Since the set of positive integers is well-ordered, S must have a least element n if S is not empty. By the definition of S, n does not factor into primes, and so in particular n is not prime. Since n is not prime it must factor into $n = rs$, with $1 < r < n$ and $1 < s < n$. Since n is the smallest element of S, r and s are not in S, so they each factor into primes. But then n factors into primes, so n cannot belong to S, and therefore S is empty. This proves Theorem 1. ●

The well-ordering property of the positive integers is sometimes called the *Minimum Principle*. The *Maximum Principle* asserts that every finite set of positive integers has a maximum element. These two principles imply that the numbers described in the next definition exist.

> **DEFINITION 2.1.2**
> **i.** If m and n are positive integers, the *greatest common divisor* g.c.d.(m, n) of m and n is the largest positive integer which divides both m and n.
> **ii.** If m and n are positive integers, the *least common multiple* l.c.m.(m, n) of m and n is the smallest positive integer which is divisible by both m and n.

The *Euclidean Algorithm* is an algorithm which computes the greatest common divisor of two positive integers. Euclid proved this in the first two Propositions of Book VII of *Elements*, using the Division Algorithm in the proofs.

THE DIVISION ALGORITHM   Let m and n be positive integers. There are non-negative integers q and r, where $0 \leq r < n$, such that

$$m = qn + r$$

The integer r is the *remainder* obtained when m is divided by n.

THE EUCLIDEAN ALGORITHM   Let m and n be positive integers, with $n \leq m$. Let

$$m = q_1 n + r_1$$
$$n = q_2 r_1 + r_2$$
$$r_1 = q_3 r_2 + r_3$$
$$\ldots$$
$$r_{N-2} = q_{N-1} r_{N-1} + r_N$$
$$r_{N-1} = q_N r_N$$

be the result of iterating the Division Algorithm, where r_N is the last non-zero remainder. Then

$$g.c.d.(m, n) = r_N.$$

By writing $r_N = r_{N-2} - q_{N-1} r_{N-1}$ and successively substituting in the previous equations we can write

$$g.c.d.(m, n) = Am + Bn$$

where A and B are integers.

The greatest common divisor and the least common multiple of two integers are related by the formula

$$l.c.m.(m, n) = \frac{m \cdot n}{g.c.d.(m, n)}$$

Thus the Euclidean Algorithm also gives an algorithmic computation of the least common multiple of two integers.

EXAMPLE 1   Use the Euclidean Algorithm to find the greatest common divisor of 8 and 13, and write $g.c.d.(13, 8)$ in the form $A \cdot 13 + B \cdot 8$.

Solution

$$13 = 1 \cdot 8 + 5$$
$$8 = 1 \cdot 5 + 3$$
$$5 = 1 \cdot 3 + 2$$
$$3 = 1 \cdot 2 + 1$$
$$2 = 2 \cdot 1$$

We see that $g.c.d.(13, 8) = 1$, and substituting back we find that $1 = g.c.d.(13, 8) = (-3) \cdot 13 + 5 \cdot 8$. ●

The following theorem is Proposition 30 of Book VII of the *Elements*.

THEOREM 2.1.2   Let m and n be positive integers, and p a prime. If p divides mn then p divides m or p divides n.

Proof   Assume that p divides mn and that p does not divide m. Since p is prime, this implies that $1 = g.c.d.(m, p) = Am + Bp$. If we multiply this equation by n we get $n = Amn + Bpn$. Since p divides both terms on the right-hand side of this equation it must divide the left-hand side, that is p divides n. ●

The theorem which we have just proved is the fundamental fact about the divisibility of the integers. It implies the *unique factorization* property of the integers.

THEOREM 2.1.3   If n is a positive integer, then the numbers of primes in any two prime factorizations of n are the same, and the primes which occur are the same except possibly for the order in which they occur.

Proof   Suppose that $n = p_1 \cdot p_2 \cdot \ldots \cdot p_s = q_1 \cdot q_2 \cdot \ldots \cdot q_t$, where the integers p_i and q_j are primes. We must prove that $s = t$, and also that the primes p_i are the same as the primes q_j except perhaps for the order in which they occur.

Clearly p_1 divides $n = q_1 \cdot (q_2 \cdot \ldots \cdot q_t)$. By the previous theorem, either p_1 divides q_1, so that $p_1 = q_1$ or, if not, p_1 divides $(q_2 \cdot \ldots \cdot q_t)$. In the latter case p_1 will either divide q_2, so that $p_1 = q_2$, or p_1 will divide $(q_3 \cdot \ldots \cdot q_t)$. Continuing this argument, we see that p_1 must be equal to one of the primes q_j.

Renumbering the primes q_j if necessary, we can assume that $p_1 = q_1$. Then $p_2 \cdot \ldots \cdot p_s = q_2 \cdot \ldots \cdot q_t$. Repeating the argument above, we see that p_2 is equal to one of the primes q_j, and after renumbering the q_j if necessary, we can assume that $p_2 = q_2$. Using the argument repeatedly, we see that $p_3 = q_3, \ldots, p_s = q_s$. It follows in particular that $s \leq t$.

Now repeat the same argument with the roles of the p_i and q_j interchanged, and conclude that $t \leq s$. Then $s = t$, so the number of primes p_i and q_j is the same, and the p_i coincide with the q_j except perhaps for the order in which they occur. This proves the theorem. ●

For example,

$$196{,}560 = 2^4 \cdot 3^3 \cdot 5 \cdot 7 \cdot 13$$

However we factor 196,560 into primes, we will find that the same primes occur, each prime occurs the same number of times, and if we arrange the prime factors in increasing order we will always obtain the factorization above.

The uniqueness of the factorization of a positive integer into primes is of interest for many reasons, one of which is the following.

A positive integer n is a code for its prime factors.

There are many applications of this principle. *Public Key Encryption* depends on it, which we shall describe at the end of Section 2.3.

If we are given a positive integer n, we can ask whether n is prime, and we can also ask for a factorization of n into primes. These problems are different, and separate algorithms have been developed to solve them.

The Primality Problem   Given a positive integer n, when is n prime?

For example, on a computer we can verify that

$$2^{61} - 1 = 2,305,843,009,213,693,951$$

is prime.

The Factorization Problem   Given a positive integer n, how can n be factored into primes?

For example, on a computer we can verify that

$$2^{67} - 1 = 147,573,952,589,676,412,927 = (193,707,721) \cdot (761,838,257,287)$$

It is believed that the Primality Problem is simpler than the Factorization Problem. There are algorithms which will check whether a 100-digit number is prime in a few seconds, while algorithms which factor 100-digit numbers, although they do terminate, usually take longer. The problem of factoring 200-digit numbers seems intractible at present.

Pierre de Fermat (1601–1665) was a French mathematician whose exchange of letters with Pascal in 1654 is often viewed as the beginning of Probability Theory. Fermat is best known for the Fermat Problem, which asserts that the equation

$$x^n + y^n = z^n$$

has no solution in non-zero integers if $n > 2$. This was proved in 1994 by Andrew Wiles.

It was Fermat who described the first interesting algorithm for factoring a positive integer.

The Fermat Factorization Algorithm   Given a positive odd integer N which is to be factored, the factorization takes the form

$$N = x^2 - y^2 = (x + y) \cdot (x - y)$$

which we can also write as

$$y^2 = x^2 - N.$$

Choose

$$x = n, n + 1, n + 2, \ldots, (N + 1)/2$$

where $n \geq \sqrt{N}$ is as small as possible, and stop when $y^2 = x^2 - N$ is the square of an integer.

The following example is due to Fermat.

EXAMPLE 2   Apply the Fermat Factorization Algorithm to $N = 2,027,651,281$.

Solution   Since $\sqrt{2,027,651,281} = 45,029.449\ldots$ we begin with $n = 45,030$ and consider the integers $x^2 - N$:

$$(45,030)^2 - 2,027,651,281 = 49,619$$
$$(45,031)^2 - 2,027,651,281 = 139,680$$
$$(45,032)^2 - 2,027,651,281 = 229,743$$
$$(45,033)^2 - 2,027,651,281 = 319,808$$
$$(45,034)^2 - 2,027,651,281 = 409,875$$
$$(45,035)^2 - 2,027,651,281 = 499,944$$
$$(45,036)^2 - 2,027,651,281 = 590,015$$
$$(45,037)^2 - 2,027,651,281 = 680,088$$
$$(45,038)^2 - 2,027,651,281 = 770,163$$
$$(45,039)^2 - 2,027,651,281 = 860,240$$
$$(45,040)^2 - 2,027,651,281 = 950,319$$
$$(45,041)^2 - 2,027,651,281 = 1,040,400 = (1,020)^2$$

This gives the factorization

$$2,027,651,281 = (45,041 + 1,020) \cdot (45,041 - 1,020) = (46,061) \cdot (44,021) \quad \bullet$$

Fermat's algorithm was developed into a powerful factorization algorithm called the Quadratic Sieve by Maurice Kraitchik, Carl Pomerance, and others. Fermat's algorithm has this interesting property:

The best known factorization algorithms run most slowly in the case of integers which factor into two primes of roughly equal size, but Fermat's algorithm runs most quickly in this case.

It is possible that this aspect of Fermat's algorithm could be exploited to improve other factorization algorithms.

Here is another basic fact about the integers (this is Proposition 20 of Book IX of *Elements*). Euclid did not state this theorem in terms of infinite sets. A rough translation of his statement is "The primes are greater than any number of them."

THEOREM 2.1.4   The set of primes is infinite.

Proof   Suppose that we have listed the first N primes $p_1, p_2, \ldots, p_N$. Notice that $p_1 \cdot p_2 \cdot \ldots \cdot p_N$ is divisible by $p_1, p_2, \ldots, p_N$, and that the multiples of each prime p_k occur at the endpoints of intervals of length p_k. Therefore $(p_1 \cdot p_2 \cdot \ldots \cdot p_N) + 1$ is not divisible by any of the primes $p_1, p_2, \ldots, p_N$, since each prime is greater than 1. Since every integer factors into primes, there must be at least one prime which divides this integer and is different from all the primes $p_1, p_2, \ldots, p_N$. Thus any finite list of prime numbers cannot be a list of all the primes, and so the set of primes must be infinite. ●

Euclid's proof that the number of primes is infinite can be viewed either as an abstract argument, or as an algorithm whose input is a given set of primes and whose output is another set of primes, each of which is different from all the primes in the input set.

EXAMPLE 3   List the first eleven numbers given by Euclid's prime-generating algorithm, and their prime factorizations.

Solution

$$
\begin{aligned}
2 + 1 &= 3 \\
2 \cdot 3 + 1 &= 7 \\
2 \cdot 3 \cdot 5 + 1 &= 31 \\
2 \cdot 3 \cdot 5 \cdot 7 + 1 &= 211 \\
2 \cdot 3 \cdot 5 \cdot 7 \cdot 11 + 1 &= 2{,}311 \\
2 \cdot \ldots \cdot 13 + 1 &= 30{,}031 &&= (59) \cdot (509) \\
2 \cdot \ldots \cdot 17 + 1 &= 510{,}511 &&= (19) \cdot (97) \cdot (277) \\
2 \cdot \ldots \cdot 19 + 1 &= 9{,}699{,}691 &&= (347) \cdot (27{,}953) \\
2 \cdot \ldots \cdot 23 + 1 &= 223{,}092{,}871 &&= (317) \cdot (703{,}763) \\
2 \cdot \ldots \cdot 29 + 1 &= 6{,}469{,}693{,}231 &&= (331) \cdot (571) \cdot (34{,}231) \\
2 \cdot \ldots \cdot 31 + 1 &= 200{,}560{,}490{,}131
\end{aligned}
$$
●

Notice that the first five numbers and the eleventh number in this sequence are prime. There are other prime numbers in this sequence. In the computer exercises we shall find three more.

There is a useful model associated with the divisors of a positive integer.

DEFINITION 2.1.3   If n is a positive integer, the set of positive integers which divide n is denoted Div(n).

There is a graph associated with Div(n), which we shall also denote Div(n). The vertices of this graph are the positive integers which divide n. Vertices labelled s and t are joined by an edge if s divides t and there is no divisor of t properly between s and t.

Figure 2.1 shows the graphs for Div(8), Div(12), and Div(30). Notice that the graph for Div(30) is a cube.

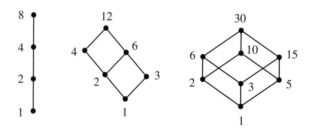

Figure 2.1   Div(8), Div(12), and Div(30)

The greatest common divisor and the least common multiple of two positive integers m and n can be described in terms of the prime factorizations of m and n, and also in terms of the graph Div(n).

Let $p_1, p_2, \ldots, p_s$ be all the primes that divide either m or n, and suppose that the prime factorizations are

$$m = p_1^{i_1} \cdot p_2^{i_2} \cdot \ldots \cdot p_s^{i_s}$$

and

$$n = p_1^{j_1} \cdot p_2^{j_2} \cdot \ldots \cdot p_s^{j_s}$$

Then

$$g.c.d.(m, n) = p_1^{h_1} \cdot p_2^{h_2} \cdot \ldots \cdot p_s^{h_s}$$

where h_k is the minimum of i_k and j_k, and

$$l.c.m.(m, n) = p_1^{l_1} \cdot p_2^{l_2} \cdot \ldots \cdot p_s^{l_s}$$

where l_k is the maximum of i_k and j_k.

We can picture two integers m and n and their greatest common divisor and least common multiple as vertices of a rectangle in Div(n). Figure 2.2 shows the graph for Div(12) again, and the two integers are $m = 4$ and $n = 6$. We have $g.c.d.(4, 6) = 2$ and $l.c.m.(4, 6) = 12$. We can realize these respectively as the vertex of Div(12) where two falling lines from the vertices 4 and 6 meet, and the vertex where two rising lines from the vertices 4 and 6 meet.

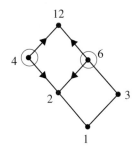

Figure 2.2   The Greatest Common Divisor and the Least Common
Multiple

APPLICATION   *When Is Div(n) a Tree?*

We shall answer this question by proving the following theorem. Theorems are often
stated in this way, as the assertion that two or more statements or conditions are logically
equivalent. One then has to prove that each statement logically implies the others.

THEOREM 2.1.5   If n is a positive integer, the following conditions are logically equivalent:

    (i)   Div(n) is a tree.

    (ii) There is a prime p and a non-negative integer k such that $n = p^k$.

Proof   To prove this, we must show that each statement logically implies the other. First assume
that Div(n) is a tree. Then n cannot be divisible by two different primes p and q, for if
it is, Div(n) will contain the cycle

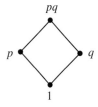

Remember that a tree contains no cycles, so n can only be divisible by a single prime
p, and hence must be a power of p.

Now assume that $n = p^k$. Then the graph of Div(n) is as follows.

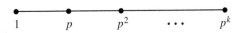

Clearly it is a tree.

EXERCISES

Use the Euclidean Algorithm to compute the greatest common divisor and the least common multiple of each of the following pairs of integers, and to write the greatest common divisor in the form $g.c.d.(m, n) = Am + Bn$.

1. $(m, n) = (64, 28)$.   **2.** $(m, n) = (59, 37)$.

3. $(m, n) = (130, 23)$.   **4.** $(m, n) = (691, 103)$.

5. $(m, n) = (6765, 4181)$.

6. $(m, n) = (10425, 2655)$.

7. Use the Euclidean Algorithm to find the greatest common divisor of the pairs of integers $(1597, 987)$ and $(1590, 997)$. Notice that these two pairs of integers are of comparable size. The algorithm terminates more quickly in one case than in the other. Can you suggest a reason for this?

Use the Euclidean Algorithm to find the greatest common divisor of each of the following pairs of integers. Find the number of steps that the algorithm requires to terminate in each case.

8. $(m, n) = (89, 55)$.   **9.** $(m, n) = (93, 47)$.

10. $(m, n) = (610, 377)$.   **11.** $(m, n) = (605, 322)$.

12. $(m, n) = (75025, 46368)$.

13. $(m, n) = (70000, 38502)$.

Factor each of the following integers n into primes.

14. $n = 7,920$.   **15.** $n = 95,040$.

16. $n = 443,520$.   **17.** $n = 10,200,960$.

18. $n = 244,823,040$.

Use the Fermat Factorization Algorithm to factor the following integers.

19. $N = 2,183$.   **20.** $N = 5,429$.

21. $N = 16,459$.   **22.** $N = 135,199$.

23. $N = 111,557$.   **24.** $N = 112,219$.

For each of the following integers n, draw the graph for $Div(n)$, and for the given pair of integers s and t indicate $g.c.d.(s, t)$ and $l.c.m.(s, t)$ on your graph.

25. $n = 20$, $s = 4$, $t = 10$.

26. $n = 45$, $s = 9$, $t = 15$.

27. $n = 105$, $s = 15$, $t = 21$.

28. $n = 105$, $s = 15$, $t = 7$.

29. $n = 210$, $s = 6$, $t = 15$.

30. $n = 210$, $s = 15$, $t = 21$.

31. $n = 210$, $s = 6$, $t = 35$.

ADVANCED EXERCISES

Marin Mersenne (1588–1648) was a monk who was a friend of Descartes, Fermat, and many others who were involved in the development of French science at the time.

In 1640 Fermat wrote Mersenne a letter containing three Propositions:

(i)   If a positive integer n is not prime, then $2^n - 1$ is not prime.

(ii)   If n is a prime then $2^n - 2$ is a multiple of $2n$.

(iii)   If n is a prime, and p is a prime divisor of $2^n - 1$ then $p - 1$ is a multiple of n.

If p is prime, the number $M_p = 2^p - 1$ is called a *Mersenne number*, and if M_p is prime it is called a *Mersenne prime*. Some Mersenne numbers are prime

and some are not. For example,

$$M_7 = 2^7 - 1 = 127$$

is prime, but

$$M_{11} = 2^{11} - 1 = 2,047 = (23) \cdot (89)$$

is not.

1. Prove Fermat's Proposition (i) above. If n is not prime, then it factors into $n = a \cdot b$. Use the identity

$$x^b = (x - 1)(x^{b-1} + x^{b-2} + \ldots + x + 1)$$

and substitute $x = 2^a$.

2. Check Fermat's Proposition (ii) for the following primes: $p = 7, 11, 17, 19$.

Fermat's Proposition (iii) can be used to find factors of Mersenne numbers if they are not prime. If n is prime and p is a prime factor of $M_n = 2^n - 1$, then $p - 1$ is a multiple of n. Thus $p = nk + 1$ for some integer k, and since p is odd we must actually have $p = 2nk + 1$.

3. Fermat found a factorization of $M_{37} = 2^{37} - 1 = 137,438,953,471$ by this method, by inspecting possible factors of the form $74k + 1$. Find this factorization yourself by dividing M_{37} by these numbers for $k = 1, 2, \ldots$.

4. Factor $M_{23} = 2^{23} - 1 = 8,388,607$ by Fermat's method.

5. Factor $M_{29} = 2^{29} - 1 = 536,870,911$ by Fermat's method.

We have seen that Mersenne numbers M_p may or may not be prime. Lucas discovered a criterion which describes exactly when a Mersenne number is prime. First define the *Lucas sequence* as follows:

$$S_0 = 4, \; S_1 = 4^2 - 2 = 14, \; S_2 = 14^2 - 2 = 194, \ldots$$

with the recursion $S_k = S_{k-1}^2 - 2$.

Lucas' Theorem   The Mersenne number $M_p = 2^p - 1$ is prime if and only if M_p divides S_{p-2}.

6. Check that $M_5 = 2^5 - 1 = 31$ is prime by using Lucas' Theorem.

COMPUTER EXERCISES

Remember that the *Mersenne numbers* are defined by the expression

$$M_p = 2^p - 1$$

where p is prime.

1. Make a list of the Mersenne numbers for the first 200 primes, identify those which are prime, and give the factorizations of those which are not prime.

You may not be able to factor all of the Mersenne numbers which are composite (the factorization algorithm may take too long in some cases), but this is part of the problem. Factor as many of the Mersenne numbers in this range as you can.

By contrast, you will be able to find *all* of the Mersenne primes in this range. The reason is that the factorization algorithm in any program that you are likely to use will identify a number as prime, if it is prime, much faster than it will factor a composite number of comparable size.

Use Lucas' Theorem, described in the Advanced Exercises, to prove that the following Mersenne numbers are prime.

2. $M_{13} = 2^{13} - 1$.   **3.** $M_{17} = 2^{17} - 1$.

4. $M_{19} = 2^{19} - 1$.   **5.** $M_{31} = 2^{31} - 1$.

6. Consider the numbers

$$N_n = p_1 \cdot p_2 \cdot \ldots \cdot p_n + 1$$

which arise in Euclid's prime-generating algorithm (here $p_1, p_2, \ldots, p_n$ are the first n primes in order).

Consider the first 200 of these numbers.

(i)   Verify that the numbers $N_1, N_2, N_3, N_4,$ $N_5, N_{11}, N_{75}, N_{171},$ and N_{172} are prime, and that no other numbers N_n are prime in the range $1 \le n \le 200$.

(ii) Factor the other integers N_n into primes, again for $1 \le n \le 200$.

As in the previous problem, you may not be able to factor all of the composite integers N_n into primes, because the factorization algorithm may take too long to run in some cases. This is part of the problem. Factor as many of them as you can.

7. Factor the following integer using the factorization algorithm in *Derive*, *Maple*, *Mathematica*, or another program. Note the amount of time that the factorization takes on the computer that you use.

$$N_{19} = 2 \cdot \ldots \cdot 67 + 1$$
$$= 7,858,321,551,080,267,055,879,091$$

When this integer has been factored, compute the number of steps that the Fermat Factorization Algorithm would require to factor it. Estimate the amount of time that would be needed by the same computer to factor this integer using the Fermat Factorization Algorithm.

8. Factor

$$n = 808,017,424,794,512,875,886,459,904,$$
$$961,710,757,005,754,368,000,000,000$$

into primes. This number is the order of the *Monster*, a set of symmetries in a space of $196, 883$ dimensions.

2.2 LAMÉ'S THEOREM

He came as a Butcher: but gravely declared,

When the ship had been sailing a week,

He could only kill Beavers. The Bellman looked scared,

And was almost too frightened to speak:

At last he explained, in a tremulous tone,

There was only one Beaver on board;

And that was a tame one he had of his own,

Whose death would be deeply deplored.

The Swiss mathematician Leonhard Euler (1707–1783) wrote the *Introductio in Analysin Infinitorum* (Introduction to the Analysis of the Infinite, published in 1748). In the *Introductio* Euler described the Lucas numbers, long before they were named for Lucas, and stated the formula which we shall give next in the case of Lucas numbers. Euler stated the formula again in a paper published in 1765, but it is now named for the French mathematician Jacques Phillipe Marie Binet.

Binet's Formula   The Fibonacci numbers and the Lucas numbers can be expressed by the formulas

$$F_n = \frac{1}{\sqrt{5}} \left(\left(\frac{1 + \sqrt{5}}{2} \right)^n - \left(\frac{1 - \sqrt{5}}{2} \right)^n \right)$$

and

$$L_n = \left(\frac{1 + \sqrt{5}}{2} \right)^n + \left(\frac{1 - \sqrt{5}}{2} \right)^n$$

The number $\phi = \frac{1+\sqrt{5}}{2}$ which occurs in these formulas is called the *Golden Ratio*, or *phi*.

Proof of Binet's Formula   We shall prove Binet's Formula in the case of the Fibonacci numbers. The proof in the case of the Lucas numbers is similar. To prove the formula, we must show that the expression in Binet's Formula gives the initial values of the Fibonacci sequence,

$$F_0 = 0 \text{ and } F_1 = 1$$

and also the recursion

$$F_n = F_{n-1} + F_{n-2}$$

if $n \geq 2$.

We have

$$F_0 = \frac{1}{\sqrt{5}}\left(\left(\frac{1+\sqrt{5}}{2}\right)^0 - \left(\frac{1-\sqrt{5}}{2}\right)^0\right) = \frac{1}{\sqrt{5}}(1-1) = 0$$

and

$$F_1 = \frac{1}{\sqrt{5}}\left(\left(\frac{1+\sqrt{5}}{2}\right)^1 - \left(\frac{1-\sqrt{5}}{2}\right)^1\right) = \frac{1}{\sqrt{5}}\left(\frac{2\sqrt{5}}{2}\right) = 1$$

To carry out the inductive step, notice first that

$$\left(\frac{1+\sqrt{5}}{2}\right)^2 = \frac{1+2\sqrt{5}+5}{4} = \frac{1+\sqrt{5}}{2} + 1$$

and similarly that

$$\left(\frac{1-\sqrt{5}}{2}\right)^2 = \frac{1-\sqrt{5}}{2} + 1$$

Now assume that $n \geq 2$. We have

$$F_{n-1} + F_{n-2} = \frac{1}{\sqrt{5}}\left(\left(\frac{1+\sqrt{5}}{2}\right)^{n-1} - \left(\frac{1-\sqrt{5}}{2}\right)^{n-1}\right)$$

$$+ \frac{1}{\sqrt{5}}\left(\left(\frac{1+\sqrt{5}}{2}\right)^{n-2} - \left(\frac{1-\sqrt{5}}{2}\right)^{n-2}\right)$$

$$= \frac{1}{\sqrt{5}}\left(\left(\frac{1+\sqrt{5}}{2}\right)^{n-1} + \left(\frac{1+\sqrt{5}}{2}\right)^{n-2}\right)$$

$$- \frac{1}{\sqrt{5}}\left(\left(\frac{1-\sqrt{5}}{2}\right)^{n-1} + \left(\frac{1-\sqrt{5}}{2}\right)^{n-2}\right)$$

$$= \frac{1}{\sqrt{5}}\left(\left(\frac{1+\sqrt{5}}{2}\right)^{n-2}\left(\frac{1+\sqrt{5}}{2} + 1\right)\right)$$

$$- \frac{1}{\sqrt{5}}\left(\left(\frac{1-\sqrt{5}}{2}\right)^{n-2}\left(\frac{1-\sqrt{5}}{2} + 1\right)\right)$$

$$= \frac{1}{\sqrt{5}}\left(\frac{1+\sqrt{5}}{2}\right)^n - \frac{1}{\sqrt{5}}\left(\frac{1-\sqrt{5}}{2}\right)^n$$

$$= \frac{1}{\sqrt{5}}\left(\left(\frac{1+\sqrt{5}}{2}\right)^n - \left(\frac{1-\sqrt{5}}{2}\right)^n\right) = F_n$$

Euler mentioned Binet's Formula in passing, as if it were well-known. It seems that Binet's Formula was known early in the eighteenth century, but the history of the formula is obscure.

The number ϕ is called the *Golden Ratio*. A rectangle such that the ratio of the length and width is ϕ is called a *Golden Rectangle*. It is believed that of all rectangular shapes, the Golden Rectangle is most pleasing to the eye.

Binet's Formula gives useful approximations to the Fibonacci numbers and the Lucas numbers. Since $\phi = 1.618033989\ldots$ and $\phi' = -.618033989\ldots$, it follows that

$$F_n \text{ is the nearest integer to } \frac{\phi^n}{\sqrt{5}}$$

and if $n \geq 3$,

$$L_n \text{ is the nearest integer to } \phi^n$$

This has the consequence that the Fibonacci numbers and the Lucas numbers grow like the powers of a real number larger than 1. That is, the Fibonacci numbers and the Lucas numbers exhibit *exponential growth*.

EXAMPLE 1   Compute F_{30} and L_{30} using Binet's Formula.

Solution   We have

$$\frac{\phi^{30}}{\sqrt{5}} = 832,040.0042\ldots$$

so $F_{30} = 832,040$, and

$$\phi^{30} = 1,860,498.009\ldots$$

so $L_{30} = 1,860,498$.   ●

We shall understand Binet's Formula better if we expand the two expressions that it contains by the Binomial Theorem. This leads to the *Second Form* of Binet's Theorem.

Binet's Formula, Second Form   The Fibonacci numbers F_n and the Lucas numbers L_n can be expressed by the formulas

$$F_n = 2^{-n+1}\left(\binom{n}{1} + \binom{n}{3} \cdot 5 + \binom{n}{5} \cdot 5^2 + \ldots\right)$$

and

$$L_n = 2^{-n+1}\left(\binom{n}{0} + \binom{n}{2} \cdot 5 + \binom{n}{4} \cdot 5^2 + \ldots\right)$$

The sums continue as long as the lower index is less than or equal to the upper index.

EXAMPLE 2   Compute $F_5 = 5$ using the Second Form of Binet's Formula.

Solution

$$F_5 = 2^{-4} \left(\binom{5}{1} + \binom{5}{3} \cdot 5 + \binom{5}{5} \cdot 5^2 \right) = \left(\frac{1}{16} \right) (5 + 10 \cdot 5 + 25)$$

$$= \left(\frac{1}{16} \right) (80) = 5$$

●

EXAMPLE 3 Compute $L_5 = 11$ using the Second Form of Binet's Formula.

Solution

$$L_5 = 2^{-4} \left(\binom{5}{0} + \binom{5}{2} \cdot 5 + \binom{5}{4} \cdot 5^2 \right) = \left(\frac{1}{16} \right) (1 + 10 \cdot 5 + 5 \cdot 5^2)$$

$$= \left(\frac{1}{16} \right) (176) = 11$$

●

Gabriel Lamé (1795–1870) was perhaps the first person to concern himself with the number of steps that an algorithm takes to terminate, or what we would now call the *complexity* of an algorithm. He proved that if the Euclidean Algorithm is applied to a pair of integers, and if N is the number of digits in the smaller of the two, then the Euclidean Algorithm will terminate in no more than $5N$ steps.

Because Lamé's Theorem is a basic example of a theorem about complexity, we shall prove it in the several forms which are to be found in Lamé's paper of 1844. Before we do this, we will give an example which illustrates an idea which is basic to the theorem.

EXAMPLE 4 Apply the Euclidean Algorithm to the pair of consecutive Fibonacci numbers $(n, m) = (34, 55)$ and also to $(n, m) = (35, 54)$.

Solution

$$55 = 1 \cdot 34 + 21$$
$$34 = 1 \cdot 21 + 13$$
$$21 = 1 \cdot 13 + 8$$
$$13 = 1 \cdot 8 + 5$$
$$8 = 1 \cdot 5 + 3$$
$$5 = 1 \cdot 3 + 2$$
$$3 = 1 \cdot 2 + 1$$
$$54 = 1 \cdot 35 + 19$$
$$35 = 1 \cdot 19 + 16$$
$$19 = 1 \cdot 16 + 3$$
$$16 = 5 \cdot 3 + 1$$

The Euclidean Algorithm computes $g.c.d.(34, 55) = 1$ in seven steps, while it computes $g.c.d.(35, 54) = 1$ in only four steps. The sequence of remainders for the pair of consecutive Fibonacci numbers is the Fibonacci sequence in reverse. It is a fact that the Euclidean Algorithm takes longest to terminate when applied to a pair of consecutive Fibonacci numbers, as compared with the number of steps required for a pair of integers of comparable size. This idea underlies the proof of Lamé's Theorem.   ●

Lamé's Theorem

(i)   If $k \geq 3$ and F_k and F_{k+1} are consecutive Fibonacci numbers, then $F_{k+1}/F_k < 2$.

(ii)   If $k \geq 1$ and F_k and F_{k+1} are consecutive Fibonacci numbers, $g.c.d.(F_k, F_{k+1}) = 1$. If the Euclidean Algorithm is applied to this pair of integers, the sequence of remainders is the Fibonacci sequence in reverse, beginning with F_{k-1}. In particular, the Euclidean Algorithm gives $g.c.d.(F_k, F_{k+1}) = 1$ in $k - 1$ steps.

(iii)   If m and n are positive integers with $n \leq m$, and if the Euclidean Algorithm computes $g.c.d.(m, n)$ in k steps, then $n \geq F_{k+1}$.

sharper ⟹ **(iv)**   With the same assumptions as in (iii), the Euclidean Algorithm will compute
bound $g.c.d.(m, n)$ in no more than $\log_\phi(n) + 1$ steps. size of m,n

↳ **(v)**   With the same assumptions as in (iii), the Euclidean Algorithm will compute $g.c.d.(m, n)$ in no more than $5N$ steps, where N is the number of digits in the decimal expansion of n. decimal expansion of m,n

Proof   Notice that $F_1 = F_2 = 1$, but if $k \geq 2$ then $F_k < F_{k+1}$. Thus if $k \geq 3$,

$$F_{k+1}/F_k = (F_k + F_{k-1})/F_k < 2F_k/F_k = 2$$

which proves (i).

(i) implies (ii), because the statement is clearly true if $k = 1$ or $k = 2$. If $k \geq 3$, we must have $q = 1$ in the equation $F_{k+1} = q F_k + r$, and then by the recursion $r = F_{k-1}$. The last non-zero remainder is $F_2 = 1$, so $g.c.d.(F_k, F_{k+1}) = 1$, and the number of steps is $k - 2$. This proves (ii).

We will prove (iii) by induction on k, using the Strong Form of the Principle of Induction. If $k = 1$, then $n \geq F_2 = 1$, and if $k = 2$ clearly $n \geq F_3 = 2$. If $k \geq 3$, write $m = q_1 n + r_1$ and $n = q_2 r_1 + r_2$. By induction, $r_1 \geq F_{k+1}$ and $r_2 \geq F_k$, so $n = q_2 r_1 + r_2 \geq r_1 + r_2 = F_{k+1} + F_k = F_{k+2}$.

To prove (iv) we will use the result in (iii) that if the Euclidean Algorithm computes $g.c.d.(m, n)$ in k steps, then

$$F_{k+1} \leq n$$

By Binet's Formula

$$\frac{1}{\sqrt{5}} \left(\phi^{k+1} - (\phi')^{k+1} \right) \leq n$$

It follows that

$$\phi^{k+1} \leq \sqrt{5}n + (\phi')^{k+1}$$

If $n = 1$, then $k = 0 = \log_\phi(1)$, so the formula holds. If $n \geq 2$, then $k \geq 1$, so $(\phi')^2 = 0.3819\ldots < (0.3)2 = 0.6 \leq (0.3)n$, and so

$$\phi^{k+1} < \sqrt{5}n + (0.3)n = \left(\sqrt{5} + 0.3\right)n$$

Then

$$k + 1 = \log_\phi\left(\phi^{k+1}\right) < \log_\phi\left(\left(\sqrt{5} + 0.3\right)n\right) = \log_\phi\left(\sqrt{5} + 0.3\right) + \log_\phi(n)$$

Finally, $\log_\phi(\sqrt{5} + 0.3) = 1.9338\ldots < 2$, so

$$k + 1 < 2 + \log_\phi(n)$$

so $k < \log_\phi(n) + 1$.

Finally, we will prove (v) by induction on N. If $N = 1$, then n has only one decimal digit and so $n \leq 9$, so by (iv) the Euclidean Algorithm will compute $g.c.d.(m, n)$ in one more step than $\log_\phi(n) \leq \log_\phi(9) = \ln(9)/\ln(\phi) = 4.566\ldots \leq 5$. Since the number of steps is an integer, it is no more than 5.

To prove the inductive step we must show that if we increase the number of decimal digits of n by 1 then the Euclidean Algorithm will compute $g.c.d.(m, n)$ in no more than 5 more steps. If $n \geq 9$ then

$$\log_\phi(10n + 9) \leq \log_\phi(11n) = \log_\phi(11) + \log_\phi(n)$$

and since $\log_\phi(11) = \ln(11)/\ln(\phi) = 4.983\ldots < 5$ this is true, and this completes the proof of Lamé's Theorem.   ●

EXAMPLE 5   Use Lamé's Theorem to estimate the number of steps that the Euclidean Algorithm will take to compute $g.c.d(9087, 120)$, and compare this with the actual number of steps.

Solution   Lamé's Theorem asserts that no more than

$$\log_\phi(120) + 1 = \ln(120)/\ln(\phi) + 1 = 10.948\ldots$$

steps will be needed, so the number of steps is no more than 10, and in fact the Euclidean Algorithm computes $g.c.d.(9087, 120) = 3$ in six steps.   ●

APPLICATION   ___Egyptian Fractions___

Here we will describe two variants of the Euclidean Algorithm, the *Egyptian Fraction Algorithm* and the *Strict Egyptian Fraction Algorithm*.

In 1858, the Scot Henry Rhind bought a papyrus, now called the *Rhind Papyrus*, which was written about 1650 B.C. by a scribe named Ahmose, who stated that he was copying material which was 400 years or more older.

For reasons which are not well understood, the ancient Egyptians manipulated fractions by first writing them as a sum of *unit fractions*, fractions of the form $\frac{1}{n}$. A positive fraction can always be written in this form, and although the result is not explicitly stated in the Rhind Papyrus, it is clear that the scribe Ahmose was aware of it.

The Egyptian Fraction Algorithm   Let $\frac{m}{n}$ be a ratio of positive integers with no common factor, with $m < n$.

(i) If the fraction $\frac{p}{q}$ has been obtained, choose the smallest positive integer k such that $\frac{p}{q} \geq \frac{1}{k}$ and compute the fraction $\frac{p}{q} - \frac{1}{k}$.

(ii) Repeat (i) until $\frac{m}{n}$ has been expressed as a sum of unit fractions.

The Egyptian Fraction is a variant of the Euclidean Algorithm, and takes approximately the same number of steps to terminate.

EXAMPLE 6   Apply the Egyptian Fraction Algorithm to $\frac{6}{7}$.

Solution   The largest unit fraction less than $\frac{6}{7}$ is $\frac{1}{2}$.

$$\frac{6}{7} - \frac{1}{2} = \frac{5}{14}, \text{ and the largest unit fraction less than } \frac{5}{14} \text{ is } \frac{1}{3}.$$

$$\frac{5}{14} - \frac{1}{3} = \frac{1}{42}.$$

The result is that $\dfrac{6}{7} = \dfrac{1}{2} + \dfrac{1}{3} + \dfrac{1}{42}$ ●

The Egyptian Fraction Algorithm is an example of a *greedy algorithm*. A greedy algorithm always selects the largest, or smallest, or nearest ... object which is being considered.

The Strict Egyptian Fraction Algorithm   Let $\frac{m}{n}$ be a ratio of positive integers.

(i) Write $\frac{m}{n} = \frac{1}{n} + \ldots + \frac{1}{n}$, where the unit fraction occurs m times.

(ii) If a sum of unit fractions has been obtained in which two denominators coincide, choose a term with the smallest denominator which occurs more than once and apply the identity

$$\frac{1}{x} = \frac{1}{x+1} + \frac{1}{x(x+1)}$$

(iii) Repeat (ii) until all the denominators of the unit fractions are distinct.

EXAMPLE 7   Apply the Strict Egyptian Fraction Algorithm to $\frac{3}{2}$.

Solution

$$\frac{3}{2} = \frac{1}{2} + \frac{1}{2} + \frac{1}{2} = \frac{1}{2} + \frac{1}{2} + \frac{1}{3} + \frac{1}{6}$$

$$= \frac{1}{2} + \frac{1}{3} + \frac{1}{6} + \frac{1}{3} + \frac{1}{6}$$

$$= \frac{1}{2} + \frac{1}{3} + \frac{1}{6} + \frac{1}{4} + \frac{1}{12} + \frac{1}{6}$$

$$= \frac{1}{2} + \frac{1}{3} + \frac{1}{4} + \frac{1}{6} + \frac{1}{7} + \frac{1}{12} + \frac{1}{42}$$ ●

The Strict Egyptian Fraction Algorithm is the first example we have seen of an algorithm for which it is far from obvious that the algorithm terminates. The point is that as the algorithm is implemented, repeated denominators keep being introduced, and it is not clear that the algorithm eventually makes all the denominators distinct. The first proof that the algorithm terminates was published in a paper by Laurent Beekmans in 1993.

If the Strict Egyptian Fraction Algorithm is applied to the fraction 5/2, it terminates with a sum of 46 unit fractions! The algorithm produces its output extremely slowly, in contrast to the Egyptian Fraction Algorithm, which terminates rather quickly. In other words, the *complexity* of the two algorithms is very different.

The Egyptian Fraction Algorithm was known a millenium and a half before Euclid stated the Euclidean Algorithm. The Strict Egyptian Fraction Algorithm was studied two millenia later. Both algorithms are of interest in computer science.

EXERCISES

Use the approximation formula given in this section to find the following Fibonacci numbers and Lucas numbers.

1. F_{10} and L_{10} **2.** F_{20} and L_{20} **3.** F_{40} and L_{40}

Use the Euclidean Algorithm to compute the greatest common divisor of each of the following pairs of integers, and verify the statement in Lamé's Theorem that no more than k steps are used, where $k \leq log_\phi(n) = ln(n)/ln(\phi)$.

4. $(m, n) = (987, 610)$.   **5.** $(m, n) = (988, 602)$.

6. $(m, n) = (3524578, 2178309)$.

7. $(m, n) = (3524580, 2178300)$.

Use the Second Form of Binet's Theorem to compute the following Fibonacci numbers and Lucas numbers.

8. $F_6 = 8$.   **9.** $F_7 = 13$.   **10.** $F_8 = 21$.

11. $L_6 = 18$.   **12.** $L_7 = 29$.   **13.** $L_8 = 47$.

14. In Book VI, Proposition 30 of Euclid's *Elements*, a construction is given to divide a line segment in *Mean and Extreme Ratio*. This means that the total length of the line segment is $x + 1$, and x satisfies the identity

$$\frac{x}{x+1} = \frac{1}{x}$$

Prove that $x = \phi = \frac{1+\sqrt{5}}{2}$.

Apply the Egyptian Fraction Algorithm to the following fractions.

15. $\dfrac{2}{3}$ **16.** $\dfrac{3}{5}$ **17.** $\dfrac{5}{7}$

18. $\dfrac{5}{11}$ **19.** $\dfrac{8}{11}$

20. Apply the Strict Egyptian Fraction Algorithm to

$$\frac{4}{2} = \frac{1}{2} + \frac{1}{2} + \frac{1}{2} + \frac{1}{2}$$

ADVANCED EXERCISES

The Fibonacci numbers and the Lucas numbers satisfy numerous identities. Listed are ten of these identities, which can be proved by induction, or by using Binet's Formula. These formulas are all valid if $n \geq 1$.

I. $\sum_{k=1}^{n} F_k = F_{n+2} - 1$

II. $\sum_{k=1}^{n} L_k = L_{n+2} - 3$

III. $F_{n+1}F_{n-1} - F_n^2 = (-1)^n$

IV. $L_{n+1}L_{n-1} - L_n^2 = 5(-1)^{n+1}$

V. $L_n = F_{n+1} + F_{n-1}$

VI. $F_{2n+1} = F_{n+1}^2 + F_n^2$

VII. $F_{2n} = F_{n+1}^2 - F_{n-1}^2$

VIII. $F_{2n} = F_n L_n$

IX. $F_{n+m+1} = F_{n+1}F_{m+1} + F_n F_m$

X. $\sum_{k=1}^{n} F_k^2 = F_n F_{n+1}$

1. Prove identity I.   **2.** Prove identity II.

3. Prove identity III.   **4.** Prove identity IV.   **7.** Prove identity VII.   **8.** Prove identity VIII.

5. Prove identity V.   **6.** Prove identity VI.   **9.** Prove identity IX.   **10.** Prove identity X.

COMPUTER EXERCISES

Verify that the ten identities in the Advanced Exercises hold if $n = 100$.

1. Show that identity I holds if $n = 100$.

2. Show that identity II holds if $n = 100$.

3. Show that identity III holds if $n = 100$.

4. Show that identity IV holds if $n = 100$.

5. Show that identity V holds if $n = 100$.

6. Show that identity VI holds if $n = 100$.

7. Show that identity VII holds if $n = 100$.

8. Show that identity VIII holds if $n = 100$.

9. Show that identity IX holds if $n = 100$.

10. Show that identity X holds if $n = 100$.

11. Apply the Strict Egyptian Fraction Algorithm to $\dfrac{5}{2}$.

12. Apply the Strict Egyptian Fraction Algorithm to $\dfrac{6}{2}$.

2.3   THE INTEGERS MOD n

The Beaver's best course was no doubt to procure

A second-hand dagger-proof coat—

So the Baker advised it—and next, to insure

Its life in some Office of note:

This the Banker suggested, and offered for hire

(On moderate terms), or for sale,

Two excellent policies, one Against Fire,

And one Against Damage From Hail.

In 1801 Carl Friedrich Gauss (1777–1855) completed the dedication of his book *Disquisitiones Arithmeticae* (Arithmetical Investigations) to his patron Charles Wilhelm Ferdinand, duke of Brunswick and Lüneburg:

> ... *YOU have never excluded from YOUR patronage those sciences which are commonly regarded as being too recondite and too removed from ordinary life. YOU YOURSELF in YOUR supreme wisdom are well aware of the intimate and necessary bond that unites all sciences among themselves and with whatever pertains to the prosperity of human society* ...

The *Disquisitiones* is a work of wonderful invention and originality. The first two sections of the *Disquisitiones* are devoted to the properties of *congruences* or, in a slightly different language, to the properties of the *Integers modulo n*, or *mod n*. Gauss hinted at their usefulness. Here are only a few of their applications:

   I. In the construction of the codes which are used in computer science.

  II. In the construction of other useful codes.

 III. In models for the genetic code.

 IV. In the definition of recursive sequences. In particular, in the definition of binary recursions which, as shift register output, occur in the codes of computers and communications machines.

  V. In the solution of combinatorial problems.

 VI. In many encryption schemes, including *Public Key Encryption*, which we shall describe at the end of this section.

DEFINITION 2.3.1   If *n* is a positive integer, then two integers *A* and *B* are *congruent modulo n*

$$A \equiv B \ (\text{mod } n)$$

if *n* divides $A - B$.

For example,

$$11 \equiv 1 \ (\text{mod } 10)$$

and

$$-3 \equiv 7 \ (\text{mod } 10)$$

DEFINITION 2.3.2   If *r* is an integer, the *congruence class* of *r* (mod *n*), denoted [*r*], is the set

$$[r] = \{r + kn : k \in Z\}$$

of integers congruent to *r* (mod *n*).

For example, if $n = 2$, there are two congruence classes. One is the congruence class of 0, the set of even integers

$$[0] = \{2k : k \in Z\}$$

and the other is the congruence class of 1, the set of odd integers

$$[1] = \{2k + 1 : k \in Z\}$$

The connection between the relation of congruence and congruence classes is explained by the following theorem.

THEOREM 2.3.1   The following conditions are equivalent:

    (i) $A \equiv B$ (mod n)

    (ii) The congruence classes of A and B (mod n) are equal: $[A] = [B]$.

There are exactly n congruence classes (mod n), namely

$$[0], [1], \ldots, [n-1]$$

Proof   If A and B are congruent (mod n) then $A = B + kn$. If $C \in [A]$ then $A \equiv C$(mod n) so $A = C + hn$. Then $B + kn = C + hn$, so $C = B + (k - h)n$, so $C \in [B]$. This proves that $[A] \subset [B]$. Similarly, $[B] \subset [A]$, so $[A] = [B]$.

If $[A] = [B]$ then $A = B + kn$, so $A \equiv B$(mod n).

Since every integer is congruent (mod n) to exactly one integer r, where $0 \le r \le n - 1$, there are exactly n congruence classes (mod n), the congruence classes of these n integers.   ●

> **DEFINITION 2.3.3**   The *Integers modulo n*, denoted Z_n, is the set of congruence classes (mod n).
>
> $$Z_n = \{[0], [1], [2], \ldots, [n-1]\}$$

There are concepts of *symmetry* associated with all the number systems which we have mentioned. We can think of the real numbers as expressing a concept of *translational symmetry* (translation to the right or left), and the integers as giving a discrete version of this concept of translational symmetry.

In a similar way, we can think of the circle as expressing a concept of *circular symmetry*, or as modelling *periodic* phenomena. The discrete version of circular symmetry is expressed by Z_n. It gives a very general model for n events which occur in a certain order and then are repeated in the same order. For example, we can picture the elements of Z_{12} as the hours on a clock (Figure 2.3).

The set of congruence classes Z_n is a number system with n elements. That is, it is possible to add and multiply congruence classes according to the rules $[r] + [s] = [r + s]$ and $[r] \cdot [s] = [rs]$. These will not be valid definitions unless we can be sure that the congruence class of the sum and product of two congruence classes do not depend on the *representatives*, r and s, of these congruence classes which we have chosen. The following theorem proves that this is true (that the addition and multiplication of congruence classes is *well-defined*).

THEOREM 2.3.2   If A, B, C, and D are integers, and $A \equiv B$(mod n) and $C \equiv D$(mod n), then

    (i) $A + C \equiv B + D$(mod n), and

    (ii) $AC \equiv BD$(mod n).

Proof   If $A - B = hn$ and $C - D = kn$, then $(A + C) - (B + D) = hn - kn = (h - k)n$, and this proves (i).

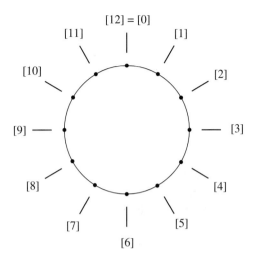

Figure 2.3 Z_{12} as a Model for a Clock

If $A - B = hn$ and $C - D = kn$, then $AC - BD = AC - BC + BC - BD = (A - B)C + B(C - D) = hnC + Bkn = (hC + Bk)n$, and this proves (*ii*). ●

DEFINITION 2.3.4   If $[r]$ and $[s]$ are congruence classed (mod *n*), then

$$[r] + [s] = [r + s]$$

and

$$[r] \cdot [s] = [rs]$$

The congruence class $[0]$ behaves like 0 in other number systems and the class $[1]$ behaves like 1.

Suppose that we add the congruence class $[1]$ to itself in Z_2:

$$[1] + [1] = [1 + 1] = [2] = [0]$$

This is binary addition, the addition that is used in binary code in a computer. The elements $[0]$ and $[1]$ of Z_2 are called *binary bits*.

In Z_3,

$$[1] + [2] = [1 + 2] = [3] = [0]$$

so $[2] = -[1]$.

In Z_4,

$$[3] \cdot [3] = [9] = [1] \text{ and } [2] \cdot [2] = [4] = [0]$$

so $[3] = [3]^{-1}$. But $[2]$ behaves differently, for although $[2] \neq [0]$ in Z_4 its square is zero. This kind of behavior does not occur in the other number systems we have considered. Also, although $[2] \neq [0]$, $[2]$ has no multiplicative inverse in Z_4.

When working with Z_n, it is tedious to keep writing square brackets for congruence classes. From now on we shall omit the square brackets if there is no danger of confusion. For example,

$$Z_2 = \{0, 1\}$$
$$Z_3 = \{0, 1, 2\}$$
$$Z_4 = \{0, 1, 2, 3\}$$

and so on.

We have seen that Z_n is a number system with n elements. What are its applications? Following are some applications for $n = 2$, $n = 3$, and $n = 4$.

APPLICATION   *Applications of Z_2*

The elements of $Z_2 = \{0, 1\}$ are binary bits, which we have used in binary vectors and binary strings.

The *machine code* of a computer is composed of binary strings. All input to the computer is converted to a binary string

$$1100010111010000101000111011001111111001 \ldots$$

which is then manipulated further. First it is processed in the computer's *compiler*, and then a program is run with the processed string as input.

We should stress the fact that

Any information can be represented as a binary string.

This principle was recognized long ago, and is at the basis of most kinds of signalling. For example, in *Morse Code*, the letters and punctuation marks of the English language are converted into sequences of dots and dashes, which could be replaced by 0s and 1s. To be more precise, Morse Code uses dots, dashes, and some other symbols. A closely related code, the *Continental Code*, uses only dots and dashes.

The problem of converting a continuous signal into a binary string, called the problem of *quantization*, can be solved by using one of several algorithms. Telephone companies use these algorithms to convert the input from a person's voice into a string of binary bits.

Binary vectors can be used to construct binary codes. An example of such a code, the Hamming Code H_8, is given at the end of Section 1.4.

Binary recursive sequences are used in most computers to generate code which is then modified for specific uses. We described these sequences, and the shift registers which generate them, at the end of Section 1.5.

Binary strings are also used to transmit visual images. The method which is usually used is to divide the visual field into small squares, and then to transmit the light intensity in each square (called a *pixel*), using a binary code. Pictures from the first Mars lander were sent to earth in this way, using a binary code called the *Reed–Muller Code*. This error-correcting code was used because of the danger that noise in the receiver would distort the very faint incoming signal.

APPLICATION   *Applications of Z_3*

A code whose letters are the elements 0, 1, 2, of Z_3 is called a *ternary code*. Although binary codes are used more often, there are also interesting ternary codes. Figure 2.4 lists six vectors of the *Ternary Golay Code G_{12}*.

$$(1, 1, 1, 1, 1, 1, 1, 1, 1, 1, 1, 1)$$
$$(0, 1, 1, 1, 0, 1, 1, 0, 1, 0, 0, 0)$$
$$(0, 1, 1, 0, 1, 1, 0, 1, 0, 0, 0, 1)$$
$$(0, 1, 0, 1, 1, 0, 1, 0, 0, 0, 1, 1)$$
$$(0, 0, 1, 1, 0, 1, 0, 0, 0, 1, 1, 1)$$
$$(0, 1, 1, 0, 1, 0, 0, 0, 1, 1, 1, 0)$$

Figure 2.4   Six Vectors of the Ternary Golay Code

G_{12} is a linear code, which means that any finite sum of vectors in G_{12} with coefficients in Z_3 is again a vector in G_{12}. There are

$$3^6 = 729$$

vectors in G_{12} altogether. All vectors in G_{12} can be obtained as sums of the six vectors in Figure 2.4, with coefficients in Z_3. Here are two examples of sums of vectors in G_{12}:

$$(0, 1, 1, 1, 0, 1, 1, 0, 1, 0, 0, 0) + (0, 1, 1, 0, 1, 1, 0, 1, 0, 0, 0, 1)$$
$$= (0, 2, 2, 1, 1, 2, 1, 1, 1, 0, 0, 1)$$

and

$$(0, 1, 1, 1, 0, 1, 1, 0, 1, 0, 0, 0) + 2 \cdot (0, 1, 1, 0, 1, 1, 0, 1, 0, 0, 0, 1)$$
$$= (0, 1, 1, 1, 0, 1, 1, 0, 1, 0, 0, 0) + (0, 2, 2, 0, 2, 2, 0, 2, 0, 0, 0, 2)$$
$$= (0, 0, 0, 1, 2, 0, 1, 2, 1, 0, 0, 2)$$

APPLICATION   *Applications of Z_4*

Z_4 can be used to give a model for the genetic code. The genetic code is stored in the chromosomes of all plants and animals, and has the form of a *double helix*. There are two strands in the double helix, each consisting of a sequence of four organic chemicals called *nucleotides*: T = Thymine, A = Adenine, C = Cytosine, and G = Guanine. These form a substance called *DNA* (*Deoxyribonucleic Acid*). Another substance called *RNA* (*Ribonucleic Acid*) transmits the information stored in *DNA* to construct the proteins which constitute all plant and animal tissue.

In thinking about the genetic code one is struck by the fact that it is structured almost entirely differently from computer codes. The genetic code has existed for almost four billion years, and it is hard to believe that in such a long period of time it has not optimized itself in some mathematical sense. The simplest aspect of this difference is striking and calls for a mathematical explanation. Almost all computer codes are binary codes, and the genetic code is always a four-letter code. What is the reason for this difference?

Why does the genetic code have four letters?

Almost nothing is known about the linear structure of the genetic code. We might ask, for example,

Does the genetic code contain recursive sequences?
Does the genetic code contain linear recursive sequences?
Does the genetic code contain codewords from known error-correcting codes?

Far from being answered, these questions have scarcely been asked. Perhaps the reason is that so far one has not used an algebraic model for the genetic code.

In 1949 the chemist Irwin Chargaff noticed that in *DNA*, the proportions of thymine (T) and adenine (A) are very nearly equal, and also that the proportions of guanine (G) and cytosine (C) are very nearly equal. Later it was realized that these proportions are exactly equal for the reason that, in the double helix, thymine is always paired with adenine (A), and guanine (G) is always paired with cytosine (C). These facts are now known as *Chargaff's Rules*.

If we wish to find an algebraic model for the genetic code, we must find a number system with four elements which we can identify with the four nucleotides T, A, G, C. It must also have some property which will allow us to respect the $T-A$ and $G-C$ pairings described by Chargaff's Rules.

The number system Z_4 satisfies these conditions. We can set $T = 0, A = 2, G = 1$, and $C = 3$. Chargaff's Rules are respected because

$$0 + 2 = 2 \text{ and } 2 + 2 = 0$$

and

$$1 + 2 = 3 \text{ and } 3 + 2 = 1$$

In other words, we get a model for the genetic code by setting $T = 0, A = 2, G = 1$, and $C = 3$ in Z_4, and interchanging the strands of the double helix by adding the vector

$$2222222222222 \ldots 222222222222$$

to either strand.

Figure 2.5 shows a segment of the double helix for the human TSH-β gene, and the model for this segment of the double helix using Z_4. By using this model, we can study the questions which we have posed above.

To illustrate computations in Z_n, we shall compute the Fibonacci numbers and the Lucas numbers in Z_n for several values of n. When we computed the Fibonacci numbers

GGTCACCACAGCATCTGCTCACCAATGCAAAGTAAG
CCAGTGGTGTCGTAGACGAGTGGTTACGTTTCATTC
11032333232132030130323322013222210221
332101101031021231210110023100032003

Figure 2.5   A Model for the Double Helix

and the Lucas numbers in the integers, we could of course not find them all. In Z_n, by contrast, we shall be able to find them all because both the Fibonacci numbers and the Lucas numbers are *periodic* in Z_n for any n. This means that the initial values recur after a finite number of steps called the *period*.

Recall that we have already computed the Fibonacci numbers in Z_2, and that the terms of the sequence were 0, 1, 1, 0, 1, 1, ..., so the period of the Fibonacci numbers in Z_2 is three. Since $2 = 0$ in Z_2, the sequence of Lucas numbers is the same as the sequence of Fibonacci numbers in Z_2, and so its period is also three.

EXAMPLE 1   Compute the Fibonacci numbers and the Lucas numbers in Z_3, and find the period of each.

Solution   For the Fibonacci numbers we have

$$0, 1, 1, 2, 0, 2, 2, 1, 0, 1, \ldots$$

so the period is eight.
For the Lucas numbers we have

$$2, 1, 0, 1, 1, 2, 0, 2, 2, 1, \ldots$$

so the period is also eight.   ●

EXAMPLE 2   Compute the Fibonacci numbers and the Lucas numbers in Z_4, and find the period of each.

Solution   For the Fibonacci numbers we have

$$0, 1, 1, 2, 3, 1, 0, 1, \ldots$$

so the period is six.
For the Lucas numbers we have

$$2, 1, 3, 0, 3, 3, 2, 1, \ldots$$

so the period is also six.   ●

The periods of the Fibonacci numbers and the Lucas numbers in Z_n are not always the same. The first time this occurs is in the next case, for $n = 5$.

EXAMPLE 3   Compute the Fibonacci numbers and the Lucas numbers in Z_5, and find the period of each.

Solution   For the Fibonacci numbers we have

$$0, 1, 1, 2, 3, 0, 3, 3, 1, 4, 0, 4, 4, 3, 2, 0, 2, 2, 4, 1, 0, 1, \ldots$$

so the period is twenty.

For the Lucas numbers we have

$$2, 1, 3, 4, 2, 1, \ldots$$

so the period is four.   ●

We now continue the discussion of the properties of Z_n. To avoid confusion, we shall again use the square bracket notation for congruence classes.

DEFINITION 2.3.5

i. A non-zero element $[r] \in Z_n$ such that there is another non-zero element $[s] \in Z_n$ with $[r] \cdot [s] = [0]$ is called a *zero-divisor*.

ii. An element $[r] \in Z_n$ which has a multiplicative inverse is called a *unit*.

THEOREM 2.3.3   The following conditions are equivalent for a congruence class $[r] \in Z_n$.

    (i) $g.c.d.(r, n) = 1$.

    (ii) $[r]$ has a multiplicative inverse.

    (iii) $[r]$ is not a zero-divisor in Z_n.

Proof   We will prove this theorem by proving a *circle of implications*: if (i) is true then (ii) is true; if (ii) is true then (iii) is true; and if (iii) is true then (i) is true.

If (i) then (ii):

If $g.c.d.(r, n) = 1$ then, by the Euclidean Algorithm, there are integers A and B such that $1 = Ar + Bn$. Then, $[1] = [Ar + Bn] = [A] \cdot [r] + [B] \cdot [n] = [A] \cdot [r]$, since $[n] = [0]$ in Z_n. Thus $[A] = [r]^{-1}$.

If (ii) then (iii):

Suppose that $[r]$ has a multiplicative inverse and $[r] \cdot [s] = [0]$ in Z_n. Then $[r]^{-1} \cdot ([r] \cdot [s]) = ([r]^{-1} \cdot [r]) \cdot [s] = [1] \cdot [s] = [s] = [r]^{-1} \cdot [0] = [0]$, so $[r]$ is not a zero-divisor.

If (iii) then (i):

Suppose that $g.c.d.(r, n) = d > 1$. We then have $n = dm$ and $r = ds$, with $1 < m < n$, so that $[m] \neq [0]$. Now $[r] \cdot [m] = [rm] = [dsm] = [sdm] = [sn] = [0]$, which contradicts the assumption that $[r]$ is not a zero-divisor. Thus, $g.c.d.(r, n) = 1$.   ●

The method that we have used to prove this theorem, proving a circle of implications, is a basic proof technique.

Notice that it follows from this theorem that every non-zero element of Z_n is either a unit or a zero-divisor, but not both. The theorem describes the cases in which the non-zero classes of Z_n are all units.

COROLLARY 2.3.1 If p is a prime, then every non-zero class $[r] \in Z_p$ has a multiplicative inverse. Conversely, if every non-zero class in Z_n is invertible, then $n = p$ is prime. ●

Because of this, Z_p resembles the number systems that we have studied more closely than Z_n does for composite integers n.

EXAMPLE 4 Find the multiplicative inverses of all the non-zero elements of Z_5.

Solution These are: $[1]^{-1} = [1]$, $[2]^{-1} = [3]$, $[3]^{-1} = [2]$, and $[4]^{-1} = [4]$. ●

EXAMPLE 5 Find the additive inverse of each element of Z_{10}. Divide the non-zero elements of Z_{10} into units and zero-divisors.

Solution The additive inverses are $-[0] = [0]$, $-[1] = [9]$, $-[2] = [8]$, $-[3] = [7]$, $-[4] = [6]$, and $-[5] = [5]$. The classes $[r]$ which have multiplicative inverses are those for which $g.c.d.(r, 10) = 1$. These are $[1]$, $[3]$, $[7]$, and $[9]$. The other non-zero classes $[2]$, $[4]$, $[5]$, $[6]$, and $[8]$ are zero-divisors, for example $[4] \cdot [5] = [0]$. The multiplicative inverses are $[1]^{-1} = [1]$, $[3]^{-1} = [7]$, $[7]^{-1} = [3]$, and $[9]^{-1} = [9]$. ●

Joseph Louis Lagrange (1736–1813) wrote several of the earliest papers about recursive sequences. He was also the first to observe and prove that the 1s digits of the Fibonacci numbers are periodic.

THEOREM 2.3.4 The period of the 1s digits of the Fibonacci numbers is 60.

This means that the 1s digits of the Fibonacci numbers repeat in intervals of length 60, but not in shorter intervals. We also prove that the period of the 1s digits of the Lucas numbers is 12.

Proof We prove this by computing the Fibonacci numbers and the Lucas numbers in Z_{10}. The results are recorded in Figure 2.6 and Figure 2.7. ●

```
2 1 3          0 1 1 2 3    5 8 3 1 4    5 9 4 3 7
4 7 1          0 7 7 4 1    5 6 1 7 8    5 3 8 1 9
8 9 7          0 9 9 8 7    5 2 7 9 6    5 1 6 7 3
6 3 9          0 3 3 6 9    5 4 9 3 2    5 7 2 9 1
2 1            0 1
```

Figure 2.7
The 1s Digits
of the Lucas
Numbers

Figure 2.6 The 1s Digits of the Fibonacci Numbers

In these arrays of 1s digits, we can see a curious pattern in the columns. The second column from the left in the table for the Fibonacci numbers is 1, 7, 9, 3, which are the units in Z_{10}. The other columns, starting from the left, are multiples of this one by 0, 1, 1, 2, 3, 5, ..., that is, by the classes of the successive Fibonacci numbers in Z_{10}. In the table for the Lucas numbers, the columns are multiples of this same column by 2, 1, 3, that is, by the classes of the first three Lucas numbers in Z_{10}.

APPLICATION   *Public Key Encryption*

There has always been a need to transmit and store data securely, and with the increasing use of computers, this need has grown. There have been many attempts to devise methods for the secure transmission and storage of data. There is some agreement that the most successful of these is a method which has come to be called Public Key Encryption. We shall describe a particular type of Public Key Encryption called *RSA Encryption*.

Here are the steps in the construction of an RSA cryptosystem.

I.   Choose two different prime numbers P and Q. In practice, these numbers will be chosen to be quite large.

II.   Choose a positive integer E such that $g.c.d.(E, (P-1)(Q-1)) = 1$. E need not be prime, but it must be odd, since $(P-1)(Q-1)$ is even.

III.   Since we have chosen E so that its greatest common divisor with $(P-1)$ $(Q-1)$ is 1, the congruence class of E will have a multiplicative inverse in the integers modulo $(P-1)(Q-1)$. In other words, there will be a positive integer D such that

$$ED \equiv 1 \;(\text{mod } (P-1)(Q-1))$$

and D can be computed from E, P and Q, as we have explained, using the Euclidean Algorithm.

IV.   An easy argument shows that there are $(P-1)(Q-1)$ elements of Z_{PQ} which have multiplicative inverses, and from this it easily follows that for any positive integer T,

$$T \equiv T^{ED} \equiv (T^E)^D \;(\text{mod } PQ)$$

Now suppose that T is the *plaintext* (the message to be sent, encoded as a positive integer). The *encryption function* is

$$T^E \;(\text{mod } PQ)$$

V.   The *decryption function*, applied to an integer C, is

$$C^D \;(\text{mod } PQ)$$

Starting with a positive integer T, if we apply the encryption function first and then the decryption function, we will obtain T again, as we have seen above.

The *public key* is the pair (PQ, E). The *private key* is the number D.

The point of this procedure is that D can be easily found if the two primes P and Q are known. If they are not, and if P and Q are large, it will take so long to factor PQ and find D that it will be impractical to decode the message.

In the Computer Exercises, we shall give examples of encoding and decoding with an RSA cryptosystem.

There are two potential problems with RSA cryptosystems:

I.   It has not been proved that factoring large integers is intrinsically difficult.

II.   It has not been proved that decoding an RSA cryptosystem depends in an intrinsic way on factoring PQ.

The goal of a *provably secure cryptosystem* has not yet been reached.

EXERCISES

Write an addition table and a multiplication table for Z_n in the following cases.

1. For Z_2.   **2.** For Z_3.   **3.** For Z_4.

4. For Z_5.   **5.** For Z_8.   **6.** For Z_{10}.

7. For Z_{12}.

Figure 2.4 shows six vectors of the ternary Golay Code G_{12}. Label these vectors in order $v_1, v_2, v_3, v_4, v_5, v_6$ and compute the following sums with coefficients in Z_3.

8. $v_1 + v_2$.   **9.** $v_1 + v_3$.   **10.** $v_2 + 2v_4$.

11. $v_3 + v_5$.   **12.** $v_2 + v_3 + v_4$.

13. $2v_4 + v_5 + 2v_6$.

In the following two problems, we will give segments from one strand in the double helix in certain animal genes. Write the other strand of the double helix, and give the model with entries from Z_4, as in Figure 2.5.

14. From the mouse Osteopontin gene:

$$ATGAGATTGGCAGTGATTT$$

$$GCTTTTGCCTGTTTGGC$$

15. From the rat Osteocalcin gene:

$$ATGAGGACCCTCTCTCTG$$

$$CTCACTCTGCTGGCCCTG$$

Find the periods of F_n and L_n in Z_k in the following cases.

16. In Z_6.   **17.** In Z_7.   **18.** In Z_8.

19. In Z_9.

Make a list of the units in Z_n in the following cases, and with each unit give its multiplicative inverse.

20. In Z_{16}.   **21.** In Z_{24}.   **22.** In Z_{32}.

In the next problems, we will be concerned with sequences of integers $a_0, a_1, a_2, \ldots$, where a_0 and a_1 are the initial values and the recursion $a_{n+2} = a_{n+1} + a_n$ is satisfied. We will compute these sequences in Z_{10}, in order to find the period of their 1s digits. These exercises are designed to illustrate the following facts:

(i)   The period of the 1s digits of every such sequence is a divisor of 120.

(ii)   Every divisor of 120 can occur as the period of the 1s digit of such a sequence.

(iii)   If two of these sequences are such that the period of the 1s digit is 120, then the two sequences of 1s digits can be exchanged by a cyclic permutation. (One can be turned into the other by successively placing the last digit in the period first.)

Find the period of the sequence of 1s digits for these sequences with the following initial values, by computing the sequences in Z_{10}. If the period is divisible by 4, arrange the classes of 1s digits in four rows as we have done in the text.

23. $a_0 = 0, a_1 = 5$.   **24.** $a_0 = 5, a_1 = 5$.

25. $a_0 = 1, a_1 = 3$.   **26.** $a_0 = 2, a_1 = 2$.

27. $a_0 = 1, a_1 = 7$.   **28.** $a_0 = 9, a_1 = 3$.

29. Prove that the sequences of 1s digits in Exercises 12 and 13 are cyclic permutations of each other, that is, that each can be turned into the other by successively placing the last term in the sequence first.

30. By computing the Fibonacci numbers in Z_{100}, prove that the period of the last two digits is 300.

31. By computing the Lucas numbers in Z_{100}, find the period of the last two digits.

ADVANCED EXERCISES

Here we shall describe a result involving computations with integers which is very useful in computer algorithms. Because it was known to Chinese astronomers it is called the *Chinese Remainder Theorem*.

The Chinese Remainder Theorem   Let a_1, a_2, $\ldots$, a_r be any integers, and let $n_1, n_2, \ldots, n_r$ be positive integers such that $g.c.d.(n_i, n_j) = 1$, if $i \neq j$. Then the system of congruences

$$X \equiv a_1 \pmod{n_1}$$
$$X \equiv a_2 \pmod{n_2}$$
$$\ldots$$
$$X \equiv a_r \pmod{n_r}$$

has a solution X, and that solution is unique modulo $n_1 n_2 \ldots n_r$.

Proof   Let N_i be the product of all the integers n_j except for n_i. It is clear that $g.c.d.(N_i, n_i) = 1$, and so the congruence class of N_i has a multiplicative inverse in Z_{n_i}, $[N_i]^{-1} = [A_i]$. Clearly,

$$N_i \equiv 0 \pmod{n_j}$$

and

$$N_i A_i \equiv 1 \pmod{n_i}$$

So, if

$$X = \sum_{i=1}^{r} N_i A_i a_i$$

then X will solve the system of congruences.   ●

For example, let us solve the system

$$X \equiv 1 \pmod 2$$
$$X \equiv 2 \pmod 3$$
$$X \equiv 3 \pmod 5$$

We have

$$X = N_1 A_1 a_1 + N_2 A_2 a_2 + N_3 A_3 a_3$$
$$= 15 \cdot 1 \cdot 1 + 10 \cdot 1 \cdot 2 + 6 \cdot 1 \cdot 3$$
$$= 53 \equiv 23 \pmod{30}$$

and this value of X solves the system of congruences.

Solve the following systems of congruences.

1. $X \equiv 0 \pmod 2$
    $X \equiv 1 \pmod 3$
    $X \equiv 4 \pmod 5$

2. $X \equiv 2 \pmod 3$
    $X \equiv 3 \pmod 5$
    $X \equiv 6 \pmod 7$

3. $X \equiv 2 \pmod 3$
    $X \equiv 3 \pmod 5$
    $X \equiv 4 \pmod 7$
    $X \equiv 5 \pmod{11}$

COMPUTER EXERCISES

1. The period of the last three digits of the Fibonacci numbers is 1500. Verify this by computing the first 1502 Fibonacci numbers in $Z_{1,000}$.

2. By computing the Lucas numbers in $Z_{1,000}$, find the period of the last three digits.

3. The period of the last four digits of the Fibonacci numbers is 15,000. Verify this by computing the first 15,002 Fibonacci numbers in $Z_{10,000}$.

4. By computing the Lucas numbers in $Z_{10,000}$, find the period of the last four digits.

5. Recall that the Tribonacci numbers are defined by the initial values $T_0 = 0$, $T_1 = 0$, $T_2 = 1$, and the recursion $T_n = T_{n-1} + T_{n-2} + T_{n-3}$ if $n \geq 3$.

    By computing the Tribonacci numbers in Z_{10}, find the period of the ones digits of the Tribonacci numbers.

6. By computing the Tribonacci numbers in Z_{100}, find the period of the first two digits of the Tribonacci numbers.

7. Prime numbers of the form

    $$2^{2^k} + 1$$

    are called *Gaussian primes*. Verify that these numbers are prime if $k = 0, 1, 2, 3, 4$.

    No other exponents k are known for which these numbers are prime. Factor them for as many values of k as you can.

In these exercises, we will ask you to go through the encoding and decoding process for RSA encryption. In each case, you will be able to carry out the decoding process because the primes P and Q involved are not very large. However, in the last examples, the decoding process will take some time. Perhaps this will illustrate how difficult decoding would be if the primes were larger.

In each case, you will be given two primes P and Q, a positive integer E, and a positive integer T. You are asked to find the integer D (recall that this can be done using the Euclidean Agorithm), and then to apply first the encryption function and then the decryption function to T, and to verify that this gives T again.

8. $P = 37$, $Q = 47$, $E = 5$, $T = 100$.
9. $P = 71$, $Q = 101$, $E = 11$, $T = 1,000$.
10. $P = 229$, $Q = 389$, $E = 23$, $T = 10,000$.
11. $P = 2003$, $Q = 4001$, $E = 1003$, $T = 100,000$.
12. $P = 54,730,729,297$, $Q = 143,581,524,529,603$, $E = 3613$, $T = 100$.

2.4   CHAPTER SUMMARY AND SUPPLEMENTARY EXERCISES

Yet still, ever after that sorrowful day,

Whenever the Butcher was by,

The Beaver kept looking the opposite way,

And appeared unaccountably shy.

Chapter Summary

Section 2.1

In this section, we discussed the basic properties of the integers. We showed that any integer can be factored into primes, and proved that this factorization is unique. We mentioned the Primality Problem and the Factorization Problem. We proved that the set of primes is infinite. We discussed the set of factors of an integer, $\text{Div}(n)$.

Section 2.2

We began this section with the statement and proof of Binet's Formula, which gives an explicit expression for the Fibonacci numbers and the Lucas numbers. We discussed the Golden Ratio ϕ, Lamé's Theorem, and two forms of the Egyptian Fraction Algorithm.

Section 2.3

In this section, we discussed the integers (mod n). We discussed the basic properties of the integers (mod n), and proved that they form a number system with n elements. We described applications of Z_n. We discussed the periodicity of the Fibonacci numbers and the Lucas numbers in Z_n. We showed how the elements of Z_n can be divided into the zero element, zero-divisors, and units. We described an application of Z_n of great practical importance, RSA Encryption.

EXERCISES

Use the Euclidean Algorithm to find the greatest common divisor of the following pairs of integers. In each case, verify the statement of Lamé's Theorem, that the number of steps is no greater than $5N$, where N is the number of digits of the smaller of the two integers.

1. $(m, n) = (100, 13)$.   **2.** $(m, n) = (750, 89)$.

3. $(m, n) = (1220, 754)$.

4. $(m, n) = (4791, 2961)$.

5. $(m, n) = (35422, 21892)$.

Use the Fermat Factorization Algorithm to factor each of the following integers n (each of these integers is a product of two primes).

6. $n = 5,893$.   **7.** $n = 37,979$.   **8.** $n = 99,301$.

9. $n = 4,284,347$.   **10.** $n = 14,022,053$.

Draw the graph of $\text{Div}(n)$ in the following cases.

11. $n = 70$.   **12.** $n = 231$.   **13.** $n = 28$.

14. $n = 125$.   **15.** $n = 462$.

ADVANCED EXERCISES

In combinatorics there are large numbers of "Pascal Triangles," that is, arrays of numbers whose entries have some kind of combinatorial significance. Here is an example, which Lucas included in his book on Number Theory.

```
                 1
             1   1   1
         1   2   3   2   1
     1   3   6   7   6   3   1
   1  4  10  16  19  16  10  4  1
```

1. Describe the recursion that generates this triangle.
   Prove by induction that the nth row of this Pascal Triangle consists of the coefficients of powers of x in

the expansion of

$$(x^2 + x + 1)^n$$

2. Now suppose that we have a chessboard with a top row and infinitely many rows and columns, and that there is a king on one of the squares in the top row. Suppose that the king moves down the chessboard, each move being diagonally to the left and downward, directly downward, or diagonally to the right and downward.
   Describe how the nth row of this Pascal Triangle gives the number of paths which the king has taken after n moves, and prove by induction that your description is correct.

COMPUTER EXERCISES

In these exercises, you are asked to carry out the RSA encryption-decryption procedure. In each exercise, you are given two primes P and Q, the number E which defines the encryption function, and the plaintext T. You must find an integer D which defines the decryption function, and verify that after encryption and decryption, you again obtain the plaintext T.

1. $P = 23$, $Q = 29$, $E = 13$, $T = 100$.

2. $P = 31$, $Q = 41$, $E = 47$, $T = 100$.

3. $P = 97$, $Q = 131$, $E = 337$, $T = 100$.

4. $P = 163$, $Q = 467$, $E = 67$, $T = 100$.

5. $P = 383$, $Q = 457$, $E = 103$, $T = 100$.

Functions, Relations, and Counting

3.1 FUNCTIONS AND RELATIONS

The Bellman himself they all praised to the skies—

Such a carriage, such ease and such grace!

Such solemnity, too! You could see he was wise,

The moment one looked in his face!

Leonhard Euler (1707–1783) made numerous contributions to combinatorics, and was the founder of graph theory. It was Euler who introduced the notation $f(x)$ for a function, $\sum$ for sum, Δ for the difference operator, i for $\sqrt{-1}$, and e for the base of the natural logarithm system.

DEFINITION 3.1.1

i. Let A and B be sets. A *function* $f : A \to B$ is a rule which assigns a unique $f(a) \in B$ to each $a \in A$. $f(a)$ is the *value* of the function f at $a \in A$. The set A is the *domain* of the function f and the set

$$f(A) = \{f(a) : a \in A\} \subset B$$

is the *range* of the function f.

ii. A function f is 1–1 or *injective* if $a_1 \in A$ and $a_2 \in A$ and $a_1 \neq a_2$ imply that $f(a_1) \neq f(a_2)$.

---

iii. A function f is *onto* or *surjective* if for each $b \in B$ there is an $a \in A$ such that $f(a) = b$.

iv. A function $f : A \rightarrow B$ which is both injective and surjective is called a *one-one correspondence* between A and B.

EXAMPLE 1 Let Z be the set of integers and define three functions $f, g, h : Z \rightarrow Z$, as follows: $f(n) = 2n$; $g(n) = n/2$ if n is even and $g(n) = (n + 1)/2$ if n is odd; and $h(n) = n + 1$. Discuss the injectivity and surjectivity of these three functions.

Solution The function f is injective, but it is not surjective because no odd integer is in the range of f.

The function g is surjective, because for any integer n, $g(2n) = n$. g is not injective because for example $g(1) = (1 + 1)/2 = 1 = g(2)$.

The function h is injective and surjective, and hence is a 1–1 correspondence between Z and Z. ●

DEFINITION 3.1.2

i. If $f : A \rightarrow B$ and $g : B \rightarrow C$ are functions, the *composite function* $g \circ f : A \rightarrow C$ is defined by $(g \circ f)(a) = g(f(a))$.

ii. If $f : A \rightarrow B$ is a function, then a function $g : B \rightarrow A$ is the *inverse* of f, in symbols $g = f^{-1}$, if $g(f(a)) = a$ for all $a \in A$ and $f(g(b)) = b$ for all $b \in B$.

A function may or may not have an inverse. Inverse functions are unique if they exist.

EXAMPLE 2 If $f(x) = x^2$ and $g(x) = x + 1$, find $f \circ g$ and $g \circ f$.

Solution

$$(f \circ g)(x) = f(g(x)) = f(x + 1) = (x + 1)^2 = x^2 + 2x + 1$$

and

$$(g \circ f)(x) = g(f(x)) = g(x^2) = x^2 + 1$$

Notice that the two composite functions are not the same. ●

EXAMPLE 3 Find the inverse of the function $f : R \rightarrow R$ defined by $f(x) = 2x + 3$.

Solution

$$g(x) = (x - 3)/2$$

because

$$g(f(x)) = g(2x + 3) = ((2x + 3) - 3)/2 = 2x/2 = x$$

and

$$f(g(x)) = f((x - 3)/2) = 2(x - 3)/2 + 3 = x \qquad \bullet$$

THEOREM 3.1.1   The following conditions are equivalent:

    (i)   The function $f : A \to B$ is a one–one correspondence.

    (ii) f has an inverse function $g : B \to A$.

Proof   Assume that $f : A \to B$ is one–one correspondence. If $b \in B$ then, since f is onto, there is an $a \in A$ such that $f(a) = b$, and since f is 1–1 this element $a \in A$ is unique. Thus we can define $g(b) = a$, and clearly $g(f(a)) = a$ for all $a \in A$, so f has an inverse function $g : B \to A$. From the definition it is clear that $f(g(b)) = b$ for all $b \in B$.

    Now assume that f has an inverse function $g : B \to A$. For any $b \in B$ we have $f(a) = b$, where $a = g(b)$, so f is onto. Suppose that we have $a_1 \in A$ and $a_2 \in A$ such that $f(a_1) = f(a_2)$. Then $a_1 = g(f(a_1)) = g(f(a_2)) = a_2$ which means that f is $1 - 1$, and therefore f is a one-one correspondence. $\bullet$

    The following are examples of the injectivity and surjectivity of linear and quadratic functions.

EXAMPLE 4   Prove that a non-constant linear function $f(x) = ax + b$ from the set of real numbers to itself is always a one–one correspondence.

Solution   If $f(x)$ is non-constant then $a \neq 0$, and $f(x)$ has the inverse function $g(x) = (x - b)/a$. $\bullet$

    Although a non-constant linear function is always a one–one correspondence from the set of real numbers to itself, this may not be true if the function is restricted to the set of integers. The function will always be injective, but it may not be surjective.

EXAMPLE 5   If a and b are integers, describe when the function $f(x) = ax + b$ is surjective as a function from the set of integers to itself.

Solution   If $a = 1$ or $a = -1$ then $f(x)$ has an inverse $g(x) = a(x - b)$ and so $f(x)$ is surjective. If a is greater than 1 in absolute value, then the integer $b + 1$ cannot be in the range of $f(x)$. This would mean that $f(x) = ax + b = b + 1$ so that $ax = 1$, which is impossible.

    Thus a linear function $f(x) = ax + b$ is a one–one correspondence from the set of integers to itself if and only if $a = 1$ or $a = -1$. $\bullet$

    A quadratic function $f(x) = ax^2 + bx + c$ from the set of real numbers to itself has a graph which is a parabola. There are numbers which are not in the range of $f(x)$, so $f(x)$ is never surjective, and there are infinitely many numbers in the range which are values of $f(x)$ for two different values of x, so $f(x)$ is never injective.

If a, b, and c are integers, the function $f(x) = ax^2 + bx + c$ as a function from the set of integers to itself can never be surjective. However, it can be injective.

EXAMPLE 6   Describe when a quadratic function $f(x) = ax^2 + bx + c$, with integer coefficients, is injective as a function from the set of integers to itself.

Solution   Suppose that m and n are different integers and $f(m) = f(n)$. This means that

$$am^2 + bm + c = an^2 + bn + c$$

so

$$a(m^2 - n^2) = a(m - n)(m + n) = b(n - m)$$

so

$$-a(m + n) = b$$

and

$$b/a = -(m + n)$$

If a does not divide b, this equation cannot hold, so $f(x)$ is injective. If a does divide b, then we can find pairs of integers which satisfy the equation, so $f(x)$ is not injective.

For example, $f(x) = 2x^2 + x$ is an injective function from the set of integers to itself. ●

The *graph* of a function $f(x)$ is the set of ordered pairs

$$\{(x, f(x))\}$$

where x is in the domain of $f(x)$. A graph, in the sense of the picture that one draws in precalculus or calculus, is also a graph in the sense of being a directed graph. We will give an example of a graph of a function from a finite set to itself.

EXAMPLE 7   Find the graph of the function $f : Z_{10} \rightarrow Z_{10}$ defined on congruence classes $[k]$ by $f([k]) = [2k]$.

Solution   The values of f are $f([0]) = [0]$, $f([1]) = [2]$, $f([2]) = [4]$, $f([3]) = [6]$, $f([4]) = [8]$, $f([5]) = [0]$, $f([6]) = [2]$, $f([7]) = [4]$, $f([8]) = [6]$, $f([9]) = [8]$. The graph of f is in Figure 3.1. Notice that the graph has two components, each with one cycle. This is always the case:

> *The graph of any function from a finite set to itself consists of finitely many components, each with one cycle.* ●

EXAMPLE 8   Let X be any set and define $f : P(X) \rightarrow P(X)$ by $f(A) = A^c = X - A$. Then f is a one–one correspondence.

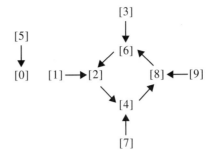

Figure 3.1   The Graph of a Function from Z_{10} to Z_{10}

Proof   To prove that f is surjective, we must show that for any subset $B \subset X$ there is a subset $A \subset X$ such that $f(A) = B$. Choose $A = B^c$. Since

$$A^c = (B^c)^c = B$$

we have $f(A) = B$ and so f is surjective.

Suppose that we have two subsets $A_1 \subset X$ and $A_2 \subset X$ such that $A_1^c = f(A_1) = f(A_2) = A_2^c$. Then

$$A_1 = (A_1^c)^c = (A_2^c)^c = A_2$$

which means that f is injective.   ●

DEFINITION 3.1.3   If G_1 and G_2 are graphs, a *graph isomorphism* $f : G_1 \to G_2$ is a one–one correspondence $f : V(G_1) \to V(G_2)$ such that the function $g : E(G_1) \to E(G_2)$ defined by

$$g(\{v_1, v_2\}) = \{f(v_1), f(v_2)\}$$

is a one–one correspondence.

EXAMPLE 9   Give an example of an isomorphism of the graph of the 3-Cube with itself.

Solution   If we regard the 3-Cube as the set of all subsets of $X = \{1, 2, 3\}$, the function $f : P(X) \to P(X)$ defined by $f(A) = A^c$ defines an isomorphism of the graph of the 3-Cube with itself. The effect of this isomorphism is shown in Figure 3.2.   ●

Since $f(f(A)) = (A^c)^c = A$, the function f is its own inverse: $f^{-1} = f$.

Given a finite set $A = \{a_1, a_2, \ldots, a_n\}$ and a subset $B \subset A$, we defined the characteristic function χ_B of B as follows: for each $a \in A$, $\chi_B(a) = 1$ if $a \in B$ and $\chi_B(a) = 0$ if $a \notin B$.

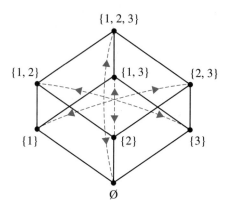

Figure 3.2   An Isomorphism of the 3-Cube with Itself

Then define a function

$$\chi : P(A) \rightarrow Z_2^n$$

by the expression

$$\chi(B) = (\chi_B(a_1), \chi_B(a_2), \ldots, \chi_B(a_n))$$

for each subset $B \subset A$.

THEOREM 3.1.2   For any finite set $A = \{a_1, a_2, \ldots, a_n\}$ the function

$$\chi : P(A) \rightarrow Z_2^n$$

is a one–one correspondence.

Proof χ *is injective.*   Suppose that $B_1 \subset A$ and $B_2 \subset A$ are subsets and $B_1 \neq B_2$. Then one of these two subsets must contain an element $a \in A$ which the other does not, say $a \in B_1 - B_2$. Then $\chi_{B_1}(a) = 1 \neq 0 = \chi_{B_2}(a)$, so $\chi(B_1) \neq \chi(B_2)$. This means that χ is injective.

χ *is surjective.*   Given a finite set $A = \{a_1, a_2, \ldots, a_n\}$ and a binary vector $(b_1, b_2, \ldots, b_n)$, define a subset $B \subset A$ as follows: given $a_k \in A$, $a_k \in B$ if $b_k = 1$ and $a_k \notin B$ if $b_k = 0$. Then $\chi(B) = (b_1, b_2, \ldots, b_n)$, so χ is surjective.   ●

Closely related to the concept of a function is the concept of a *relation*.

DEFINITION 3.1.4

i. A *relation* R on a set X is a subset of the product

$$R \subset X \times X.$$

> **ii.** A relation $R \subset X \times X$ is *reflexive* if $(a, a) \in R$ for all $a \in X$.
> **iii.** A relation $R \subset X \times X$ is *symmetric* if $(a, b) \in R$ implies $(b, a) \in R$ for all $a \in X$ and $b \in X$.
> **iv.** A relation $R \subset X \times X$ is *antisymmetric* if $(a, b) \in R$ and $a \neq b$ imply $(b, a) \notin R$ for all $a \in X$ and $b \in X$.
> **v.** A relation $R \subset X \times X$ is *transitive* if $(a, b) \in R$ and $(b, c) \in R$ imply $(a, c) \in R$ for all $a \in X$, $b \in X$, and $c \in X$.
> **vi.** A relation $R \subset X \times X$ is a *partial ordering* if R is reflexive, antisymmetric, and transitive.
> **vii.** A pair (X, R) where X is a set and R is a partial ordering on X is a *partially ordered set* or a *poset*.
> **viii.** A relation $R \subset X \times X$ is an *equivalence relation* if R is reflexive, symmetric, and transitive.

We will use the notation aRb for $(a, b) \in R$, that is, aRb means that "a stands in the relation R to b."

Examples of Partial Orderings

(i)   If X is any set and $P(X)$ is the set of all subsets of X, then the relation of inclusion $\subset$ is a partial ordering on $P(X)$.

(ii) In particular, if $X = \{1, 2, \ldots, n\}$, the poset $(P(X), \subset)$ is called the *n-Cube*.

(iii) $\mathrm{Div}(n)$ is the set of positive integers which divide n. The relation of divisibility is a partial ordering on the set $X = \mathrm{Div}(n)$.

> **DEFINITION 3.1.5**   The *Hasse diagram* of a poset (X, R) is the following graph G:
> **i.** $V(G) = X$.
> **ii.** If $a \in X$ and $b \in X$, then $\{a, b\} \in E(G)$ if aRb, $a \neq b$, and there is no element $c \in X$ with $a \neq c$, $b \neq c$, and aRc and cRb (we say that b is an *immediate successor* of a).

Figure 3.3 shows the Hasse diagram of the poset $\mathrm{Div}(36)$.
We can also consider relations which relate the elements of different sets.

> **DEFINITION 3.1.6**
> **i.** A *binary relation* between the elements of two sets X and Y is a subset
> $$R \subset X \times Y.$$

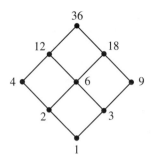

Figure 3.3   The Hasse Diagram of Div(36)

ii. An *n-ary relation* between the elements of the sets $X_1, X_2, \ldots, X_n$ is a subset
$$R \subset X_1 \times X_2 \times \ldots X_n.$$

Examples of Equivalence Relations

  (i)   If X is any set, the relation "$a = b$" is an equivalence relation on X.

 (ii)   If Z is the set of integers and n is any positive integer, the relation "a is congruent to b (mod n)" is an equivalence relation.

(iii)   If X is any set and $\{X_\alpha\}_{\alpha \in A}$ is any partition of X (this means that $X = \cup_{\alpha \in A} X_\alpha$ and $X_\alpha \cap X_\beta = \emptyset$ if $\alpha \neq \beta$), then the relation "aRb if there is an $\alpha \in A$ such that $a \in X_\alpha$ and $b \in X_\alpha$" is an equivalence relation on X.

In mathematics we often want to consider objects to be "the same" if they are equivalent in some sense. This idea is expressed by the following definition.

DEFINITION 3.1.7   Let X be a set, R an equivalence relation on X, and $a \in X$. The *equivalence class* of a is the set
$$R[a] = \{b \in X : aRb\}$$
that is, the set of elements $b \in X$ which stand in the relation R to $a \in X$.

Using the concept of an equivalence class, we can prove that every equivalence relation arises from a partition.

The subsets $X_\alpha \subset X$ in a partition are often called the *cells* of the partition.

THEOREM 3.1.3   **The Partition Theorem**   Each of the following items of data determines the other:

  (i)   A partition $\{X_\alpha\}_{\alpha \in A}$ of a set X.

 (ii)   An equivalence relation R on X.

Proof   ***(i) implies (ii).***   Given a partition $\{X_\alpha\}_{\alpha\in A}$ of X, we define the relation R as follows: given $a \in X$ and $b \in X$, aRb if there is an $\alpha \in A$ such that $a \in X_\alpha$ and $b \in X_\alpha$. That is, a and b are in the same cell of the partition.

Given $a \in X$ there is an $\alpha \in A$ such that $a \in X_\alpha$, which means that aRa, so R is reflexive. Given $a \in X$ and $b \in X$ such that aRb there is an $\alpha \in X$ such that $a \in X_\alpha$ and $b \in X_\alpha$. This also implies that bRa, so R is symmetric. Given $a \in X, b \in X$, and $c \in X$ such that aRb and bRc, there must be an $\alpha \in X$ and a $\beta \in X$ such that $a \in X_\alpha$ and $b \in X_\alpha$, and also such that $b \in X_\beta$ and $c \in X_\beta$. Then we must have $\alpha = \beta$, so aRc and R is transitive. Since R is reflexive, symmetric, and transitive, it is an equivalence relation.

(ii) implies (i).   Given an equivalence relation R on X, we must find a partition $\{X_\alpha\}_{\alpha\in A}$ of X. We define the cells X_α of the partition to be the *different* equivalence classes $R[a]$ of R, and we must prove that these form a partition of X. We must prove that every element $a \in X$ is in some equivalence class of R, and that different equivalence classes have empty intersection.

First, since R is reflexive, we have aRa for any $a \in X$, so $a \in R[a]$. Thus every $a \in X$ is in some equivalence class of R. Now suppose that $R[a] \neq R[b]$, we must prove that $R[a] \cap R[b] = \emptyset$. It is more convenient to prove the logically equivalent statement: if $R[a] \cap R[b] \neq \emptyset$ then $R[a] = R[b]$. It is sufficient to prove that $R[a] \subset R[b]$, since an exactly similar argument will prove that $R[b] \subset R[a]$, and hence that $R[a] = R[b]$.

If $R[a] \cap R[b] \neq \emptyset$ then there must be an element $c \in R[a] \cap R[b]$. Suppose that $d \in R[a]$. We must prove that $d \in R[b]$. Altogether we are assuming that aRd, aRc, and bRc. By the symmetry of R we have cRb. By transitivity, aRb. By symmetry, dRa. By transitivity, dRb. By symmetry again, bRd, so $d \in R[b]$. We have proved that $R[a] \subset R[b]$, and this completes the proof.   ●

Just as any function has a graph which can be drawn, at least in principle, the graph of any relation is a directed graph. An equivalence relation has a graph with a loop at each node (reflexivity); each edge is double, with arrows in opposite directions (symmetry); and if three nodes are joined by two double edges, the triangle that they form must be closed on the third side (transitivity). Figure 3.4 shows an equivalence relation with two equivalence classes.

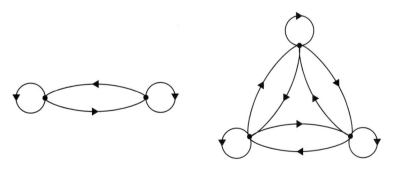

Figure 3.4   An Equivalence Relation with Two Equivalence Classes

The digraph of any equivalence relation R breaks up into components which correspond to the equivalence classes of R.

The Prüfer Correspondence

Let T_n be the set of labelled trees with n vertices, that is, the set of trees with n vertices whose vertices are labelled with the integers $1, 2, \ldots, n$. We shall describe a 1–1 correspondence P between T_n and the set $[n]^{n-2}$ of $(n-2)$-tuples of integers $1, 2, \ldots, n$. This 1–1 correspondence is called the *Prüfer Correspondence* because it was described by the German mathematician Ernst Paul Heinz Prüfer (1896–1934) in a paper published in 1918.

We shall define the function $P : T_n \to [n]^{n-2}$ and its inverse, but we shall omit the proof that the composition of these two functions is the identity. In what follows, we assume that $n \geq 3$.

Definition of P   Given a labelled tree T with n vertices, define $P(T)$ as follows:

   (i)   Choose the vertex of degree one of T with the smallest label (every tree T has at least one vertex of degree one).

  (ii)   Take the (unique) vertex of T adjacent to the vertex in (i) and make its label the first term in the sequence $P(T)$.

 (iii)   Remove the vertex in (i) and its adjacent edge from T.

  (iv)   Repeat these steps until there are only two vertices left in T. The resulting ordered $(n-2)$-tuple $P(T)$ is called the *Prüfer Code* of T.

Definition of P^{-1}   Given a sequence $S \in [n]^{n-2}$, define the labelled tree $P^{-1}(T)$ as follows:

   (i)   Draw n vertices, labelled $1, 2, \ldots, n$, and also make a list $1, 2, \ldots, n$.

  (ii)   Find the smallest number which is in the list $1, 2, \ldots, n$ but not in S, and join the vertex with this label to the vertex whose label is the first number in S, by an edge.

 (iii)   Remove the first number in S from S and the number in (ii) from the list.

  (iv)   Repeat these steps until there are only two numbers left on the list and then join the vertices with these two labels, by an edge.

EXAMPLE 10   Prove that the Prüfer code of the tree T in Figure 3.5 is $P(T) = (2, 3, 2)$ and also that $P^{-1}((2, 3, 2)) = T$.

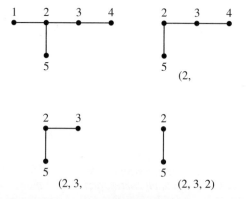

Figure 3.5   The Prüfer Code of a Labelled Tree

Solution   We carry out these steps in Figure 3.5 and Figure 3.6.

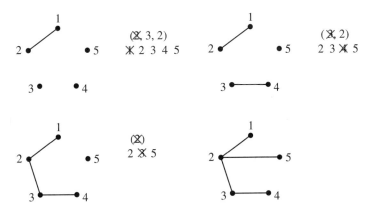

Figure 3.6   The Inverse Prüfer Correspondence

EXERCISES

In the next six exercises, prove that each function, with domain and range the set of real numbers R, is injective and surjective. Hence each of these functions is a one–one correspondence. Find the inverse of each function.

1. $f(x) = 2x + 3$

2. $g(x) = 3x + 2$

3. $h(x) = 3x - 1$

4. $k(x) = 5x + 3$

5. $l(x) = 23x - 13$

6. $m(x) = 47x - 71$

In the next four exercises, find the composite functions $f \circ g$ and $g \circ f$. Are the two composites ever equal?

7. $f(x) = x^2$ and $g(x) = x + 2$

8. $f(x) = x^2 + 1$ and $g(x) = 2x$

9. $f(x) = x^3$ and $g(x) = x + 1$

10. $f(x) = x^4$ and $g(x) = x^3$

11. Let $X = \{1, 2, 3, 4\}$, and define $f : P(X) \to P(X)$ by $f(A) = A^c$ ($f(A)$ is the complement of $A \subset X$). Draw the graph of the 4-Cube labelled with the subsets of $\{1, 2, 3, 4\}$. The mapping f gives an isomorphism of this graph with itself. Indicate the effect of this graph isomorphism in your diagram, as in Figure 3.2.

12. List all subsets of $X = \{1, 2, 3\}$. Next to this, list all binary vectors of length three in such a way that for each subset A in the first list, the binary vector in the second list is $\chi(A)$.

13. List all subsets of $X = \{1, 2, 3, 4\}$. Next to this, list all binary vectors of length four in such a way that for each subset A in the first list, the binary vector in the second list is $\chi(A)$.

For each of the following relations R on the set X, state whether the relation is or is not reflexive, symmetric, antisymmetric, transitive, a partial ordering, or an equivalence relation.

14. $X = \{1, 2, 3, 4\}$ and $R = \{(1, 1), (2, 2), (3, 3), (4, 4), (1, 2), (1, 3), (2, 4), (3, 4), (1, 4), \}$

15. $X = \{1, 2, 3, 4\}$ and $R = \{(1, 1), (2, 2), (3, 3), (1, 2), (2, 4)\}$

16. $X = \{1, 2, 3, 4, 5\}$ and $R = \{(1, 1), (2, 2), (3, 3), (4, 4), (5, 5), (1, 2), (1, 4), (1, 5), (2, 4), (2, 5), (1, 3), (3, 4), (3, 5), (4, 5)\}$

17. $X = \{1, 2, 3, 4\}$ and $R = \{(1, 1), (2, 2), (3, 3), (4, 4), (1, 2), (2, 1)\}$

18. $X = \{1, 2, 3, 4, 5, 6\}$ and $R = \{(1, 1), (2, 2), (3, 3), (4, 4), (5, 5), (6, 6), (1, 2), (2, 1), (1, 3), (3, 1), (2, 3), (3, 2), (4, 5), (5, 4)\}$

For each of the following partitions of a set X, write the corresponding equivalence relation $R \subset X \times X$.

19. $X = \{1, 2, 3, 4\}$, $X_1 = \{1, 2\}$, and $X_2 = \{3, 4\}$

20. $X = \{1, 2, 3, 4, 5, 6\}$, $X_1 = \{1, 2\}$, $X_2 = \{3, 4\}$, and $X_3 = \{5, 6\}$

21. $X = \{1, 2, 3, 4, 5, 6\}$, $X_1 = \{1, 2, 3\}$, $X_2 = \{4, 5\}$, and $X_3 = \{6\}$

22. Find the partition of $X = \{1, 2, 3, 4\}$ corresponding to the equivalence relation $R = \{(1, 1), (2, 2), (3, 3), (4, 4), (1, 2), (2, 1)\}$

23. Give the Prüfer code for the labelled tree

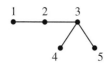

24. Give the Prüfer code for the labelled tree

25. Give the Prüfer code for the labelled tree

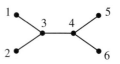

26. Find the labelled tree with five vertices whose Prüfer code is $(2, 3, 4)$.

27. Find the labelled tree with five vertices whose Prüfer code is $(4, 3, 2)$.

28. Find the labelled tree with five vertices whose Prüfer code is $(5, 5, 5)$.

ADVANCED EXERCISES

A positive integer n is *square-free* if n is not divisible by a square. Equivalently, each prime divisor of n occurs exactly once. In these exercises we shall illustrate the following fact, which we shall prove in Chapter 7.

> *The graph of $Div(n)$ is a cube if and only if n is square-free.*

Draw the graph of $Div(n)$ for the following values of n.

1. $n = 27$
2. $n = 45$
3. $n = 105$
4. $n = 16$
5. $n = 24$
6. $n = 36$
7. $n = 60$
8. $n = 210$

3.2 A SURVEY OF COUNTING RULES

> *He was thoughtful and grave—but the orders he gave*
> *Were enough to bewilder a crew.*
> *When he cried "Steer to starboard, but keep her head larboard!"*
> *What on earth was the helmsman to do?*

Gottfried Wilhelm Leibniz (1646–1716) published his *Dissertatio de arte combinatoria* in 1666. The word *combinatorics*, the name for the science of counting, comes from the title of this work. In the *Dissertatio*, Leibniz proposed a unified theory of logic, combinatorics, and probability theory. What we now call discrete mathematics still appears to be made up of different topics, but much progress has resulted from attempts to unify it.

In this section we shall describe some elementary counting rules.

If X is a finite set, we will denote the number of elements of X by $|X|$.

Recall that a (finite) partition of a set X is a family of subsets $X_1, X_2, \ldots, X_r$ of X such that

$$X = X_1 \cup X_2 \cup \ldots \cup X_r$$

and

$$X_i \cap X_j = \emptyset$$

if $1 \leq i, j \leq r$, and $i \neq j$. The subsets X_i are sometimes called the cells of the partition.

The Sum Rule

If X is a finite set and $X_1, X_2, \ldots, X_r$ is a partition of X, then

$$|X| = |X_1| + |X_2| + \ldots + |X_r|$$

The Sum Rule can be proved by induction. We shall leave this as an exercise.

The Product Rule

   i. If X and Y are finite sets then

$$|X \times Y| = |X| \cdot |Y|$$

   ii. If $X_1, X_2, \ldots, X_r$ are all finite sets then

$$|X_1 \times X_2 \times \ldots \times X_r| = |X_1| \cdot |X_2| \cdot \ldots \cdot |X_r|$$

Proof of (i)   The sets

$$\{X \times \{y\}\}_{y \in Y}$$

form a partition of $X \times Y$ with $|Y|$ cells. Since $|X \times \{y\}| = |X|$ for each $y \in Y$, $|X \times Y| = |X| \cdot |Y|$ by the Sum Rule. (ii) follows from (i) by induction.   ●

The Surjective Function Rule

If X and Y are finite sets and $f : X \to Y$ is a surjective function, then if we define

$$f^{-1}(y) = \{x \in X : f(x) = y\}$$

then

$$\{f^{-1}(y)\}_{y \in Y}$$

is a partition of X, and then

$$|X| = \sum_{y \in Y} |f^{-1}(Y)|$$

In particular, if the sets $f^{-1}(y)$ all have the same number of elements N, then

$$|X| = N \cdot |Y|$$

Proof   Let $x \in X$. Since the domain of f is X, there is a $y \in Y$ such that $f(x) = y$, so that $x \in f^{-1}(y)$. This implies that $X = \cup_{y \in Y} f^{-1}(y)$.

If y_1 and y_2 are two different elements of Y and if $x \in f^{-1}(y_1)$, then $f(x) = y_1$. Since $f(x)$ is a unique element of Y (by definition of a function), $f(x) \neq y_2$, and so $x \notin f^{-1}(y_2)$. This means that $f^{-1}(y_1) \cap f^{-1}(y_2) = \emptyset$, so the sets

$$\{f^{-1}(y)\}_{y \in Y}$$

form a partition of X.   The Surjective Function Rule now follows from the Sum Rule.   ●

The Pigeonhole Principle

   i. If $r + 1$ balls are placed in r slots, then some slot must contain at least two balls.

   ii. If $rs + 1$ balls are placed in r slots, then some slot must contain at least $s + 1$ balls.

Proof of (i)   If each of the r slots contains no more than one ball, then by the Sum Rule the total number of balls is no more than r, which contradicts the assumption that there are $r + 1$ balls.   ●

Proof of (ii)   If each of the r slots contains no more than s balls, then by the Sum Rule the total number of balls is no more than rs, which contradicts the assumption that there are $rs + 1$ balls.   ●

Following are three applications of the Pigeonhole Principle.

EXAMPLE 1   Prove that in any group of thirteen people, at least two must have birthdays in the same month.

Solution   We may think of the twelve months in which the birthdays may occur as the slots, and the thirteen or more people as the balls. Then, by the Pigeonhole Principle, there must be at least two balls in some slot. That is, at least two of the people must have birthdays in the same month.   ●

EXAMPLE 2   Each week a man goes to a shopping center where there are seven stores, and shops at two of the stores. If he goes to the shopping center for 43 weeks, prove that he must shop at some pair of stores at least three times.

Solution   The number of pairs of stores is $\binom{7}{2} = 21$. Think of the 43 visits to the shopping center as balls, and the 21 pairs of stores as slots. Since $43 > 2 \cdot 21 = 42$, some slot must contain at least 3 balls by the Pigeonhole Principle. In other words, the man must visit some pair of stores at least three times.   ●

Suppose that X and Y are finite sets and $f : X \to Y$ is a function from X to Y. If f is injective, then $|X| \le |Y|$ and if f is surjective, then $|X| \ge |Y|$, so if f is a one–one correspondence, $|X| = |Y|$.

THEOREM 3.2.1   Suppose that $f : X \to Y$ is a function from one finite set to another, and $|X| = |Y|$. Prove that if f is injective, then f is surjective, and that if f is surjective, then f is injective.

Proof   Suppose that f is surjective. Then by the Surjective Function Rule and the Pigeonhole Principle, f must be injective. If f is injective, then $|f(X)| = |Y|$, so f must be surjective.   ●

The theorem implies that if we want to check whether a function f, from one finite set to another, with the same number of elements, is a one–one correspondence, it is sufficient to verify that f is injective or that f is surjective. In particular, this is true if $f : X \to X$ is a function from a finite set to itself.

Applied Form of the Product Rule   Suppose that a process is carried out in r steps. The first step involves n_1 choices, the second step involves n_2 choices, and so on, and the rth step involves n_r choices. Suppose also that the choices are independent (the way in which one is carried out does not affect the way in which any other is carried out).

In this situation, the process can be carried out in

$$N = n_1 \cdot n_2 \cdot \ldots \cdot n_r$$

ways in all.

EXAMPLE 3   In a certain company there are seven people in Division A, ten people in Division B, and nineteen people in Division C. A committee is composed of two people from Division A, three people from Division B, and two people from Division C. How many committees are possible?

Solution   There are $\binom{7}{2} = 21$ pairs of people from Division A, $\binom{10}{3} = 120$ sets of three people from Division B, and $\binom{19}{2} = 171$ pairs of people from Division C. By the Product Rule there are

$$21 \cdot 120 \cdot 171 = 430,920$$

committees in all.   ●

EXAMPLE 4   If A is a set with r elements, count the strings of length n of elements of A.

Solution   The strings of length n of elements of A can be thought of as elements of the set

$$A \times A \times \ldots \times A$$

where the set A occurs n times. By the Product Rule the number of elements of this set is

$$|A|^n = r^n \qquad ●$$

Thus the number of binary strings of length n is 2^n, the number of ternary strings of length n is 3^n, and so on.

EXAMPLE 5   If X and Y are finite sets, and if Y^X is the set of all functions from X to Y, then

$$|Y^X| = |Y|^{|X|}$$

Solution   If $X = \{x_1, x_2, \ldots, x_n\}$ and $f : X \to Y$ is a function, define

$$F(f) = (f(x_1), f(x_2), \ldots, f(x_n)) \in Y \times Y \times \ldots \times Y$$

where the set Y occurs $|X| = n$ times in the product. The mapping

$$F : Y^X \to Y \times Y \times \ldots \times Y$$

is a one–one correspondence, so the formula follows from the Product Rule.   ●

Sometimes problems involving product sets are best solved by representing the product set directly.

EXAMPLE 6   Two dice are thrown. In how many ways can the sum of the numbers on the dice be a seven or eleven?

Solution   The set of outcomes when two dice are thrown can be pictured as a six-by-six grid (the product set $A \times A$, where $A = \{1, 2, 3, 4, 5, 6\}$), where the first number is the number which comes up on the first die and the second number is the number which comes up on the second die.

$$
\begin{array}{cccccc}
(1,1) & (1,2) & (1,3) & (1,4) & (1,5) & (1,6) \\
(2,1) & (2,2) & (2,3) & (2,4) & (2,5) & (2,6) \\
(3,1) & (3,2) & (3,3) & (3,4) & (3,5) & (3,6) \\
(4,1) & (4,2) & (4,3) & (4,4) & (4,5) & (4,6) \\
(5,1) & (5,2) & (5,3) & (5,4) & (5,5) & (5,6) \\
(6,1) & (6,2) & (6,3) & (6,4) & (6,5) & (6,6)
\end{array}
$$

Inspecting this grid, we see that there are six cases where the sum of the numbers on the dice is seven, and two cases where the sum is eleven. So there are eight cases, out of thirty-six cases in all, in which a seven or eleven occurs.   ●

DEFINITION 3.2.1

i. If $A = \{a_1, a_2, \ldots, a_n\}$ is a set with n elements, a *k-permutation* of the elements of A is an ordered k-tuple of different elements of A. The number of k-permutations of the elements of A will be denoted $P(n, k)$. If $k = n$, n-permutations are called simply *permutations* of the elements of A.

ii. A *k-combination* of the elements of A is an unordered k-element subset of A. The number of k-combinations of the elements of A will be denoted $C(n, k)$.

We shall use round brackets for ordered sets and curly brackets for unordered sets.

EXAMPLE 7   List the permutations of the elements of $\{1, 2, 3\}$.

Solution   These are:

$$(1, 2, 3)$$
$$(1, 3, 2)$$
$$(2, 1, 3)$$
$$(2, 3, 1)$$
$$(3, 1, 2)$$
$$(3, 2, 1)$$

●

THEOREM 3.2.2   If $1 \leq k \leq n$, then

$$P(n, k) = \frac{n!}{(n-k)!} = n \cdot (n-1) \cdot \ldots \cdot (n-k+1)$$

In particular,

$$P(n, n) = n!$$

Proof   The proof is by induction on n. If $n = 1$, then the formula is $P(1, 1) = \frac{1!}{(1-1)!} = 1$. Since the number of ordered one-tuples of a one-element set is clearly 1, this starts the induction. Now assume by induction that the formula holds for any positive integers n and k with $1 \leq k \leq n$, and suppose that $|A| = n + 1$. Define a mapping f from ordered k-tuples of elements of A to A by setting

$$f((a_1, a_2, \ldots, a_k)) = a_k$$

The mapping f is clearly surjective. For $a \in A$, each set $f^{-1}(a)$ has $P(n, k-1)$ elements. The Surjective Function Rule implies that $P(n+1, k) = (n+1)P(n, k-1)$.

Since this is clearly true for the expression above for $P(n, k)$, the proof by induction is complete. ●

In many books it is stated that the formula for $P(n, k)$ follows from the Product Rule. This argument is not correct: the correct argument uses the Surjective Function Rule, as we have done here.

We have already proved Pascal's Formula

$$C(n, k) = \binom{n}{k} = \frac{n!}{k!(n-k)!}$$

and it follows that

$$C(n, k) = P(n, k)/k!$$

EXAMPLE 8   In how many ways can a 5-card hand be dealt from a deck of 52 cards? How many 5-card hands are there?

Solution   In counting the number of ways to deal a 5-card hand, order matters, so the answer is

$$P(52, 5) = 52 \cdot 51 \cdot 50 \cdot 49 \cdot 48 = 311, 875, 200$$

In counting the number of 5-card hands, order does not matter, so the answer is

$$C(52, 5) = \binom{52}{5} = 311, 875, 200/5! = 311, 875, 200/120 = 2, 598, 960$$ ●

If n is a non-negative integer and $n_1, n_2, \ldots, n_r$ are non-negative integers such that $n = n_1 + n_2 + \ldots + n_r$, the *multinomial coefficient* is defined by the expression

$$\left(\begin{array}{cccc} & & n & \\ n_1 & n_2 & \ldots & n_r \end{array} \right) = \frac{n!}{n_1! n_2! \ldots n_r!}$$

Clearly, the multinomial coefficient is a generalization of the binomial coefficient, which we obtain from the expression above if $r = 2$.

Like the binomial coefficients, multinomial coefficients have many interpretations in combinatorics. Two of these are

(i)   The multinomial coefficient above counts the number of ordered partitions of a set X with n elements into r numbered cells $X_1, X_2, \ldots, X_r$, where the cell X_i has n_i elements.

(ii)   The multinomial coefficient above counts the number of strings of length n on an r-letter alphabet in which the ith letter occurs n_i times.

EXAMPLE 9   How many ordered partitions (X_1, X_2, X_3) are there of the set $\{1, 2, 3, 4\}$ into three cells with 2, 1, and 1 elements respectively?

Solution   These partitions are counted by the multinomial coefficient above with $n = 4$, $n_1 = 2$, $n_2 = 1$, and $n_3 = 1$. There are

$$4!/2!1!1! = 24/2 = 12$$

of these. The partitions are

$$(\{1, 2\}, \{3\}, \{4\}), (\{1, 2\}, \{4\}, \{3\})$$
$$(\{1, 3\}, \{2\}, \{4\}), (\{1, 3\}, \{4\}, \{2\})$$
$$(\{1, 4\}, \{2\}, \{3\}), (\{1, 4\}, \{3\}, \{2\})$$
$$(\{2, 3\}, \{1\}, \{4\}), (\{2, 3\}, \{4\}, \{1\})$$
$$(\{2, 4\}, \{1\}, \{3\}), (\{2, 4\}, \{3\}, \{1\})$$
$$(\{3, 4\}, \{1\}, \{2\}), (\{3, 4\}, \{2\}, \{1\})$$

●

EXAMPLE 10   Remember that DNA consists of a sequence of four nucleotides T = Thymine, A = Adenine, G = Guanine, and C = Cytosine. How many DNA sequences of length twelve are there in which thymine, adenine, guanine, and cytosine each occur three times?

Solution   By the formula that we have given, the number of sequences of this kind is

$$\begin{pmatrix} 12 \\ 3 \ 3 \ 3 \ 3 \end{pmatrix} = 12!/3!3!3!3! = 369,600$$

●

THEOREM 3.2.3   Suppose that we have an $n_1 \times n_2$ rectangular grid which is divided into squares, with a vertex A at one corner and a vertex B at the opposite corner as in Figure 3.7.

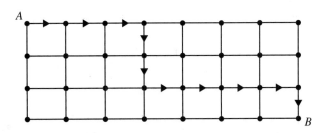

Figure 3.7   Paths Through a Rectangular Grid

The number of paths from vertex A to vertex B, always moving from A to B, is

$$\begin{pmatrix} n_1 + n_2 \\ n_1 \end{pmatrix}$$

EXAMPLE 11   In how many ways can we spell the word ABRACADABRA by drawing a path downward through the following array, starting at the top and ending at the bottom, like the following path?

Solution   The problem is to count the number of paths through this 5×5 grid. By the formula, the number of paths is $\binom{10}{5} = 252$, so the word ABRACADABRA can be spelled in 252 ways by paths of this kind.   ●

DEFINITION 3.2.2   A *multiset* is an unordered set whose elements may be repeated.

For example, $\{1, 1, 1\}$, $\{1, 1, 2\}$, and $\{1, 2, 3\}$ are all multisets with three elements.

THEOREM 3.2.4   The number of multisets with k elements whose elements are taken from a set with n elements is

$$\binom{n + k - 1}{k}$$

EXAMPLE 12   How many 3-element multisets are there whose elements are taken from the set $\{1, 2\}$?

Solution   By the theorem, there are

$$\binom{2 + 3 - 1}{3} = \binom{4}{3} = 4$$

such multisets. They are

$$\{1, 1, 1\}, \{1, 1, 2\}, \{1, 2, 2\}, \{2, 2, 2\}$$

●

APPLICATION   ***Counting Boolean Functions***

A *Boolean function* of n variables is a function

$$f : Z_2^n \to Z_2$$

In other words, it is a function $f(x_1, x_2, \ldots, x_n)$ where the variables x_i and the values of f are equal to 0 or 1.

We have proved that there are 2^n binary vectors of length n, so the result in Example 4 of this section implies that the number of Boolean functions of n variables is

$$2^{2^n}$$

For 1, 2, 3, 4, 5, or 6 variables we have

$$2^{2^1} = 4, \ 2^{2^2} = 16, \ 2^{2^3} = 256, \ 2^{2^4} = 65,536$$

$$2^{2^5} = 4,294,967,296$$

$$2^{2^6} = 18,446,744,073,709,551,616$$

Boolean functions.

EXERCISES

Solve the following problems using the Pigeonhole Principle.

1. Each month a coach chooses two rowers from a team of five to man the team's two-man shell. If the rowing team goes to one meet a month for eleven months, prove that the coach must choose the same pair of rowers more than once.

2. Prove that in any group of six people there are at least three people who mutually know each other, or at least three people who mutually do not know each other.

Solve the following four problems using the Product Rule.

3. In a certain university, there are twenty faculty in Department A, seventeen faculty in Department B, and twenty-three faculty in Department C. The dean of the college forms a committee by choosing three of the faculty from Department A, two of the faculty from Department B, and four of the faculty from Department C. How many committees are possible?

4. The genetic code, which contains the genetic material for all plants and animals, can be viewed as a sequence of four letters T, A, G, C. How many six-letter sequences are there? How many seven-letter sequences are there? How many eight-letter sequences are there?

5. A sequence is *palindromic* if it is unaltered when read in reverse order. How many palindromic sequences of length six are there in the four-letter code T, A, G, C?

6. Biochemists use *cutting sequences* to analyse DNA by cutting the DNA sequence at specific points. Some cutting sequences have length six. For example, AAGCTT is a cutting sequence. Cutting sequences are usually not palindromic, but they have a palindromic property. If we read the sequence in reverse order, and exchange T and A and also exchange G and C, the sequence is unaltered. Notice that the sequence AAGCTT has this property. How many sequences of length six in the letters T, A, G, C have this property?

Use a product set model to solve the following two problems.

7. The ancient Romans played the game of Odds and Evens. A player would toss a certain number of dice and bet with one or more other players that the sum of the numbers on the dice would be odd or even. Prove that this is a fair game if played with two fair dice. That is, that the number of cases in which the sum of the numbers on the two dice is odd is the same as the number of cases in which the sum of the numbers on the dice is even.

8. Prove that the game of Odds and Evens is fair if it is played with three fair dice.

Compute the following values of $P(n, k)$ and $C(n, k)$.

9. $P(6, 3)$ **10.** $C(6, 3)$ **11.** $P(7, 4)$

12. $C(7, 4)$ **13.** $P(10, 3)$ **14.** $C(10, 3)$

15. $P(14, 6)$ **16.** $C(14, 6)$ **17.** $P(20, 10)$

18. In how many ways can a 13-card bridge hand be dealt?

19. How many 13-card bridge hands are there?

20. Find the number of ordered partitions of the set $\{1, 2, 3, 4, 5\}$ into three cells with 3 elements, 1 element, and 1 element, respectively. List these partitions.

21. How many strings of length 20 on the letters A, B, C are there if the letter A occurs 10 times, the letter B occurs 5 times, and the letter C occurs 5 times?

22. How many strings of length 30 on the letters A, B, C, D are there if the letter A occurs 12 times, the letter B occurs 7 times, the letter C occurs 5 times, and the letter D occurs 6 times?

23. In how many ways can we spell the word ALGORITHM by a downward path starting at the top of the following array, like the path below?

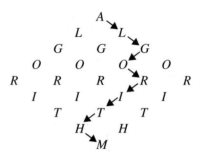

ADVANCED EXERCISES

1. Count and list the 3-element multisets whose elements are taken from $\{1, 2, 3\}$.

2. Count and list the 3-element multisets whose elements are taken from $\{1, 2, 3, 4\}$.

3. Count and list the 4-element multisets whose elements are taken from $\{1, 2, 3, 4\}$.

COMPUTER EXERCISES

1. Count the number of Boolean functions of seven variables.

2. Count the number of Boolean functions of eight variables.

3. Count the number of Boolean functions of nine variables.

3.3   THREE COUNTING TECHNIQUES

Then the bowsprit got mixed with the rudder sometimes:

A thing, as the Bellman remarked,

That frequently happens in tropical climes,

When a vessel is, so to speak, "snarked."

But the principal failing occurred in the sailing,

And the Bellman, perplexed and distressed,

Said he had hoped, at least, when the wind blew due East,

That the ship would not travel due West!

Early in his career, the Belgian mathematician Eugène-Charles Catalan (1814–1894) improved Euler's solution to an interesting counting problem, using what are now called

Catalan numbers. The use of Catalan numbers is part of a larger theme in combinatorics, that of counting with binomial coefficients, and using *binomial coefficient identities.* We shall describe a few of these identities, and show how Catalan numbers can be used to solve Euler's problem, and also to count trees of a certain type.

Next we shall use the *Principle of Inclusion-Exclusion* in counting problems, and show how it can be used to count prime numbers.

Some problems which seem as if they could be solved only by a slow algorithm can in fact be solved by using a suitable counting formula. An example of this is finding the sum of the first n integers, or the sum of the first n squares, the first n cubes, and so on. This problem was solved by Jakob Bernoulli in the *Ars Conjectandi.* We shall describe Bernoulli's solution of the problem and use it, as Bernoulli did, to find the sum of the first thousand tenth powers.

The binomial coefficients satisfy a large number of identities, many of which are useful in counting problems. Here are some of these identities.

I.   Symmetry:

$$\binom{n}{k} = \binom{n}{n-k}$$

II.   Extraction:

$$\binom{n}{k} = \frac{n}{k}\binom{n-1}{k-1}$$

III.   Trinomial Revision:

$$\binom{n}{k}\binom{k}{r} = \binom{n}{r}\binom{n-r}{k-r}$$

IV.   Binomial Theorem:

$$(X+Y)^n = \sum_{k=0}^{n}\binom{n}{k}X^{n-k}Y^k$$

V.   Middle Coefficient Expansion:

$$\sum_{k=0}^{n}\binom{n}{k}^2 = \binom{2n}{n}$$

VI.   Parallel Summation:

$$\sum_{k=0}^{n}\binom{r+k}{k} = \binom{r+n+1}{n}$$

VII.   Inverse Expansion:

$$\sum_{k=0}^{n-r}(-1)^k\binom{n-r}{k}\frac{n+1}{r+k+1} = \binom{n}{r}^{-1}$$

EXAMPLE 1   Check the Middle Coefficient Expansion identity for $n = 4$.

Solution

$$\binom{4}{0}^2 + \binom{4}{1}^2 + \binom{4}{2}^2 + \binom{4}{3}^2 + \binom{4}{4}^2 = 1^2 + 4^2 + 6^2 + 4^2 + 1^2$$

$$= 1 + 16 + 36 + 16 + 1 = 70 = \binom{8}{4}$$

so the identity holds in this case. ●

EXAMPLE 2 Check the Parallel Summation identity for $n = 2$ and $r = 2$.

Solution

$$\binom{2}{0} + \binom{3}{1} + \binom{4}{2} = 1 + 3 + 6 = 10 = \binom{5}{2}$$

so the identity holds in this case. ●

EXAMPLE 3 Check the Inverse Expansion identity for $n = 4$ and $r = 2$.

Solution

$$\binom{2}{0}\frac{5}{3} - \binom{2}{1}\frac{5}{4} + \binom{2}{2}\frac{5}{5} = \frac{5}{3} - \frac{5}{2} + 1 = \frac{1}{6} = \frac{1}{\binom{4}{2}}$$

so the identity holds in this case. ●

Binomial coefficients and the identities that relate them are very useful in solving counting problems. For example, several interesting counting problems can be solved using the Catalan numbers, which can be defined by either of the two expressions

$$C_n = \frac{1}{n+1}\binom{2n}{n} = \binom{2n}{n} - \binom{2n}{n-1}$$

The first few Catalan numbers are

$$C_0 = 1, C_1 = 1, C_2 = 2, C_3 = 5, C_4 = 14, C_5 = 42$$

The Catalan numbers were first studied by Euler. Euler was studying the problem of counting the number of triangulations of a regular polygon with n sides (the number of ways of dividing such a polygon into triangles, by straight lines joining the vertices of the polygon, in such a way that these lines do not cross).

Figure 3.8 shows the triangulations of a hexagon. There are 14 of these, the fourth Catalan number. If T_n is the number of triangulations of a regular polygon with n sides, Euler proved that

$$T_n = C_{n-2}$$

A contemporary of Euler's named von Segner found that the Catalan numbers satisfy the recursion

$$C_{n+1} = C_0 C_n + C_1 C_{n-1} + \ldots + C_n C_0$$

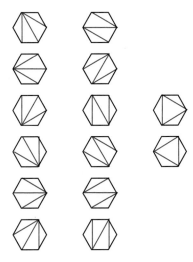

Figure 3.8 Triangulations of a Hexagon

For example,

$$C_5 = 42 = 1 \cdot 14 + 1 \cdot 5 + 2 \cdot 2 + 5 \cdot 1 + 14 \cdot 1$$

One of the basic problems in combinatorics is to count trees of various types. Rooted trees are particularly important, and we may ask for the number of rooted trees with a given number of nodes. For example, there are nine rooted trees with five nodes, shown in Figure 3.9.

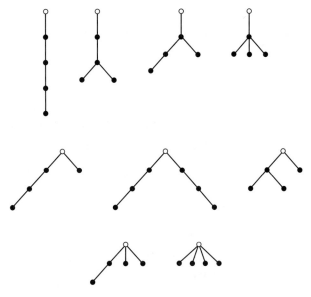

Figure 3.9 Rooted Trees with Five Nodes

The problem of counting rooted trees with a given number of nodes is rather difficult. However, a closely related type of tree can be counted with the Catalan numbers.

A rooted tree is *planted* if the root has degree one, and *trivalent* if all nodes have degree one or three. The terminal nodes of such a tree are labelled in order $1, 2, 3, \ldots$. These trees can all be obtained from triangulations of a regular polygon. Figure 3.10 shows five planted trivalent trees with four terminal nodes, and how one can be obtained from a triangulation of a pentagon.

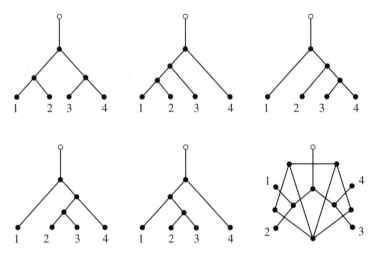

Figure 3.10   Five Planted Trivalent Trees

In general,

The number of planted trivalent trees with n terminal nodes is C_{n-1}.

The *Principle of Inclusion-Exclusion* was first stated clearly by Abraham DeMoivre in his book *The Doctrine of Chances*, which appeared in several editions in 1718, 1738, and 1756. This principle is a formula which counts the number of elements in the union of finitely many finite sets. For two or three sets the formula states that

$$|A \cup B| = |A| + |B| - |A \cap B|$$

and

$$|A \cup B \cup C| = |A| + |B| + |C| - |A \cap B| - |A \cap C| - |B \cap C| + |A \cap B \cap C|$$

These formulas compute the number of elements in the union by first adding the number of elements in each set, then correcting by subtracting the number of elements in the pairwise intersections, then correcting that correction, and so on. If we have n finite sets

$$A_1, A_2, \ldots, A_n$$

let S_k be the sum of the numbers of elements in all the k-fold intersections

$$A_{i_1} \cap A_{i_2} \cap \ldots \cap A_{i_k}$$

The Principle of Inclusion-Exclusion   The number of elements in the union of the sets A_i is given by the formula

$$|A_1 \cup A_2 \cup \ldots \cup A_n| = S_1 - S_2 + \ldots + (-1)^n S_n$$

This formula can be proved by induction, but we will leave the proof as an exercise.

Suppose that the sets $A_1, A_2, \ldots, A_n$ are all subsets of a set X with N elements. Suppose that the members of the set A_1 have some property, the members of A_2 have another property, and so on, and the members of A_n have the nth property on some list of properties. The Principle of Inclusion-Exclusion can be used to count the elements of X which have one or more of these properties, and also to count the elements of X which have none of the properties.

> *The number of elements of X which have one or more of the n properties is $|A_1 \cup A_2 \cup \ldots \cup A_n|$.*

> *The number of elements of X which have none of the n properties is $N - |A_1 \cup A_2 \cup \ldots \cup A_n|$.*

The Principle of Inclusion-Exclusion can be used to count prime numbers. The technique used to do this is called the *Sieve of Eratosthenes*. Because of this, methods which use the Principle of Inclusion-Exclusion are sometimes called *sieve methods*. Sieve methods can be used to determine when a positive integer is prime, and also to factor a positive integer into primes.

The Sieve of Eratosthenes   A positive integer n is prime if and only if it is not divisible by any prime p with $p \leq \sqrt{n}$.

Proof   If n is not prime, it factors properly as $n = rs$, and either r or s must be less than or equal to $\sqrt{n}$. Since every integer factors into primes, n must be divisible by a prime which is less than or equal to $\sqrt{n}$.   ●

Now suppose that the properties in the Principle of Inclusion-Exclusion are divisibility by certain prime numbers. The number of integers less than or equal to n which are divisible by a prime p is

$$\lfloor n/p \rfloor$$

(the greatest integer less than or equal to n/p), and similarly the number of integers less than or equal to n which are divisible by two primes p and q is

$$\lfloor n/pq \rfloor$$

and so on. Recall that $\pi(x)$ is the number of primes less than or equal to x.

EXAMPLE 4   Find $\pi(120)$.

Solution $\sqrt{120} = 10.95\ldots$, so the primes less than or equal to $\sqrt{120}$ are 2, 3, 5, and 7. We shall use the Principle of Inclusion-Exclusion with four sets A_1, A_2, A_3, A_4, where these are the sets of integers which are less than or equal to 120 and also divisible by 2, 3, 5, or 7 respectively. The numbers of elements in these sets and their multiple intersections are

$$\lfloor 120/2 \rfloor = 60, \lfloor 120/3 \rfloor = 40, \lfloor 120/5 \rfloor = 24$$
$$\lfloor 120/7 \rfloor = 17, \lfloor 120/6 \rfloor = 20, \lfloor 120/10 \rfloor = 10$$
$$\lfloor 120/14 \rfloor = 8, \lfloor 120/15 \rfloor = 8, \lfloor 120/21 \rfloor = 5$$
$$\lfloor 120/35 \rfloor = 3, \lfloor 120/30 \rfloor = 4, \lfloor 120/42 \rfloor = 2$$
$$\lfloor 120/70 \rfloor = 1, \lfloor 120/105 \rfloor = 1, \lfloor 120/210 \rfloor = 0$$

By the Principle of Inclusion-Exclusion, the number of integers less than or equal to 120 which are not divisible by 2, 3, 5, or 7 is

$$120 - 60 - 40 - 24 - 17 + 20 + 10 + 8 + 8 + 5 + 3 - 4 - 2 - 1 - 1 + 0 = 25$$

The four primes used in the test are not counted, and 1 has been counted as a prime which it is not, so

$$\pi(120) = 25 + 4 - 1 = 28$$

that is, 28 prime numbers are less than or equal to 120.   ●

In the *Ars conjectandi*, Bernoulli gave a formula for the sum of the kth powers of the first N integers. The Scottish mathematician Wallis had given an approximate formula for this sum by observing that

$$P_k(N) = 1^k + 2^k + \ldots + N^k$$

is an approximating sum for the integral

$$\int_0^N x^k \, dx = \frac{N^{k+1}}{k+1}$$

Thus this integral gives an approximation to the sum. Bernoulli replaced Wallis' approximation by an exact formula. His idea was that the sum should be a polynomial of degree $k + 1$, whose highest degree term is given by Wallis' formula. Recall that we defined the Bernoulli numbers B_i in Section 1.5 by taking $B_0 = 1$ and using the recursion

$$\binom{n+1}{0} B_0 + \binom{n+1}{1} B_1 + \ldots + \binom{n+1}{n} B_n = 0$$

The first few Bernoulli numbers are

$$B_0 = 1, B_1 = -\frac{1}{2}, B_2 = \frac{1}{6}, B_3 = 0, B_4 = -\frac{1}{30}, B_5 = 0,$$

$$B_6 = \frac{1}{42}, B_7 = 0, B_8 = -\frac{1}{30}, B_9 = 0, B_{10} = \frac{5}{66}, B_{11} = 0$$

The Bernoulli numbers with odd subscripts after B_1 are all equal to 0.

The nth *Bernoulli polynomial* is

$$B_n(X) = \sum_0^n \binom{n}{k} B_k X^{n-k}$$

The first few Bernoulli polynomials are

$$B_0(X) = 1, \ B_1(X) = X - \frac{1}{2}, \ B_2(X) = X^2 - X + \frac{1}{6},$$

$$B_3(X) = X^3 - \frac{3}{2}X^2 + \frac{1}{2}X$$

$B_4(x) = x^4 - 2x^3 + \frac{2}{2}x^2 -$

Bernoulli's Formula   The sum of the kth powers of the first N integers is

$$P_k(N) = \left(\frac{1}{k+1}\right)(B_{k+1}(N+1) - B_{k+1})$$

EXAMPLE 5   Compute

$$P_2(N) = 1^2 + 2^2 + \ldots + N^2$$

using Bernoulli's Formula.

Solution

$$P_2(N) = \left(\frac{1}{2+1}\right)(B_3(N+1) - B_3)$$

$$= \left(\frac{1}{3}\right)\left((N+1)^3 - \left(\frac{3}{2}\right)(N+1)^2 + \left(\frac{1}{2}\right)(N+1)\right)$$

$$= \left(\frac{(N+1)}{6}\right)(2(N+1)^2 - 3(N+1) + 1)$$

$$= \frac{N(N+1)(2N+1)}{6} \qquad \bullet$$

APPLICATION   ***The First Thousand Tenth Powers***

In the *Ars conjectandi*, Bernoulli used the formula to compute the sum of the first thousand tenth powers:

$$1^{10} + 2^{10} + 3^{10} + \ldots + 1000^{10} = 91,409,924,241,424,243,424,241,924,242,500$$

Bernoulli wrote that he carried out this calculation in "less than half of a quarter of an hour." Follow the steps given here, and see if you can do this!

Write the sum as

$$1^{10} + 2^{10} + \ldots + 999^{10} + 1000^{10}$$

because it is simpler to apply Bernoulli's Formula to $N = 999$.

That is,

$$P_{10}(1000) = P_{10}(999) + 1000^{10}$$

The Bernoulli polynomial $B_{11}(X)$ is

$$B_{11}(X) = X^{11} + 11\left(-\frac{1}{2}\right)X^{10} + 55\left(\frac{1}{6}\right)X^9 + 330\left(-\frac{1}{30}\right)x^7$$

$$+ 462\left(\frac{1}{42}\right)X^5 + 165\left(-\frac{1}{30}\right)X^3 + 11\left(\frac{5}{66}\right)X$$

Then

$$P_{10}(1000) = \left(\frac{1}{11}\right)B_{11}(1000) + 1000^{10}$$

$$= 91,409,924,241,424,243,424,241,924,242,500$$

EXERCISES

Verify that each of the following binomial coefficient identities holds for the given values of the parameters.

1. The Middle Coefficient Expansion identity for $n = 5$.

2. The Middle Coefficient Expansion identity for $n = 6$.

3. The Middle Coefficient Expansion identity for $n = 7$.

4. The Parallel Summation identity for $n = 3$ and $r = 2$.

5. The Parallel Summation identity for $n = 4$ and $r = 2$.

6. The Parallel Summation identity for $n = 5$ and $r = 2$.

7. The Inverse Expansion identity for $n = 5$ and $r = 2$.

8. The Inverse Expansion identity for $n = 6$ and $r = 2$.

9. List the 14 planted trivalent trees with five terminal vertices. You can do this directly, or use the 14 triangulations of a hexagon to construct the trees.

Express the following Catalan numbers, using each of Euler's two expressions:

10. C_6 11. C_7 12. C_8

Use von Segner's recursion for the Catalan numbers to express

13. C_6 14. C_7 15. C_8

Use the Principle of Inclusion-Exclusion to find the number of prime numbers which are less than or equal to

16. $N = 38$ 17. $N = 47$ 18. $N = 62$

19. $N = 82$ 20. $N = 105$

Use Bernoulli's Formula to compute the following sums:

21. $1^3 + 2^3 + \ldots + N^3$ 22. $1^4 + 2^4 + \ldots + 100^4$

23. $1^5 + 2^5 \ldots + 100^5$

ADVANCED EXERCISES

1. Prove the Extraction identity using Pascal's Formula.

2. Prove the Trinomial Revision identity using Pascal's Formula.

3. Prove the Parallel Summation identity by induction.

4. Prove the Inverse Expansion identity by induction. *Hint:* Use induction on $n - r$.

COMPUTER EXERCISES

1. Verify that the Middle Coefficient Expansion identity holds if $n = 10$.

2. Verify that the Middle Coefficient Expansion identity holds if $n = 15$.

3. Verify that the Parallel Summation identity holds if $n = 10$ and $r = 3$.

4. Verify that the Parallel Summation identity holds if $n = 20$ and $r = 4$.

5. Verify that the Inverse Expansion identity holds if $n = 20$ and $r = 3$.

6. Verify that the Inverse Expansion identity holds if $n = 20$ and $r = 10$.

3.4   CHAPTER SUMMARY AND SUPPLEMENTARY EXERCISES

But the danger was past—they had landed at last,

With their boxes, portmanteaus, and bags:

But at first sight the crew was not pleased with the view,

Which consisted of chasms and crags.

The Bellman perceived that their spirits were low,

And repeated in musical tone

Some jokes he had kept for a season of woe—

But the crew would do nothing but groan.

Chapter Summary

Section 3.1

This section began with a discussion of the properties of functions. We discussed one–one correspondences, graph isomorphisms, and one–one correspondences between subsets of a set and binary vectors. We discussed the basic properties of relations, and defined partial orderings and equivalence relations. The application was the Prüfer code of a labelled tree.

Section 3.2

This section covered the basic counting rules: the Sum Rule, the Product Rule, the Surjective Function Rule, and the Pigeonhole Principle. We showed how to count permutations and combinations, and stated several other counting principles, including how to count Boolean functions.

Section 3.3

This section explained three important counting techniques. First we discussed binomial coefficient identities, which play a fundamental role in counting problems. An example is the identity for the Catalan numbers, which have many interesting interpretations. We described how the Catalan numbers can be used to count a certain type of tree. We used the Principle of Inclusion-Exclusion to count prime numbers. Finally, we described Bernoulli's method for evaluating sums of powers.

EXERCISES

Prove that each of the following functions is a one–one correspondence, and find its inverse.

1. $f(x) = 2x + 4$ **2.** $g(x) = 4x + 8$

3. $h(x) = 8x + 16$

4. Let G be the graph of the 3-Cube. The vertices are the subsets of $\{1, 2, 3\}$, and the edges are pairs of subsets which differ in a single element.

Write a list of the vertices of G (you could use the Standard Gray Code), and also the edge list of G. Describe the effect of the one–one correspondence $f(A) = A^c$ on the list of vertices of G, and on the edge list of G.

Give the Hasse diagram of the following posets:

5. Div(273)   **6.** Div(385)   **7.** Div(3, 003)

8. Div(325)   **9.** Div(343)

Apply the Prüfer Correspondence to each of the following labelled trees, and then apply the inverse Prüfer Correspondence.

10.

11.

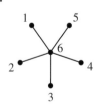

12. Two dice are thrown. In how many ways can the sum be eight?

13. Two dice are thrown. In how many ways can the sum be eight or more?

14. Two dice are thrown. In how many ways can the sum be five?

15. Two dice are thrown. In how many ways can the sum be five or less?

16. Three dice are thrown. In how many ways can the sum be seven?

17. Three dice are thrown. In how many ways can the sum be seven or less?

18. In how many ways can a hand of seven cards be dealt from a deck of 52 cards? How many seven-card hands are there?

19. How many ordered partitions (X_1, X_2, X_3) are there of the set $\{1, 2, 3, 4, 5\}$ into three cells with 2, 2, and 1 elements respectively? List these partitions.

20. Consider a 3-by-7 rectangular grid. How many paths are there from one vertex to the opposite vertex, if the paths always move down or to the right?

21. Find the number of multisets with three elements, whose elements are taken from a set with five elements, and list these multisets.

Use the Principle of Inclusion-Exclusion to count the prime numbers less than the following integers N:

22. $N = 100$ **23.** $N = 130$ **24.** $N = 168$

Use Bernoulli's Formula to compute the following sums:

25. $1^4 + 2^4 + \ldots + N^4$ **26.** $1^5 + 2^5 + \ldots + N^5$

27. $1^6 + 2^6 + \ldots + 100^6$

Graphs

4.1 GRAPHS

He had bought a large map representing the sea,

Without the least vestige of land:

And the crew were much pleased when they found it to be

A map they could all understand.

The English mathematician James Joseph Sylvester (1814–1897) founded the *American Journal of Mathematics*. A paper of Sylvester's titled *Chemistry and Algebra* appeared in Volume 17 of the journal *Nature* for the years 1877–1878. In it the word *graph* was used in the modern sense for the first time.

DEFINITION 4.1.1   A *graph* G is a set $V(G)$ of *nodes*, and a set $E(G)$ of unordered pairs of nodes called *edges*.

The nodes of a graph are often pictured as points and the edges as lines, as in the graph K_4 in Figure 4.1.

Graphs are often drawn with arrows on the edges to indicate direction.

DEFINITION 4.1.2   A *directed graph*, or *digraph* D is a set $V(D)$ of nodes and a set $E(D)$ of ordered pairs of nodes called *arcs*.

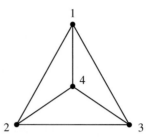

Figure 4.1   The Complete Graph K_4

If v_1 and v_2 are nodes of a digraph D, an arc from v_1 to v_2 is denoted by $(v_1, v_2) \in E(D)$. Figure 4.2 shows K_4 as a directed graph. Notice that there are many ways in which the undirected graph K_4 can be made into a directed graph.

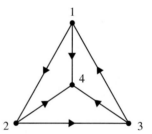

Figure 4.2 K_4 as a Directed Graph

Two nodes of a graph may be connected by more than one edge. Such edges are said to be *multiple*, and a graph with multiple edges is called a *multigraph*. An edge connecting a node to itself is called a *loop*, and a graph which contains loops is called a *pseudograph*. A graph with no multiple edges or loops is called a *simple graph*. Figure 4.3 shows a multigraph and Figure 4.4 shows a pseudograph.

Figure 4.3   A Multigraph

Figure 4.4   A Pseudograph

A graph G is completely described by its edge set, except for *isolated nodes* of G (nodes which are not joined to any others, and at which there are no loops). Thus we can describe a graph without isolated nodes by giving its edges, and these can be arranged in a list called the *edge list* of G. For example, the edge list of the graph K_4 is

$$\{1, 2\}$$
$$\{1, 3\}$$
$$\{1, 4\}$$
$$\{2, 3\}$$
$$\{2, 4\}$$
$$\{3, 4\}$$

We have arranged the edges of K_4 in *lex order*. This is like the ordering of words in a dictionary, but instead of $a, b, c, \ldots$ we have $1, 2, 3, \ldots$.

The edge list of the digraph in Figure 4.2 is

$$(1, 2)$$
$$(1, 4)$$
$$(2, 3)$$
$$(2, 4)$$
$$(3, 1)$$
$$(3, 4)$$

Notice that making the graph K_4 directed causes the lex ordering of the edges to be different.

Any graph can be described completely by a matrix called the *adjacency matrix*.

DEFINITION 4.1.3

i. The *adjacency matrix* A of a graph G with nodes $v_1, v_2, \ldots, v_n$ is the matrix $A = [a_{ij}]$ whose entries are $a_{ij} = 1$ if there is an edge connecting v_i and v_j, and $a_{ij} = 0$ if there is no edge connecting v_i and v_j.

ii. The *adjacency matrix* A of a digraph D with nodes $v_1, v_2, \ldots, v_n$ is the matrix $A = [a_{ij}]$ whose entries are $a_{ij} = 1$ if there is a directed edge from v_i to v_j and $a_{ij} = 0$ if there is no directed edge from v_i to v_j.

The adjacency matrix of a graph G is *symmetric* ($a_{ij} = a_{ji}$ for all $1 \leq i, j \leq n$). The adjacency matrix of a digraph D need not be symmetric. We will illustrate this by computing the adjacency matrices of the graph K_4 and the directed version of K_4. The adjacency matrix of K_4 (as shown in Figure 4.1) is

$$\begin{bmatrix} 0 & 1 & 1 & 1 \\ 1 & 0 & 1 & 1 \\ 1 & 1 & 0 & 1 \\ 1 & 1 & 1 & 0 \end{bmatrix}$$

The adjacency matrix of the directed graph (as shown in Figure 4.2) is

$$\begin{bmatrix} 0 & 1 & 0 & 1 \\ 0 & 0 & 1 & 1 \\ 1 & 0 & 0 & 1 \\ 0 & 0 & 0 & 0 \end{bmatrix}$$

Notice that the adjacency matrix of K_4 is symmetric, but the adjacency matrix of the directed form of K_4 is not.

DEFINITION 4.1.4

i. A *walk* on a graph G is an alternating sequence of nodes and edges

$$v_0, x_1, v_2, x_2, \ldots, x_n, v_n$$

beginning and ending with a node, such that each edge is immediately preceded and followed by the two nodes which lie on it.
ii. A *trail* is a walk all of whose edges are distinct.
iii. A *path* is a walk all of whose nodes are distinct.
iv. A *closed walk* is a walk such that $v_0 = v_n$.
v. A *cycle* is a closed path such that $n \geq 3$.
vi. A *directed path* in a digraph D is a path such that the edge x_i is the directed edge from the node v_i to the node v_{i+1}.

Figure 4.5 shows a walk which is not a trail, a trail which is not a path, a path, a closed walk which is not a cycle, and a cycle.

In Figure 4.6 shows the *Petersen graph*. This graph is named for the Danish mathematician Julius Peter Christian Petersen (1839–1910).

DEFINITION 4.1.5

i. A graph G is *connected* if there is a path from any node in G to any other node in G; otherwise G is *disconnected*.

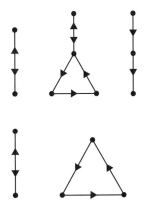

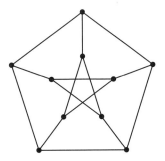

Figure 4.5   Walk, Trail, Path, Closed Walk, and Cycle

Figure 4.6   The Petersen Graph

ii. An edge x of G is a *bridge* if G becomes disconnected when x is removed.

iii. A graph G is *biconnected* if G is connected and does not have a bridge.

iv. A graph G is *k-edge connected* if G is still connected when any k edges are removed from G.

v. A graph G is *k-node connected* if G is still connected when any k nodes are removed from G.

The graph K_4 is connected, 1-edge connected, and 2-edge connected, but not 3-edge connected.

Notice that connectivity is the same as 0-edge connectivity, and biconnectivity is the same as 1-edge connectivity. When we speak of k-connectivity, we shall always mean k-edge connectivity unless we explicitly mention that we are considering node connectivity.

A finite graph G is always a finite union of maximal connected graphs which are called the *connected components* of G.

> ### DEFINITION 4.1.6
>
> **i.** A digraph D is *strongly connected* if there is a directed path from any node of D to any other node of D.
> **ii.** A *directed cycle* in D is a closed directed path.
> **iii.** D is a *directed acyclic graph* if it has no directed cycles.

The graph in Figure 4.2 is not strongly connected.

As in the case of graphs, every digraph D is a union of maximal strongly connected digraphs, its *strong components*. A directed acyclic graph is the analog, for directed graphs, of a forest, and a connected directed acyclic graph is the analog of a tree.

> ### DEFINITION 4.1.7   If D is a finite digraph, then D^*, the *condensation* of D, is the digraph whose nodes are the strong components $D_1, D_2, \ldots, D_k$ of D, and such that there is a directed edge from the ith node to the jth node if there is a directed path from some node of D_i to some node of D_j.

THEOREM 4.1.1   Let D be any directed graph. Then the condensation D^* is a directed acyclic graph.

We will always refer to a directed acyclic graph as a DAG. Figure 4.7 shows a directed graph D and its condensation D^*.

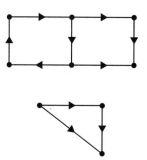

Figure 4.7   A Digraph D and Its Condensation D^*

> ### DEFINITION 4.1.8
>
> **i.** The *degree* of a node v of a graph G is the number of edges of G which contain v. Each loop at v contributes 2 to the degree of v.

ii. If G is a graph with n nodes, the *degree sequence* $(d_1, d_2, \ldots, d_n)$ of G is the sequence of degrees of nodes of G, listed in increasing order: $d_1 \leq d_2 \leq \ldots \leq d_n$.

For example, the degree sequence of the complete graph K_4 is $(3, 3, 3, 3)$.

The Handshake Lemma   If G is a graph with n nodes and e edges, then

$$\sum_{k=1}^{n} d_k = 2e$$

Proof   Each edge of G contains two nodes.   ●

COROLLARY 4.1.1   In any graph, the number of vertices of odd degree is even.

EXAMPLE 1   Is $(1, 1, 1, 1, 1, 3, 3, 5, 5)$ the degree sequence of a graph?

Solution   It is not, because the number of vertices of odd degree is odd.   ●

Following are instructions for drawing certain basic types of graphs.

The Complete Graph K_n

To draw the complete graph K_n

(i)   Draw a regular polygon with n sides.

(ii)   Join each node to every other node by an edge.

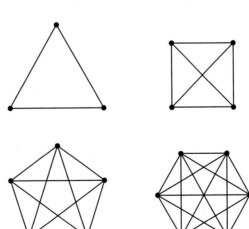

Figure 4.8   The Complete Graphs K_n, for $n = 3, 4, 5, 6$

The *n*-Cube

To draw the *n*-Cube

(i) Draw $n + 1$ rows of nodes with $\binom{n}{k}$ nodes in the kth row.

(ii) Label the nodes in the kth row with k-element subsets of $[n] = \{1, 2, \ldots, n\}$, in lex order from left to right; or with binary vectors of length n, with k 1s, in lex order for the ordering $1 < 0$, from left to right.

(iii) Join two nodes by an edge if they are labelled by subsets which differ in exactly one element, or by binary vectors of length n which differ in exactly one entry.

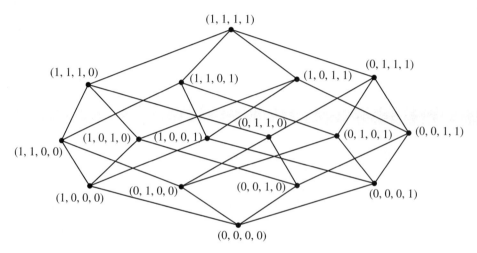

Figure 4.9   The 4-Cube Labelled with Binary Vectors

The *n*-Dimensional Octahedron

To draw the *n*-dimensional octahedron

(i) Draw a regular polygon with $2n$ sides.

(ii) Join two nodes by an edge if they are not directly opposite each other.

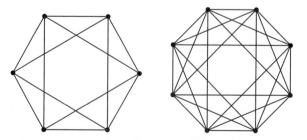

Figure 4.10   The Three- and Four-Dimensional Octahedra

Complete Multipartite Graphs

To draw the graph $K_{n_1, n_2, \ldots, n_r}$

(i)   Draw r rows of vertices with n_i nodes in the ith row.

(ii)   Join each pair of nodes by an edge, except for the nodes in the same row.

If $r = 2$, complete multipartite graphs are called *complete bipartite graphs*. Figure 4.11 shows the graph $K_{3,3}$, in which there are two rows with three nodes each.

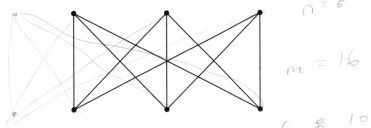

Figure 4.11   The Complete Bipartite Graph $K_{3,3}$

Notice that the graph of the n-dimensional octahedron is the same as the graph $K_{2,2,\ldots,2}$, where the 2 occurs n times.

Figure 4.12 shows that the Petersen graph contains a *homeomorph* of $K_{3,3}$ (this means that the nodes of $K_{3,3}$ are identified with nodes of the Petersen graph, and the edges of $K_{3,3}$ are identified with disjoint paths in the Petersen graph).

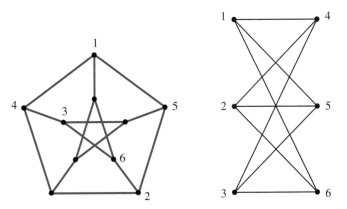

Figure 4.12   The Petersen Graph Contains a Homeomorph of $K_{3,3}$

Part of solving a problem with graphs often involves labelling the nodes or edges. The next example is a problem which can be solved by coloring the edges of a graph.

EXAMPLE 2   If the edges of the complete graph K_6 are all colored black or blue, then the resulting graph must contain a black triangle or a blue triangle.

Proof   The degree of each vertex of K_6 is five. Choose any node of K_6, and consider the five edges incident on it. At least three must be colored blue or black. Suppose that three are colored blue, and consider the nodes at the other end of these three edges. If any pair of them are joined by a blue edge, then the graph will contain a blue triangle. If they are all joined by black edges, then the graph will contain a black triangle.

The statement that we have just proved can be reformulated as a statement about relations. For example, among any six people, either three people mutually know each other or three people mutually do not know each other.

> **DEFINITION 4.1.9**   A graph is *planar* if it can be drawn in the plane in such a way that the edges do not intersect.

For example, the graph K_4 in Figure 4.1 is planar.

Five Points in the Plane   Can five points in the plane be joined by lines in such a way that the lines do not cross?

This is the same as asking if the graph K_5 planar. This is almost true, but not quite. If we remove a single edge from K_5, it becomes planar (Figure 4.13), but however we try to draw the last edge it will cross another edge.

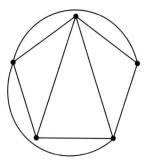

Figure 4.13 K_5 Minus an Edge is Planar

In the next section we will prove that

The complete graph K_5 is not planar.

Water, Gas, and Electricity   Lines from the water, gas, and electric utilities are to be connected to three houses A, B, and C. Can this be done in such a way that the lines do not cross?

This is the same as asking if the graph $K_{3,3}$ is planar. This is almost true, but not quite. If we remove a single edge from $K_{3,3}$ it becomes planar (Figure 4.14), but however we try to draw the last edge it will cross another edge.

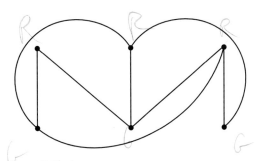

Figure 4.14 $K_{3,3}$ Minus an Edge is Planar

In the next section we will prove that

The complete bipartite graph $K_{3,3}$ is not planar.

In every dimension there is an *n-simplex*, an *n-cube*, and an *n-dimensional octa-hedron*. These figures are called *regular solids* (figures with congruent faces which are regular *n*-gons or higher-dimensional analogs of these). The graphs K_n, the *n*-cube, and the *n*-dimensional octahedron, which we described earlier, are the graphs of these figures, that is, projections of their one-dimensional parts.

Figure 4.15 shows another way to draw the graphs of the 3-simplex, the 3-cube, and the (3-dimensional) octahedron, by projecting these three figures into the plane.

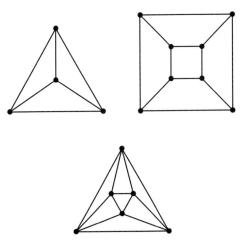

Figure 4.15   The Graphs of the 3-Simplex, the 3-Cube, and the Octahedron

There are two more regular solids in dimension three, the *icosahedron* and the *dodecahedron*. The icosahedron has twenty faces which are equilateral triangles, and the dodecahedron has twelve faces which are regular pentagons. The plane projections of these figures are shown in Figure 4.16. Cutout constructions for the icosahedron and the dodecahedron are shown in Figure 4.17. The icosahedron can be formed by fold-ing the array of twenty triangles, in Figure 4.17. The dodecahedron can be formed by folding the array of twelve pentagons in the same figure.

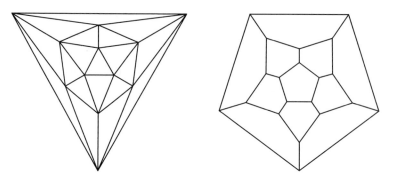

Figure 4.16   The Icosahedron and the Dodecahedron

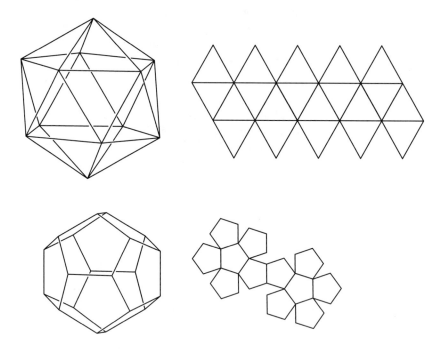

Figure 4.17   Constructions for the Icosahedron and the Dodecahedron

It can be proved that, in all dimensions, the only regular solids are the n-simplex, the n-cube, the n-dimensional octahedron, the icosahedron, the dodecahedron, and three regular solids in dimension four: the 24-cell, the 120-cell, and the 600-cell. The 24-cell has twenty-four faces which are all 3-simplices. The 120-cell has 120 faces which are all dodecahedra. The 600-cell has six hundred faces which are all 3-simplices.

There are Celtic models of the dodecahedron which date from about 1000 B.C., and which may have been used as dice. The Romans made brass models of the icosahedron and the dodecahedron, the purpose of which is not known. Many viruses have an icosahedral shape, as do some radiolarians (microscopic animals which in an aggregate form sponges). Other radiolarians have a dodecahedral shape.

The theorem that there are only five regular solids in dimension three is one of the great achievements of Greek mathematics. It is usually attributed to Theatetus, a contemporary of Plato. The thirteenth book of Euclid's *Elements* is devoted to this topic. In the nineteenth century, it was discovered that the symmetries of the icosahedron and the dodecahedron are very important in the study of the solutions of equations of the fifth degree.

APPLICATION   *The Icosahedron and the Dodecahedron*

The graphs K_n, the n-cube, and the graph of the n-dimensional octahedron are easy to draw. The graphs of the icosahedron and the dodecahedron are perhaps a little more complicated. We will simplify them by forming new graphs whose vertices are pairs of vertices and whose edges are pairs of edges.

Instead of drawing a graph on a flat surface, we may draw graphs on a sphere. The *antipodal map* on the sphere is the function which maps each point on the sphere to the symmetrically opposite point.

> **DEFINITION 4.1.10**   Let G be a graph drawn on the sphere which is *symmetric* (such that there is a node of G opposite to each node of G on the sphere, and an edge of G opposite to each edge of G). The *quotient of G by the antipodal map* is the graph whose nodes are pairs of opposite nodes of G and whose edges are pairs of opposite edges of G.

We will compute the quotients of the graphs of the icosahedron and the dodecahedron by the antipodal map using the following cutout constructions.

We can find the quotients of the graphs of the icosahedron and the dodecahedron by the antipodal map, taking half of each cutout figure and identifying opposite vertices and edges, as in Figure 4.18 and Figure 4.19.

We see that the graph of the icosahedron is related to the much simpler graph K_6:

The quotient of the graph of the icosahedron by the antipodal map is the complete graph K_6.

The Greeks regarded the dodecahedron as the most interesting of the regular solids. The thirteenth and last book of Euclid ends with a discussion of its properties. Combinatorists regard the Petersen graph as one of the most important examples of a graph. There is a relation between the two:

The quotient of the graph of the dodecahedron by the antipodal map is the Petersen graph.

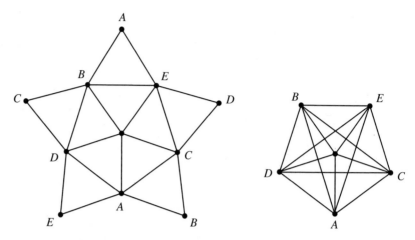

Figure 4.18   The Quotient of the Icosahedron by the Antipodal Map

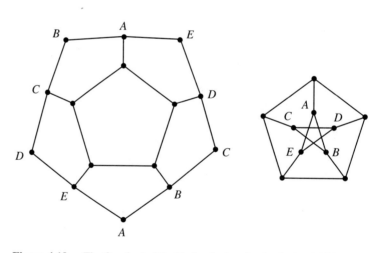

Figure 4.19   The Quotient of the Dodecahedron by the Antipodal Map

EXERCISES

Draw each of the following graphs, label the vertices $1, 2, 3, \ldots, n$, and give the edge list in lex order. Put arrows on the edges to make each graph a directed graph, and again give the edge list in lex order.

1. The complete graph K_5
2. The complete graph K_6
3. The complete bipartite graph $K_{3,3}$

4. The 3-cube   5. The 4-cube
6. The 3-dimensional octahedron
7. The 4-dimensional octahedron

In Exercises 8–14, use the labelling of the vertices from the preceding problems to write the adjacency matrix of the graph, and also the adjacency matrix of the directed form of the graph.

8. The complete graph K_5

9. The complete graph K_6

10. The complete bipartite graph $K_{3,3}$

11. The 3-cube   **12.** The 4-cube

13. The 3-dimensional octahedron

14. The 4-dimensional octahedron

15. Find the condensation D^* of the digraph

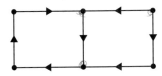

16. Find the condensation D^* of the digraph

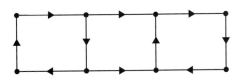

17. Find the condensation D^* of the digraph

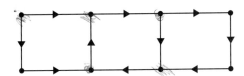

18. Is $(1, 1, 1, 2, 2, 2, 3, 3, 3, 3)$ the degree sequence of a graph?

19. Is $(1, 1, 3, 3)$ the degree sequence of a graph? If so, draw such a graph. Is $(1, 1, 3, 3)$ the degree sequence of a tree?

20. Is $(1, 1, 1, 5, 5, 5)$ the degree sequence of a graph? If so, draw such a graph. Is $(1, 1, 1, 5, 5, 5)$ the degree sequence of a tree?

21. Is $(1, 1, 1, 1, 1, 1, 1, 1, 1, 3, 3, 4, 5)$ the degree sequence of a graph? Is it the degree sequence of a tree?

22. By drawing a diagram, prove that the graph of the 3-cube contains a homeomorph of the complete graph K_4.

23. By drawing a diagram, prove that the graph of the octahedron contains a homeomorph of the complete graph K_4.

24. Does the graph of the 4-cube contain a homeomorph of the complete graph K_5?

ADVANCED EXERCISES

1. Count the number of edges of the complete graph K_n.

2. Count the number of edges of the n-cube (*Hint:* the number of vertices is 2^n and the degree of each vertex is n. Use the Handshake Lemma).

3. Count the number of edges of the n-dimensional octahedron.

4. Give a formula for the number of simple graphs with n vertices.

5. Make a list of all simple graphs with four vertices.

COMPUTER EXERCISES

Write a program which gives the edge list of the complete graph K_n in lex order, and print the edge list in the following cases.

1. For K_{20} **2.** For K_{30} **3.** For K_{40}

Write a program which gives the edge list of the graph of the n-dimensional octahedron, and print the edge list in the following cases.

4. For $n = 15$ **5.** For $n = 20$ **6.** For $n = 25$

Write a program which gives the edge list of the graph of the n-cube in lex order and print the edge list in the following cases. (You must first order the vertices of the n-cube. Can you find a good way to do this?)

7. For $n = 6$ **8.** For $n = 7$ **9.** For $n = 8$

A number of properties of graphs can be understood better if we have examples of graphs in which the edges can be chosen at random. We will construct a graph $G(n, m)$ (this graph has n vertices and m edges) and a graph $SG(n, m)$ (this is a simple graph with n vertices and no more than m edges), in such a way that the edges are chosen "at random." This will allow us, for example, to study the "evolution" of these graphs as n is held fixed and m increases.

The vertex set of both $G(n, m)$ and $SG(n, m)$ is $\{1, 2, \ldots, n\}$.

The edge set of $G(n, m)$ consists of m edges $\{A(k), B(k)\}$, where $1 \leq k \leq m$ and $A(k)$ and $B(k)$ are defined as follows:

(i) Let $C(k)$ be the remainder when $\lfloor (kn/2)\sqrt{2} \rfloor$ is divided by n, and set $A(k) = C(k) + 1$.

(ii) Let $D(k)$ be the remainder when $\lfloor (kn/2)\sqrt{3} \rfloor$ is divided by n, and set $B(k) = D(k) + 1$.

The edge set of $SG(n, m)$ is obtained from the edge set of $G(n, m)$ by omitting any loops, and including multiple edges only once. Thus $SG(n, m)$ will be a simple graph.

Write a program which generates the edge set of the graph $SG(n, m)$. Can you tell whether $SG(n, m)$ is connected? Use your program to determine whether the following graphs are connected.

10. $SG(100, 120)$

11. $SG(100, 200)$

12. $SG(500, 600)$

13. $SG(500, 800)$

14. $SG(1000, 1200)$

15. $SG(1000, 1600)$

4.2 LEONHARD EULER AND GRAPH THEORY

"What's the good of Mercator's North Poles and Equators,

Tropics, Zones, and Meridian Lines?"

So the Bellman would cry; and the crew would reply

"They are merely conventional signs!"

In 1736, Leonhard Euler gave a talk about the *Königsberg Bridge Problem*. The Pregel River runs through the city of Königsberg in Prussia, and in the river there is an island called the *Kneiphof* which is connected to the surrounding land by seven bridges:

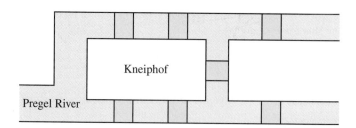

Is there a route which crosses each of the seven bridges exactly once? Euler proved that such a route does not exist, and solved the more general problem of determining when such a route exists for any configuration of islands and bridges. This is viewed as the origin of *graph theory*. It is also viewed as the first publication in topology (the study of spaces). Euler solved the problem by using what he called *analysis situs* (the geometry of position). His talk was published as a paper in the *Commentarii Academiae*

Scientiarum Imperialis Petropolitanae, also in 1736, with the title *Solutio problematis ad geometriam situs pertinentis*.

If we add one bridge to the configuration of Königsberg bridges, then there *is* a route which crosses each bridge exactly once (Figure 4.20). If we add two bridges, then we can find such a route whose starting point may be chosen arbitrarily (Figure 4.21).

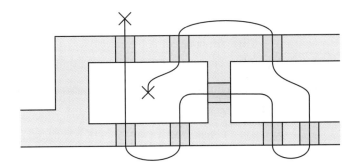

Figure 4.20   The Königsberg Bridges with One Added Bridge

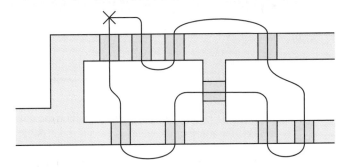

Figure 4.21   The Königsberg Bridges with Two Added Bridges

DEFINITION 4.2.1

i. An *Eulerian cycle* of a finite graph G is a closed trail which contains each edge of G exactly once.

ii. An *Eulerian circuit* of G is a trail of G which contains each edge of G exactly once.

iii. A *Hamiltonian circuit* of G is a path of G which contains each node of G exactly once.

Notice that an Eulerian cycle is an Eulerian circuit. An Eulerian circuit may have different endpoints, in which case it is not an Eulerian cycle.

A finite graph G which has an Eulerian cycle is called *Eulerian*. A graph with a Hamiltonian circuit is called *Hamiltonian*. Hamiltonian graphs are named for the Irish mathematician William Rowan Hamilton (1805–1865).

There is a simple theoretical description of Eulerian graphs in the next theorem, but there is no simple rule for drawing an arbitrary Eulerian graph. By contrast, there is no known theoretical description of Hamiltonian graphs. However, there is a simple rule for drawing any Hamiltonian graph.

THEOREM 4.2.1   If G is a connected multigraph, then the following conditions are equivalent:

    (i) G has an Eulerian cycle.

    (ii) All nodes of G have even degree.

    (iii) The edge set of G can be partitioned into cycles.

The following two conditions are also equivalent:

    (iv) G has an Eulerian circuit which is not an Eulerian cycle.

    (v) G has exactly two nodes of odd degree, and the Eulerian circuit begins at one of these nodes and ends at the other.

This theorem is attributed to Euler, although Euler actually only proved the implication $(i) \rightarrow (ii)$. The proof is a combination of arguments of Euler, C. Hierholzer, and O. Veblen.

Proof

(i) $\rightarrow$ **(ii).**   If G has an Eulerian cycle, then the edges which meet each node must be divided into pairs, along which the cycle enters and leaves the node. Thus the degree of each node must be even.

(ii) $\rightarrow$ **(iii).**   The proof is by induction on the number of cycles in G. Clearly the result holds if the number of cycles is zero or one. Given a positive integer n, assume that the result holds if G has fewer than n cycles, and assume that G has exactly n cycles. Delete one cycle from G. The resulting graph G' may be disconnected, but by induction, the edge set of each component of G' may be partitioned into cycles. The edge set of G' may be partioned into cycles by combining these partitions. Adding the cycle which was removed, we obtain the desired partition of the edges of G into cycles.

(iii) $\rightarrow$ **(i).**   Suppose that the edges of G can be partitioned into n cycles. We will prove the implication by induction on n. The result is clear if $n = 1$, and this starts the induction. Remove one cycle from the edge set of G. By induction, there is an Eulerian circuit on each component of the resulting graph G'. Now add the cycle which was removed, and traverse first this cycle and then the cycles in the components of G'. This gives an Eulerian cycle for G.

The equivalence of (iv) and (v) follows in a similar way, since by the Handshake Lemma, the number of vertices of G of odd degree is even.   ●

In the Königsberg Bridge Problem, we can associate a graph to each configuration of islands and bridges. Let each land mass be represented by a node, and each bridge by an edge. The graphs for the Königsberg bridges and the configurations in Figure 4.20 and Figure 4.21 are given in Figure 4.22.

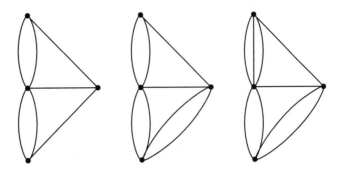

Figure 4.22   Three Graphs for the Königsberg Bridges

The degree sequences of the graphs in Figure 4.22 are $(3, 3, 3, 5)$, $(3, 4, 4, 5)$, and $(4, 4, 4, 6)$, respectively. So, according to Euler's Theorem, there is no Eulerian cycle or circuit on the first; there is an Eulerian circuit which is not an Eulerian cycle on the second (corresponding to the path in Figure 4.20); and there is an Eulerian cycle on the last (corresponding to the closed path in Figure 4.21).

Figure 4.23 shows an Eulerian graph (the graph of the 3-dimensional octahedron), and a decomposition of its edge set into disjoint cycles.

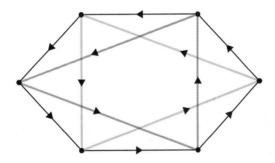

Figure 4.23   The Edge Set of the 3-Dimensional Octahedron

There is an algorithm for finding Eulerian cycles and circuits. Remember that a bridge of a graph G is an edge of G whose removal disconnects G.

FLEURY'S ALGORITHM   This algorithm will find an Eulerian cycle or circuit on a finite graph G, if such a cycle or circuit exists. If the algorithm terminates without producing an Eulerian cycle or circuit, then G does not have an Eulerian cycle or circuit.

i. Beginning with any edge, choose edges so as to give a trail in G. Erase edges as they are chosen, and also erase any isolated nodes which may occur.

ii. Never choose an edge which is a bridge unless there is no alternative.

The graph of the 3-dimensional octahedron is both Eulerian and Hamiltonian. The graph is drawn in two ways in Figure 4.24 (the two graphs are isomorphic). The left-hand graph shows an Eulerian cycle, found using Fleury's Algorithm, and the right-hand graph shows a Hamiltonian cycle.

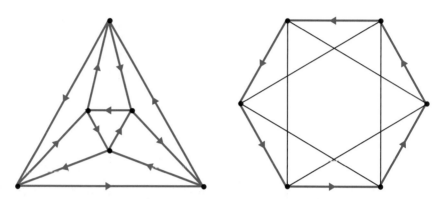

Figure 4.24   The Octahedron is Eulerian and Hamiltonian

A Hamiltonian graph must, by definition, contain a Hamiltonain circuit (a path in G which contains each vertex exactly once). Such a circuit may or may not be contained in a cycle.

THE STRUCTURE OF HAMILTONIAN GRAPHS   Any Hamiltonian graph G can be drawn in the following way:
    Draw a regular polygon with n sides, omit one edge, and then connect pairs of nodes by edges in any way.

For example, the graph of the 3-cube in Figure 4.25 was drawn in this manner.
    The complete graph K_n, the graph of the n-cube, and the graph of the n-dimensional octahedron are Eulerian in the following cases:

I.   The complete graph K_n is Eulerian if and only if n is odd (because the degree of each node of K_n is $n - 1$).

II.   The graph of the n-cube is Eulerian if and only if n is even (because the degree of each node of the graph of the n-cube is n).

III.   The graph of the n-dimensional octahedron is always Eulerian (because the degree of each node of this graph is $2n - 2$, which is always even).

The complete graph K_n, the graph of the n-cube, and the graph of the n-dimensional octahedron are always Hamiltonian.

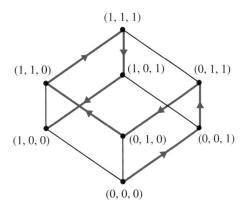

Figure 4.25   The 3-Cube

I.   The complete graph K_n is always Hamiltonian (because this graph may be drawn by drawing a regular polygon with n sides, and connecting all pairs of nodes—the regular polygon, minus an edge, gives a Hamiltonain circuit).

II.   The graph of the n-cube is always Hamiltonian (if we label the vertices with binary vectors of length n, the Standard Gray Code gives a Hamiltonian circuit).

III.   The graph of the n-dimensional octahedron is always Hamiltonian (remember that we may draw this graph by drawing a regular polygon with $2n$ sides, and connecting all pairs of vertices by an edge except those which are directly opposite—the regular polygon, minus an edge, gives a Hamiltonian circuit).

Following are three problems involving various kinds of paths and trails on graphs.

The Two-Way Street Problem   Consider any connected array of streets (that is, one in which it is possible to pass from any street corner to any other). Construct an associated graph by letting each street corner correspond to a node and each street correspond to an edge.

Double each edge. Clearly the resulting graph is Eulerian, and therefore has an Eulerian cycle. This cycle gives a route on the array of streets which passes up and down each street exactly once.

EXAMPLE 1   Solve the Two-Way Street Problem for the array of streets in Figure 4.26.

Solution   The array of streets, the graph, the graph with doubled edges, and the desired route are given in Figure 4.26.   ●

The following problem gets its name from the fact that it was studied by the Chinese mathematician Mei-ko Kwan.

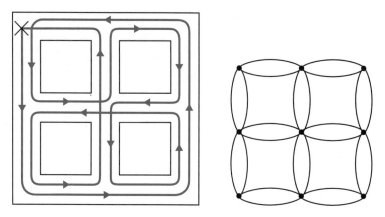

Figure 4.26   The Two-Way Street Problem

The Chinese Postman Problem   A postman must cover a certain route, passing along all streets of the route at least once and returning to his starting point. He wishes to do this in such a way that the total distance travelled is a minimum.

If the graph corresponding to the array of streets is Eulerian, then any Eulerian cycle on the graph gives a solution. If the graph is not Eulerian, then some retracing of streets is necessary and the problem is more difficult.

The Travelling Salesman Problem   A travelling salesman must visit n cities, starting at one of the cities and returning to it. If the distances between all the cities is known, what is the shortest possible route?

This problem can be modelled by the complete graph K_n, where the edges are weighted by the distances between the cities. It can be solved by finding a cycle of minimal weight which contains a Hamiltonian circuit. The problem of finding such cycles becomes very difficult as n increases, since the number of Hamiltonian circuits increases rapidly with n.

Remember that a graph G is planar if it can be drawn in the plane or on the sphere.

If a finite graph G is planar, it will have V nodes, E edges, and a certain number of *faces* F (the faces are the regions enclosed by the edges. If G is drawn in the plane, the region outside G is counted as a face).

Euler remarked that the following theorem is true. The same observation had been made earlier by Réné Descartes (1596–1650).

THEOREM 4.2.2   If a graph G is ^{connected} planar, then

$$V - E + F = 2$$

DEFINITION 4.2.2   The quantity $V - E + F$ is called the *Euler characteristic* of G.

Thus Descartes and Euler both observed that what is now called the Euler characteristic of a planar graph is always equal to 2.

EXAMPLE 2 Verify the formula $V - E + F = 2$ in the case of the icosahedron.

Solution The icosahedron has 12 nodes, 30 edges, and 20 faces, so

$$V - E + F = 12 - 30 + 20 = 2$$

●

We can use the previous theorem to prove that certain graphs *are not* planar, as in the next four examples.

First notice that if every cycle of a finite planar graph G contains at least k edges, then since each edge occurs on exactly two faces, we have the inequality

$$kF \leq 2E$$

EXAMPLE 3 The complete graph K_5 is not planar.

Proof Notice first that for this graph, $V = 5$ and $E = 10$. Each cycle of K_5 contains at least three edges. Since $V - E + F = 2$ we have $F = 7$. By the inequality

$$21 = 3F \leq 2E = 20$$

which is a contradiction, K_5 is not planar.

●

EXAMPLE 4 The complete bipartite graph $K_{3,3}$ is not planar.

Proof Notice that $V = 6$ and $E = 9$, so using Euler's formula $F = 5$ if $K_{3,3}$ is planar. Now each cycle in $K_{3,3}$ contains at least 4 edges. By the inequality

$$20 = 4F \leq 2E = 18$$

which is a contradiction, $K_{3,3}$ is not planar.

●

EXAMPLE 5 The Petersen graph is not planar.

Proof Here $V = 10$ and $E = 15$, so using Euler's formula $F = 7$ if the Petersen graph is planar. Each cycle in the Petersen graph contains at least 5 edges. By the inequality

$$35 = 5F \leq 2E = 30$$

which is a contradiction, the Petersen graph is not planar.

●

The following theorem occurs in a paper by the Polish mathematician Kazimierz Kuratowski (1896–1980). "Sur le problème des courbes gauches en topologie" (On the problem of space curves in topology) was published in 1930. Remember that a graph G contains a homeomorph of a graph H if $V(H) \subset V(G)$ and the edges of H can be identified with disjoint paths in G.

Kuratowski's Theorem If G is a finite graph, then the following conditions are equivalent:

   (i) G is not planar.

   (ii) G contains a homeomorph of K_5 or $K_{3,3}$.

In the previous section we showed that the Petersen graph contains the graph $K_{3,3}$ as a homeomorph, and this gives another proof that the Petersen graph is not planar.

Kuratowski's Theorem was proved independently by the American mathematician P. A. Smith.

APPLICATION   *The One-Way Street Problem*

In 1939 H. E. Robbins published a paper with the title "A Theorem on Graphs, with an Application to a Problem of Traffic Control." In this paper he solved what has come to be called the *One-Way Street Problem*.

If G is a connected graph, and if we make the edges of G directed, we say that we have chosen an *orientation* of G. A *strongly connected orientation* of G is an orientation such that the directed graph G is strongly connected (any node of G can be reached from any other node by a directed path).

When does a connected graph G have a strongly connected orientation? Clearly G cannot have a bridge. If, for example, G is the graph in Figure 4.27, we clearly cannot pass from any node in the right-hand component to any node in the left-hand component. Remember that a graph which is connected and does not have a bridge is called biconnected. In other words, a graph G with a strongly connected orientation must be biconnected. It was Robbins' remarkable discovery that the opposite implication also holds.

Figure 4.27   A Graph with a Bridge

Robbins' Theorem   Let G be a finite graph. Then the following conditions are equivalent:

   (i) G has a strongly connected orientation.

   (ii) G is biconnected.

The point of the theorem is that any biconnected graph has a strongly connected orientation, as shown in Figure 4.28. Given any connected array of streets such that the associated graph is biconnected, Robbins' Theorem says that there is a way to make the streets one-way such that any corner can be reached from any other corner.

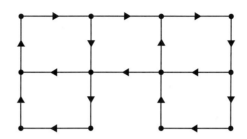

Figure 4.28   A Strongly Connected Orientation of a Biconnected Graph

EXERCISES

In the following two problems, draw the graph associated to the given array of islands and bridges. You will find that each of these graphs has four vertices of odd degree, and so has no Eulerian cycle or circuit.

   Add one bridge to each array in such a way that it then has an Eulerian circuit, and draw this circuit. Then add another bridge in such a way that the array has an Eulerian cycle, and draw this cycle.

1. The array in Figure 4.29

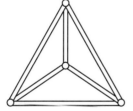

Figure 4.29
An Array of Bridges

2. The array in Figure 4.30

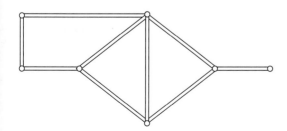

Figure 4.30
An Array of Bridges

3. Find an Eulerian cycle on the complete graph K_5.

4. Find an Eulerian cycle on the graph of the 4-cube.

5. Find an Eulerian cycle on the graph of the 4-dimensional octahedron.

6. Find a Hamiltonian cycle on the complete graph K_6.

7. Find a Hamiltonian cycle on the graph of the 5-cube.

8. Find a Hamiltonian cycle on the graph of the 5-dimensional octahedron.

9. Draw the graph of the 4-cube in the following way. Draw a regular 16-gon, and label the vertices clockwise with binary vectors of length four in lex order. Then join vertices whose labels differ in only one entry. Figure 4.25 is the graph of the 3-cube drawn in this way.

10. Use the method of the previous problem to draw the graph of the 5-cube.

11. Solve the Two-Way Street Problem for the array in Figure 4.31.

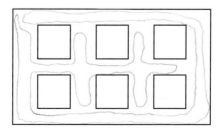

Figure 4.31
The Two-Way Street Problem

12. Solve the Chinese Postman Problem for the array in Figure 4.32. *Hint:* Notice that the associated graph is Eulerian.

Figure 4.32
The Chinese Postman Problem

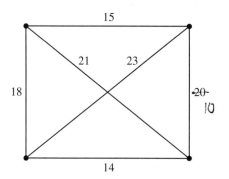

Figure 4.33
The Travelling Salesman Problem

13. Solve the Travelling Salesman Problem for the array in Figure 4.33. *Hint:* Remember that this means finding a cycle, containing a Hamiltonian circuit, of minimal weight.

14. Prove that the graph of the 4-cube is not planar, by using the Euler characteristic and the inequality in the examples at the end of this section.

    Hint: The number of vertices of this graph is $2^4 = 16$. You will need to determine the number of edges. You can draw the graph and count them,

or you may use the fact that the number of edges of the graph of the n-cube is $n \cdot 2^{n-1}$. Then find the number of faces, assuming that the graph of the 4-cube is planar. Observe that every cycle in the graph of the 4-cube has length at least 4. Finally, use the inequality $4F \leq 2E$ to derive a contradiction.

15. Prove in a similar way that the graph of the 4-dimensional octahedron is not planar.

16. Prove in a similar way that the graph $K_{2,2,3}$ is not planar.

ADVANCED EXERCISES

1. Prove that the graph of the dodecahedron is Hamiltonian.

2. Prove that the graph of the icosahedron is not Hamiltonian.

3. A monk wishes to return to his monastery, passing over each of five bridges exactly once. Find one such path (Figure 4.34).

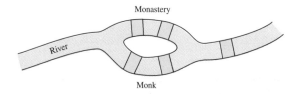

Figure 4.34
The Monk and the Bridges

4. Find *all* paths by which the monk may return to his monastery, crossing each of the five bridges exactly once (Figure 4.34).

5. The following problem is an example of the Chinese Postman Problem. The graph associated to the problem is not Eulerian, so any path which gives a solution involves retracing of segments. The problem is to minimize the amount of retracing.

   A mine inspector descends into a mine at point A. He must inspect all passages of the mine, and return to point A. Each passage is 100 yards long. What is the shortest possible route (Figure 4.35)?

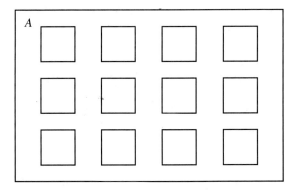

A

Figure 4.35
The Mine Inspector's Problem

Kuratowski's Theorem says that any non-planar graph must contain a homeomorph of K_5 or $K_{3,3}$. Each of the following graphs is non-planar. Draw each one, and find a homeomorph of K_5 or $K_{3,3}$ in each.

6. The graph of the 4-cube

7. $K_{2,2,2,2}$ **8.** $K_{2,2,3}$

COMPUTER EXERCISES

Write a computer program to list all the Eulerian circuits of an Eulerian graph. To write this program, label the nodes of the graph $1, 2, \ldots, n$, and begin an Eulerian cycle at vertex 1. List the Eulerian cycles by enumerating the choices at each vertex as Fleury's Algorithm is used to produce the Eulerian cycle.

Use your program to print a list of all Eulerian circuits of each of the following graphs.

1. K_5 **2.** $K_{2,2,2}$ **3.** The 4-cube

4. K_7 **5.** $K_{2,2,2,2}$

Write a program to list Hamiltonian circuits on the complete graph K_n, and print a list of Hamiltonian circuits for he following graphs:

6. K_4 **7.** K_5 **8.** K_6

4.3   TREES

"Other maps are such shapes, with their islands and capes!

But we've got our brave Captain to thank"

(So the crew would protest) "that he's bought us the best—

A perfect and absolute blank!"

On July 29, 1654, Pascal wrote to Fermat:

Sir:
I am seized with impatience as much as you and, although I am still in bed, I cannot restrain myself from telling you that yesterday evening I received, from Monsieur de Carcavi, your letter on the Problem of Points, which I admire more than I can say . . .

The English mathematician Arthur Cayley (1821–1895) founded the theory of trees in two papers in the *Philosophical Magazine: On the theory of the analytical forms called trees*, published in 1857, and *On the analytical forms called trees. Second Part*, published in 1859. He later gave an interesting application of trees to chemistry in another paper, *On the mathematical theory of isomers*, also published in the *Philosophical Magazine*, in 1874.

The Problem of Points, which we will describe in this section, was studied intensively from the middle of the seventeenth century until the early nineteenth century. We shall describe Fermat's solution to the problem, which involves the use of *event trees*.

DEFINITION 4.3.1   A *forest F* is a graph which has no cycles. A *tree T* is a connected graph which has no cycles.

There are many types of trees. We have already given a number of examples of trees and their applications:

    I.   The Full Binary Tree, in Section 1.1.

    II.   The Fibonacci Tree, in Section 1.5.

    III.   The Collatz Tree, in Section 1.5.

    IV.   The Tree Div(p^n), in Section 2.1.

DEFINITION 4.3.2

  i. An *unlabelled tree* is a tree whose nodes and edges are not labelled in any way.

  ii. A *rooted tree* is a tree with one distinguished node, called the root.

  iii. A *binary tree* is a tree each of whose nodes has degree one, two, or three.

Rooted trees are useful as models for many types of processes, where the root corresponds to the beginning of the process. Rooted binary trees are particularly important.

It is interesting to list and count trees of various types. For example, Figure 4.36 shows the unlabelled trees with six nodes.

We can list the labelled trees with n nodes (recall that the nodes of these trees are labelled $1, 2, \ldots, n$) by taking the Prüfer codes of the trees and listing them in lex order. Figure 4.37 shows the first four trees in this list if $n = 4$.

The labelled trees with n nodes are counted by the following theorem of Arthur Cayley.

Cayley's Theorem   There are n^{n-2} labelled trees with n nodes.

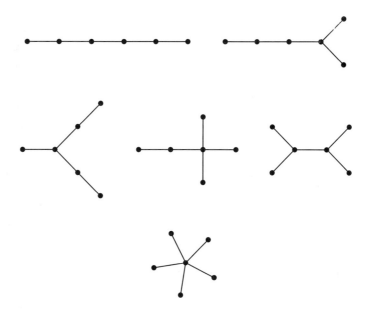

Figure 4.36   Unlabelled Trees with Six Nodes

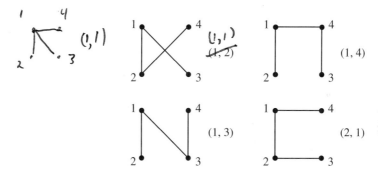

Figure 4.37   Listing Labelled Trees with the Prüfer Code

Proof   Using the Prüfer Correspondence, we see that the set of these trees is in one–one correspondence with the set of $(n-2)$-tuples of elements of $\{1, 2, \ldots, n\}$, so the theorem follows by the Product Rule.   ●

   The following theorem expresses a basic property of connected graphs. Recall that a bridge in a connected graph G is an edge whose removal disconnects G.

THEOREM 4.3.1   If G is a connected graph with n nodes then G has at least $n - 1$ edges.

Proof   The proof is by induction on n. The statement is trivially true if $n = 1$, so assume that $n > 1$. Since G is connected, it must have at least one edge. Start by removing edges from G. Eventually G must become disconnected. Suppose that removing d edges disconnects G, but removing $d-1$ edges does not. The dth edge must be a bridge. When it is removed, G will split into two connected components with n_1 nodes and n_2 nodes respectively, where $1 \le n_1 < n$ and $1 \le n_2 < n$. By induction, these connected components must have at least $n_1 - 1$ edges and $n_2 - 1$ edges respectively. Thus G must have at least $d + (n_1 - 1) + (n_2 - 1) = (n_1 + n_2) + (d - 2) = n + d - 2 \ge n - 1$ edges. This completes the proof by induction. ●

Figure 4.38 shows a disconnected graph with eight nodes and $6 < 8 - 1 = 7$ edges.

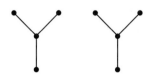

Figure 4.38   A Graph with 8 Nodes and 6 Edges

It is convenient to be able to recognize a tree in as many different ways as possible. The following theorem gives a number of different ways to recognize a tree. The proof of this theorem will illustrate a basic proof technique, proving a *circle of implications*.

THEOREM 4.3.2   Let T be a finite graph, and let n be the number of nodes of T. The following conditions are equivalent:

   (i) T is a tree.
   (ii) T has no cycles and T has $n - 1$ edges.
   (iii) T is connected and T has $n - 1$ edges.
   (iv) T is connected and every edge is a bridge.
   (v)   Any two nodes of T are connected by a unique path.
   (vi) T has no cycles but the addition of one edge to T creates a cycle.

Proof   The proof is by a circle of implications. All the statements are clearly true for $n = 1$, so we assume that $n > 1$.

(i) $\rightarrow$ **(ii)**   Since T is a tree, it contains no cycles. Thus removing any edge will decompose T into two trees, with n_1 nodes and n_2 nodes respectively. By induction these trees have $n_1 - 1$ edges and $n_2 - 1$ edges respectively. Thus T must have $1 + (n_1 - 1) + (n_2 - 1) = n - 1$ edges.

(ii) $\rightarrow$ **(iii)**   Suppose that T has r connected components. Each of these must be a tree, since T has no cycles. By induction each will have a number of edges one fewer than the number of nodes. Thus T will have $n - r$ edges. Since $n - r = n - 1$, we conclude that $r = 1$, so T is connected.

(iii) $\rightarrow$ **(iv)**   If we remove any edge of T, we obtain a graph with n nodes and $n - 2$ edges, which must be disconnected, by the previous theorem.

(iv) → (v)   Since T is connected, each pair of nodes is connected by at least one path. If any pair of vertices is connected by two different paths, then the union of these two paths will contain a cycle. An edge contained in a cycle is not a bridge, contradicting our assumption.

(v) → (vi) T cannot contain a cycle because any two nodes in a cycle are connected by two different paths. If an edge $e = \{v, w\}$ is added to T, then since v and w are already connected by a path in T, a cycle will be created.

(vi) → (i)   We must prove that T is connected. If it is not, adding an edge which joins two connected components of T will not create a cycle.   ●

A *spanning tree* T of C is a graph of C such that T is a tree and $V(T) = V(G)$. We will give a formula for the number of spanning trees of a finite connected graph G. We will assume that G is labelled, with nodes $v_1, v_2, \ldots, v_n$. Recall that the adjacency matrix $A = [a_{ij}]$ of G is the matrix whose entries are $a_{ij} = 1$ if v_i and v_j are joined by an edge, and $a_{ij} = 0$ if v_i and v_j are not joined by an edge.

An n-by-n matrix has n^2 *cofactors*. The (i, j)-th cofactor of the matrix is obtained by striking out the ith row and the jth column and multiplying the determinant of the resulting matrix by $(-1)^{i+j}$.

The following theorem was proved by the German physicist Gustav Robert Kirchhoff (1824–1887), as a part of his study of the laws of electric circuits, in a paper published in 1847.

The Matrix-Tree Theorem   Let G be a connected labelled graph with adjacency matrix A. Let B be the matrix obtained from $-A$ by replacing the ith diagonal entry by $\deg(v_i)$. Then the cofactors of B are all equal, and their common value is the number of spanning trees of G.

EXAMPLE 1   Use the Matrix-Tree Theorem to find the number of spanning trees of the complete graph K_4.

NOT oʳ TᴇST

Solution   The adjacency matrix of the graph K_4 (however the nodes are labelled) is

$$A = \begin{bmatrix} 0 & 1 & 1 & 1 \\ 1 & 0 & 1 & 1 \\ 1 & 1 & 0 & 1 \\ 1 & 1 & 1 & 0 \end{bmatrix}$$

The degree of each vertex of K_4 is 3, so when we replace the diagonal elements of $-A$ by 3 to obtain the matrix B we have

$$B = \begin{bmatrix} 3 & -1 & -1 & -1 \\ -1 & 3 & -1 & -1 \\ -1 & -1 & 3 & -1 \\ -1 & -1 & -1 & 3 \end{bmatrix}$$

The (1, 1)-th cofactor (the cofactor of the upper left-hand entry) is the determinant

$$\begin{vmatrix} 3 & -1 & -1 \\ -1 & 3 & -1 \\ -1 & -1 & 3 \end{vmatrix}$$

Subtracting the third row from the second row and adding 3 times the third row to the first row, we obtain

$$\begin{vmatrix} 0 & -4 & 8 \\ 0 & 4 & -4 \\ -1 & -1 & 3 \end{vmatrix}$$

Developing by the first column this determinant is

$$(-1)\begin{vmatrix} -4 & 8 \\ 4 & -4 \end{vmatrix} = (-1)(16 - 32) = 16$$

so we conclude that K_4 has 16 spanning trees. Cayley's Theorem says that there are $4^{4-2} = 16$ of these trees.   ●

An algorithm for finding a minimum-weight spanning tree of a weighted graph was described by R. C. Prim in a paper published in 1957.

Prim's Algorithm   This algorithm finds a minimum-weight spanning tree T of a connected weighted graph G.

i. First choose any node of G.

ii. If the tree T has been constructed, add a new edge e to T such that one node of e belongs to T, the other node does not, and the weight $w(e)$ is minimal, subject to this condition.

iii. Continue until (ii) is no longer possible.

Figure 4.39 shows an example of the implementation of Prim's algorithm.

Prim's algorithm gives a minimum-weight spanning tree for a connected graph. There are two additional basic algorithms which give spanning trees for a connected graph G with ordered nodes. The first algorithm gives a *breadth-first spanning tree* for G, and the second algorithm gives a *depth-first spanning tree* for G.

A Breadth-First Spanning Tree   Given a connected simple graph G with n vertices, this algorithm constructs a *breadth-first spanning tree* T.

(i)   Label the nodes of G

$$v_1, v_2, \ldots, v_n$$

and let v_1 be the root of a tree T.

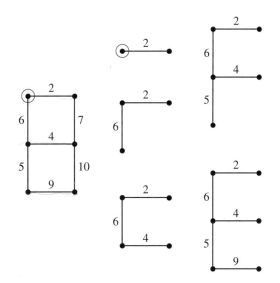

Figure 4.39   The Implementation of Prim's Algorithm

(ii)   Consider each edge $\{v_i, v_j\} \in E(G)$ with $v_i \in V(T)$ and $v_j \notin V(T)$, where first i and then j are chosen to be minimal. Add the edge to $E(T)$ and the node to $V(T)$ if doing so does not create a cycle.

(iii)   Repeat (ii) until $V(T) = V(G)$.

A Depth-First Spanning Tree   Given a connected simple graph G with n vertices, this algorithm constructs a *depth-first spanning tree T*.

(i)   Label the nodes of G

$$v_1, v_2, \ldots, v_n$$

and let v_1 be the root of a tree T.

(ii)   Choose $\{v_i, v_j\} \in E(G)$ with $v_i \in V(T)$ and $v_j \notin V(T)$, where first i and then j are chosen to be minimal. Then choose $\{v_j, v_k\} \in (G)$ with k minimal and $v_k \notin V(T)$, and so on, obtaining a path in G. Add the edges of this path to $E(T)$ in order, and their nodes to $V(T)$, but stop if the addition of an edge would create a cycle.

(iii)   Backtrack along the path defined in (ii) until a node not in $V(T)$ is found adjacent to a node in $V(T)$ and repeat (ii).

(iv)   Continue (ii) and (iii) until $V(T) = V(G)$.

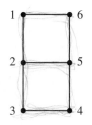

Figure 4.40

An Ordered
Graph G

Notice that the nodes of the graph G must be ordered before we can find a breadth-first or depth-first spanning tree. The structure of these trees is heavily dependent on the ordering of the nodes of G. The following example will demonstrate this.

EXAMPLE 2   Find breadth-first and depth-first spanning trees for the ordered graph G in Figure 4.40. Then change the ordering of the nodes of G in such a way that the new breadth-first and depth-first spanning trees have different degree sequences.

Solution   The construction of the breadth-first and depth-first spanning trees of G is shown in Figure 4.41. Figure 4.42 shows the graph G with a different ordering of the nodes. Figure 4.43 shows the construction of the breadth-first and depth-first spanning trees for the ordered graph in Figure 4.42. Notice that the degree sequences of these new trees are different from the degree sequences of the previous ones.   ●

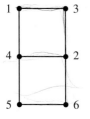

Figure 4.42
The Graph G with
a Different Ordering
of the Nodes

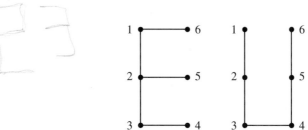

Figure 4.41   The Breadth-First and Depth-First Spanning Trees of G

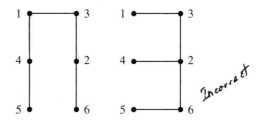

Figure 4.43   The Breadth-First and Depth-First Spanning Trees of G

Given a graph G which arises in a problem, it is a standard practice to study the problem by studying the spanning trees of the connected components of G (the spanning trees of G if G is connected). Trees themselves arise as models in the study of many problems, as we have seen. The full binary tree is a model for the coin-tossing problem, the Fibonacci tree is a model for the Fibonacci sequence, and so on.

Suppose that a tree T arises in the study of a problem. How should we continue once we have found T? Since the nodes of T represent aspects of the problem or stages in its evolution, we want to *visit* the nodes of T in order, or to *traverse* the nodes of T.

We shall assume that T is a binary tree, drawn with the root at the top, and with left- and right-subtrees of each node indicated by the way in which they are drawn on the page (this really corresponds to an ordering of the nodes of T). We shall describe three algorithms for traversing the nodes of T. Because T is assumed to have a root, $V(T)$ is non-empty.

Inorder Traversal of a Binary Tree

(i)   At the current node, pass to the left subtree of T (if this is non-empty).

(ii)   Visit the current node.

(iii)   Pass to the right subtree of T (if this is non-empty).

Preorder and postorder traversal differ from inorder traversal only in the order in which the above steps are applied.

Preorder Traversal of a Binary Tree

(i) Visit the current node.

(ii) Pass to the left subtree (if this is non-empty).

(iii) Pass to the right subtree (if this is non-empty).

Postorder Traversal of a Binary Tree

(i) Pass to the left subtree (if this is non-empty).

(ii) Pass to the right subtree (if this is non-empty).

(iii) Visit the current node.

Figure 4.44 shows a binary tree and the inorder traversal of its nodes. Figure 4.45 shows the preorder traversal of the same tree, and Figure 4.46 the postorder traversal of the same tree.

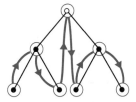

Figure 4.44   The Inorder Traversal of a Binary Tree

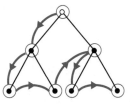

Figure 4.45   The Preorder Traversal of the Same Tree

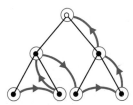

Figure 4.46   The Postorder Traversal of the Same Tree

We began this section with a quote from a letter from Pascal to Fermat which was written in the summer of 1654: they were discussing the Problem of Points. We will give an example of this problem and Fermat's method for solving it, which involves the use of event trees.

EXAMPLE 3   Player A and Player B are playing a game in which a fair coin is tossed repeatedly. Player A wins a point if heads comes up, Player B wins a point if tails comes up, and the first player to win a certain number of points will win eight dollars.

The game is interrupted when Player A needs one point to win and Player B needs three points to win. How should the eight dollars be divided between Player A and Player B?

Solution   The event tree in Figure 4.47 can be used to solve this problem. An edge to the left means a point for Player A and an edge to the right means a point for Player B. The root represents the state of the game at the time it is interrupted, and the paths in the tree represent all possible continuations of the game. The edges of the tree are each weighted with 1/2, the probability of getting heads or tails each time the coin is tossed.

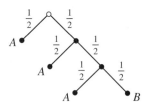

Figure 4.47   An Event Tree for the Problem of Points

The terminal vertices are labelled with an A if Player A wins and with a B if Player B wins. We compute the *expectation* of each player by taking the product of the weights of the edges along each path which leads to a win for that player, summing these products, and multiplying by 8. Thus

$$\text{Player A's expectation} = \left(\frac{1}{2} + \frac{1}{4} + \frac{1}{8}\right) \cdot 8 = 7$$

$$\text{Player B's expectation} = \left(\frac{1}{8}\right) \cdot 8 = 1$$

Player A should receive seven dollars and Player B should receive one.   ●

APPLICATION   *Structural Induction*

In the first chapter of this book, we gave a careful treatment of induction and recursion. Induction and recursion go together. Most processes in a computer are either recursive or closely connected with a recursion of some kind. Induction is a proof technique which is well adapted to proving results about recursions.

Trees can be used as models whose effect is to bring about a closer connection between induction and recursion. For example, the fact that the level sequence of the Fibonacci tree is the Fibonacci sequence gives a better understanding of the recursion which generates the Fibonacci sequence.

There are many forms of the Principle of Induction. *Structural induction* is a type of induction in which the underlying set is a binary tree (we will assume that it is the full binary tree) instead of the set of integers.

Let's compare these two types of induction (in computer science, it is usual to refer to the *base case*, instead of "starting the induction").

For the usual Principle of Induction, we have:

The base case: $P(1)$ is true.

A rule: If $P(n)$ is true, then $P(n + 1)$ is true.

Conclusion: $P(n)$ is true for all $n = 1, 2, 3, \ldots$.

For structural induction, we have:

The base case: P(the root) is true.

A rule: If T is a finite full binary tree and L and R are the left and right subtrees of T, and if $P(L)$ and $P(R)$ are true, then $P(T)$ is true.

Conclusion: $P(T)$ is true for all finite full binary trees T.

The following theorem will be proved using structural induction.

THEOREM 4.3.3   In a finite full binary tree T, the number of terminal nodes $l(T)$ is equal to the number of internal nodes $i(T)$ plus 1.

Notice that this is true for the tree in Figure 4.48.

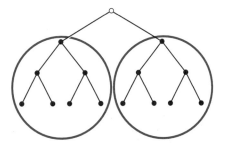

Figure 4.48   A Full Binary Tree and Its Left and Right Subtrees

Proof   If T consists only of one node (the root), then $l(T) = 1$ and $i(T) = 0$, so the statement of the theorem is true.

In general, we have $l(T) = l(L) + l(R)$, and $i(T) = i(L) + i(R) + 1$. By our inductive assumption, $l(L) = i(L) + 1$ and $l(R) = i(R) + 1$, so $l(T) = l(L) + l(R) = (i(L) + 1) + (i(R) + 1) = (i(L) + i(R) + 1) + 1 = i(T) + 1$.

EXERCISES

1. Make a list of the unlabelled trees with seven nodes. *Hint:* There are eleven of these trees.

2. Make a list of the rooted trees with six nodes. *Hint:* There are twenty of these trees.

3. Make a list of the rooted binary trees with five nodes. (Here we understand that the vertices of the trees are ordered, so that two of these trees whose left and right subtrees are interchanged are understood to be different.)

4. Using the Prüfer Code in lex order, make a list of all sixteen spanning trees of the complete graph K_4.

Draw each of the following graphs, and make a list of five of its spanning trees.

5. The graph of the octahedron

6. The graph of the 3-cube

7. The graph of the icosahedron

8. The graph of the dodecahedron

Use the Matrix-Tree Theorem to find the number of spanning trees of the following graphs:

9. The ladder graph

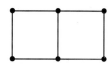

10. The 4-wheel

11. The 5-wheel

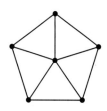

12. $K_{2,2,2}$.

13. The 3-cube. *Hint:* This graph has 384 spanning trees.

Use Prim's Algorithm to find a minimum-weight spanning tree for each of the following graphs.

14.

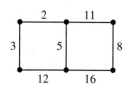

15.

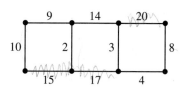

Find a breadth-first and a depth-first spanning tree for each of the following ordered graphs.

16.

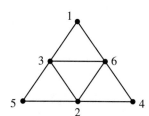

17.

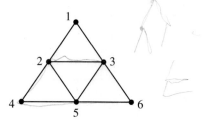

For each of the following binary trees, give a path for in-order traversal, preorder traversal, and postorder traversal.

18.

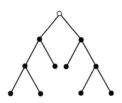

19.

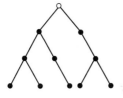

20. Player A and Player B are playing a game in which a fair coin is tossed. If the coin comes up heads, Player A wins a point. If the coin comes up tails, Player B wins a point. The first player to win a certain number of points will win sixteen dollars.

The game is interrupted when Player A needs two points to win, and Player B needs three points to win. How should the sixteen dollars be divided?

ADVANCED EXERCISES

1. Make a list of the unlabelled trees with eight nodes. *Hint:* There are twenty-three of these trees.

2. Make a list of the rooted trees with seven nodes. *Hint:* There are forty-eight of these trees.

3. Make a list of the rooted binary trees with seven nodes. (Here we understand that the vertices of the trees are ordered, so that two of these trees whose left and right subtrees are interchanged are understood to be different.)

Use the Matrix-Tree Theorem to find the number of spanning trees of each of the following graphs:

4. The 6-wheel

5. The 7-wheel

6. The ladder graph

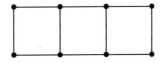

7. The ladder graph

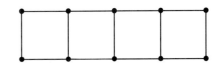

8. $K_{4,4}$

9. $K_{2,2,2,2}$

10. Player A, Player B, and Player C are playing a game in which a single fair die is tossed repeatedly. Player A wins a point if the die comes up 1 or 2. Player B wins a point if the die comes up 3 or 4. Player C wins a point if the die comes up 5 or 6. The first player to win a certain number of points will win twenty-seven dollars.

The game is interrupted when Player A needs one point to win, and Player B and Player C each need two points to win. How should the twenty-seven dollars be divided?

COMPUTER EXERCISES

Use the Matrix-Tree Theorem to count the number of spanning trees of the following graphs.

1. $K_{5,5}$
2. $K_{2,2,2,2,2}$
3. The 8-wheel

4. The 9-wheel

5. The ladder graph

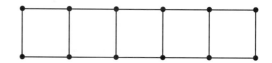

6. The ladder graph

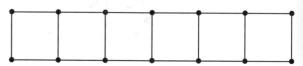

7. The graph of the icosahedron
   Hint: There are 5,184,000 of these spanning trees.

8. The graph of the dodecahedron
   Hint: There are 5,184,000 of these spanning trees.

9. Can you explain why the answers in Exercise 7 and Exercise 8 are the same?

10. The 4-cube

11. We saw earlier that the 3-cube has 384 spanning trees. Make a list of these trees (you can represent the trees either by their edge lists or their Prüfer Codes).

4.4   CHAPTER SUMMARY AND SUPPLEMENTARY EXERCISES

"Friends, Romans, and countrymen, lend me your ears!"
(They were all of them fond of quotations:
So they drank to his health, and they gave him three cheers,
While he served out additional rations).

Chapter Summary

Section 4.1

In this section we began a systematic study of graphs and their properties. We defined simple graphs, multigraphs, and pseudographs, as well as the adjacency matrix of a graph. We defined walks, trails, paths, closed walks, and cycles on a graph. We described connectivity, gave several examples of graphs, and showed how to draw certain classes of graphs. We ended the section by simplifying the graphs of the icosahedron and the dodecahedron by taking the quotient of each by the antipodal map.

Section 4.2

We began this section with a discussion of the origin of graph theory in Euler's solution of the Königsberg Bridge Problem. We defined Eulerian and Hamiltonian graphs. We described the Two-Way Street Problem, the Chinese Postman Problem, and the Travelling Salesman Problem. We defined the Euler characteristic of a planar graph, and used its properties to prove that several graphs are not planar. We ended the section with the One-Way Street Problem.

Section 4.3

We began this section by mentioning the Problem of Points, which is also discussed at the end of the section. We gave a characterization of trees, and described the adjacency matrix of a graph and the Matrix-Tree Theorem, which counts the number of spanning trees of a connected graph. We described Prim's algorithm, and breadth-first and depth-first spanning trees. We gave the basic algorithms for traversal of a binary tree: inorder traversal, preorder traversal, and postorder traversal.

EXERCISES

Use Prim's algorithm to find a minimum-weight spanning tree for each of the following graphs.

1.

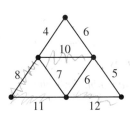

2.

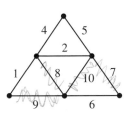

3.

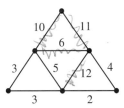

Give paths indicating the breadth-first and depth-first traversal of each of the following trees.

4.

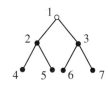

5.

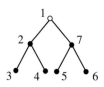

6.

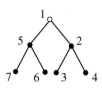

Find the inorder, preorder, and postorder traversal for each of the following binary trees.

7.

8.

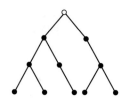

9.

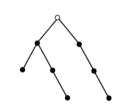

10.

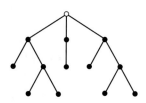

5
Proof Techniques and Logic

5.1 PROOF TECHNIQUES

> *"Come, listen, my men, while I tell you again*
>
> *The five unmistakable marks*
>
> *By which you may know, wheresoever you go,*
>
> *The genuine warranted Snarks.*
>
> *"Let us take them in order. The first is the taste,*
>
> *Which is meagre and hollow, but crisp:*
>
> *Like a coat which is rather too tight in the waist,*
>
> *With a flavour of Will-o-the-wisp."*

The Greek poet and prophet Epimenides lived in the seventh and sixth centuries B.C. Very little is known about his life, and he may have been a legendary figure. Epimenides' name is associated with a paradox called the *Epimenides Paradox*, or the *Liar Paradox*.

The Liar Paradox   Someone says: "I am lying." Is this sentence true or false?

If the sentence is true, then the person is lying, so the sentence must be false. If the sentence is false, then the person is not lying, and so the sentence must be true. Therefore the sentence cannot be either true or false.

Notice how the paradox arises through a combination of self-reference and negation. A paradox of this kind is called a *paradox of self-reference*.

Russell's Paradox, which we discussed earlier, is an example of a paradox of self-reference. The paradoxes of self-reference seem like puzzles, but they are examples of an important logical theme, which we shall develop gradually.

159

In this chapter and the next, we shall explain some basic results about logic and sets. We shall begin, in this section, by explaining and giving examples of certain basic *proof techniques*. At the end of this section, we shall test our logical skills by solving the *Liar Problem*.

In the next section, we shall give a brief description of Aristotelian logic. Aristotelian logic was the dominant form of logic for more than two millenia. We shall describe a code for the arguments of Aristotelian logic, and also an analogy between arguments in Aristotelian logic and the genetic code.

In the final section, we shall describe the *propositional calculus*, the simplest logical language.

The most important proof techniques are the following.

Proof Techniques

   I. Proofs by Induction
  II. Direct Proofs
       A. Proving that a statement is true by a direct argument
       B. Proving that a statement is false by finding a *counterexample*
 III. Indirect Proofs
       A. Proving the *contrapositive* of a statement
       B. Proving a statement by *contradiction*
  IV. Logical Equivalence
       A. Proving a *logical equivalence*
       B. Proving a *circle of implications*

We have described proofs by induction in detail. Next we shall give examples of direct proofs that a statement is true or false. First we shall give direct proofs of two identities for binomial coefficients.

A Direct Proof That a Statement Is True

We previously mentioned the Catalan numbers

$$C_n = \frac{1}{n+1}\binom{2n}{n}$$

The Catalan numbers are whole numbers, meaning that $n + 1$ must always divide the middle binomial coefficient $\binom{2n}{n}$. Why is this? We can prove that this is true if we prove that each Catalan number is a difference of binomial coefficients, and therefore is a whole number.

EXAMPLE 1   Prove that

$$C_n = \frac{1}{n+1}\binom{2n}{n} = \binom{2n}{n} - \binom{2n}{n-1}$$

Solution   By using Pascal's Formula, we will show that the right-hand side of this equation is equal to the left-hand side.

$$\binom{2n}{n} - \binom{2n}{n-1} = \frac{(2n)!}{n!n!} - \frac{(2n)!}{(n-1)!(n+1)!}$$

$$= \frac{(2n)!(n+1)}{n!n!(n+1)} - \frac{(2n)!(n)}{(n-1)!(n+1)!(n)}$$

$$= \frac{1}{n+1}\left(\frac{(2n)!(n+1) - (2n)!(n)}{n!n!}\right)$$

$$= \frac{1}{n+1}\left(\frac{(2n)!}{n!n!}\right) = \frac{1}{n+1}\binom{2n}{n}$$

   ●

EXAMPLE 2   Prove that

$$\binom{2n}{n} = \sum_{k=0}^{n}\binom{n}{k}^2 = \binom{n}{0}^2 + \binom{n}{1}^2 + \ldots + \binom{n}{n}^2$$

Solution   We shall prove this by using the Binomial Theorem and the equation

$$(X + Y)^{2n} = ((X + Y)^n)^2$$

By the Binomial Theorem, the coefficient of $X^n Y^n$ in the left-hand side of the equation is

$$\binom{2n}{n}$$

By the Binomial Theorem, the right-hand side can be expressed as

$$\left(\binom{n}{0}X^n + \ldots + \binom{n}{n}Y^n\right)^2$$

Expanding this square, the coefficient of $X^n Y^n$ is

$$\binom{n}{0}\binom{n}{n} + \ldots + \binom{n}{k}\binom{n}{n-k} + \ldots + \binom{n}{n}\binom{n}{0}$$

Since $\binom{n}{k} = \binom{n}{n-k}$, the proof is complete.

   ●

Finding a Counterexample

We shall give two examples of statements which can be shown to be false by finding a case in which they are not true, that is, by finding a counterexample. The following example was studied by Leonhard Euler.

EXAMPLE 3   Is it true that

$$n^2 - n + 41$$

is a prime number for all positive integers n?

Solution   In order to get an idea as to whether this statement is true or false, we list some of the values of this expression:

$$1^2 - 1 + 41 = 41$$
$$2^2 - 2 + 41 = 43$$
$$3^2 - 3 + 41 = 47$$
$$4^2 - 4 + 41 = 53$$
$$5^2 - 5 + 41 = 61$$
$$6^2 - 6 + 41 = 71$$
$$7^2 - 7 + 41 = 83$$
$$8^2 - 8 + 41 = 97$$

$$\cdots$$

$$40^2 - 40 + 41 = 1601$$

All of these forty numbers are prime. We can check this for the first eight numbers on the list by checking that they are not divisible by 2, 3, 5, or 7. We might think that this is evidence for the assertion that all the numbers on the list are prime.

However, the next number on the list is not prime:

$$41^2 - 41 + 41 = 1681 = 41^2$$

This is a counterexample to the statement that all numbers on this list are prime, which proves that the statement is false.   ●

For the next example, notice that every prime number except 2 is odd, and that every odd number is either of the form $4k + 1$ or $4k + 3$. If x is a positive real number, let $\pi_1(x)$ be the number of odd prime numbers of the form $4k + 1$ which are less than or equal to x, and let $\pi_3(x)$ be the number of odd prime numbers of the form $4k + 3$ which are less than or equal to x.

EXAMPLE 4   Is it always true that $\pi_1(x) \le \pi_3(x)$? In other words, are the prime numbers of the form $4k + 3$ always at least as numerous as the prime numbers of the form $4k + 1$?

Solution   To see why this question is interesting, we will make a table of the first few primes, and the values of $\pi_1(x)$ and $\pi_3(x)$:

x	$\pi_1(x)$	$\pi_3(x)$	x	$\pi_1(x)$	$\pi_3(x)$
3	0	1	29	4	5
5	1	1	31	4	6
7	1	2	37	5	6
11	1	3	41	6	6
13	2	3	43	6	7
17	3	3	47	6	8
19	3	4	$\cdots$		
23	3	5			

The pattern that we see in this table indicates that $\pi_1(x) \leq \pi_3(x)$, and in fact that $\pi_1(x) < \pi_3(x)$ except for a few values of x. This pattern continues for a long time. However, the English mathematician J. E. Littlewood proved that the two inequalities

$$\pi_1(x) < \pi_3(x) \qquad \text{and} \qquad \pi_3(x) < \pi_1(x)$$

occur infinitely often. Later, J. Leech found that for $x = 26,861$ we have

$$\pi_3(x) = 1,472 < \pi_1(x) = 1,473$$

which gives an explicit counterexample to our statement.   ●

This example shows that patterns which persist, even for a very long time, may nevertheless change. It shows that data must always be examined with great care, and that we must never forget the importance of proofs.

Before going on with our discussion of proof techniques, it will be convenient to introduce some logical notation.

When we try to prove a statement or understand a proof that someone else has given, it is often useful to break the statement up into simpler statements by picking out the *logical connectives* in the statement. These are the words "not," "and," "or," and the phrases "if-then" and "if and only if". These occur so often that there are symbols for each of them. The letters $p, q, r, \ldots$ are often used for the sentences related by these logical connectives.

> *"not p" is symbolized by "$\sim p$."*
> *"p or q" is symbolized by "$p \vee q$."*
> *"p and q" is symbolized by "$p \wedge q$."*
> *"if p then q" is symbolized by "$p \rightarrow q$."*
> *"p if and only if q" is symbolized by "$p \leftrightarrow q$."*

Let p, q, r be the following sentences:

> p: *"John is at the office."*
> q: *"Joan is at the office."*
> r: *"Laura is at the office."*

EXAMPLE 5   Use logical connectives to express the following three sentences:

> *"John is not at the office."*
> *"If Joan and Laura are at the office then John is at the office."*
> *"If John is at the office then either Joan or Laura is at the office."*

Solution

$$(\sim p)$$
$$((q \wedge r) \rightarrow p)$$
$$(p \rightarrow (q \vee r))$$ ●

EXAMPLE 6   Use logical connectives to express the following three sentences:

> *"John, Joan, and Laura are all at the office."*
> *"Joan is not at the office and either John or Laura are at the office."*
> *"If Laura is not at the office then John and Joan are both at the office."*

Solution

$$((p \wedge q) \wedge r) \; or \; (p \wedge (q \wedge r))$$
$$((\sim q) \wedge (p \vee r))$$
$$((\sim r) \rightarrow (p \wedge q)) \qquad\qquad \bullet$$

The sentence "If p then q" is often expressed as "p implies q."

Logical Equivalence   Two sentences p and q are said to be *logically equivalent* if p implies q and q implies p.

If two sentences p and q are logically equivalent we write

$$(p \leftrightarrow q)$$

which is read "p if and only if q."

Sentences which are logically equivalent are considered to be interchangeable, from a logical point of view. Logical equivalences of sentences with a different form are expressed by *logical identities*. Many proof techniques can be expressd by logical identities which replace a difficult argument by another which is easier. A basic example of a logical identity, and of a proof technique, is the *contrapositive*.

The Contrapositive   This proof technique depends on the identity

$$((p \rightarrow q) \leftrightarrow ((\sim q) \rightarrow (\sim p)))$$

for any sentences p and q.

The meaning of this identity is that if we want to prove that the sentence p implies the sentence q, it is exactly the same thing to prove that the sentence $(\sim q)$ implies the sentence $(\sim p)$.

Proving the Contrapositive of a Statement

EXAMPLE 7   Prove that if the square of an integer m is even, then the integer m must be even.

Proof   We must prove that if m^2 is even, then m is even. The contrapositive of this statement is that if m is not even, then m^2 is not even. In other words, if m is odd then m^2 is odd. If m is odd then $m = 2k + 1$ for some integer k. Then $m^2 = (2k + 1)^2 = 4k^2 + 4k + 1$, which is an odd number. $\qquad\qquad \bullet$

Proving a Statement by Contradiction

DEFINITION 5.1.1   A *contradiction* is a sentence which is always false. If p is any sentence, the sentence

$$(p \wedge (\sim p))$$

is a contradiction, since no sentence can be simultaneously true and false. Any contradiction is logically equivalent to a sentence of the form $(p \wedge (\sim p))$.

A *proof by contradiction* depends on the following identity:

$$(((\sim p) \rightarrow (q \wedge (\sim q))) \rightarrow p)$$

In other words, to prove that a sentence p is true, assume that it is false (that is, that $(\sim p)$ is true). If the assumption that $(\sim p)$ is true implies a contradiction, then $(\sim p)$ cannot be true, so p must be true.

An example of a proof by contradiction is the proof that $\sqrt{2}$ is irrational, which is due to the Pythagoreans, some time in the sixth century B.C. Aristotle mentioned the fact that the Pythagoreans had proved this, and said that he recalled that the proof was "by contradiction."

Recall that a number is rational if it is a ratio of integers, and *irrational* if it is not.

EXAMPLE 8   Prove that $\sqrt{2}$ is irrational.

Proof   To prove this statement by contradiction, assume that it is false, namely that $\sqrt{2}$ is rational. This means that

$$\sqrt{2} = \frac{m}{n}$$

where m and n are integers, and $n \neq 0$. We may also assume that m and n have no common factor, since any common factor of m and n may be factored out.

Multiplying both sides of this equation by n and squaring both sides, we have

$$(\sqrt{2}n)^2 = 2n^2 = m^2$$

which means that m^2 is even. By Example 7, this implies that m is even, say $m = 2r$. Therefore

$$2n^2 = (2r)^2 = 4r^2, \text{ so } n^2 = 2r^2$$

Thus n^2 is even. By Example 7, n must be even, say $n = 2s$. We have proved that m and n have a common factor, namely 2, which contradicts the assumption that m and n have no common factor. Therefore $\sqrt{2}$ must be irrational.   ●

The German mathematician Georg Cantor (1845–1918) is regarded as the founder of the theory of sets. Set theory began in an exchange of letters between Cantor and his friend Richard Dedekind in the fall of 1873. Cantor proved the following theorem, which he mentioned in a letter to Dedekind dated August 31, 1899.

THEOREM 5.1.1   Let A be any set, and $P(A)$ the set of all subsets of A. There cannot be a one–one correspondence

$$f : A \rightarrow P(A)$$

Proof   We will prove this by contradiction. Suppose that such a one–one correspondence $f : A \rightarrow P(A)$ exists. Define the subset $B \subset A$:

$$B = \{a \in A : a \notin f(a)\}$$

Since $f : A \rightarrow P(A)$ is a one–one correspondence, there must be a $b \in A$ such that $f(b) = B$. If $b \in B$, then by the definition of B, $b \notin B$. If $b \notin B$, then by the definition of B, $b \in B$. This contradiction proves that f cannot exist, and proves the theorem. ●

Notice the close resemblance between this argument and the argument in Russell's Paradox. Cantor found his argument shortly before Russell found his. It does not seem to be known whether Russell's argument was influenced by Cantor's.

Proving a Logical Equivalence

Remember that a logical equivalence is a statement of the form

$$(p \leftrightarrow q)$$

which means that $(p \rightarrow q)$ and $(q \rightarrow p)$.

Theorems are often stated in the form of logical equivalences, that is, assertions that two or more sentences are logically equivalent. Here is an example.

EXAMPLE 9   If m is a positive integer, the following two statements are logically equivalent:

    (i) m is even.

    (ii) m^2 is even.

Proof   **(i)** $\rightarrow$ **(ii)**   If m is even, then $m = 2r$ for some positive integer r. Then $m^2 = (2r)^2 = 4r^2$ is even.

   (ii) $\rightarrow$ **(i)**   We proved in Example 7 that if m^2 is even, then m is even.   ●

If we wish to prove that two sentences p and q are logically equivalent, we must prove two implications, $p \rightarrow q$ and $q \rightarrow p$. Suppose that we must prove that n sentences are all logically equivalent. What is the most efficient way to do this?

Proving a Circle of Implications

Suppose that we must prove that n sentences

$$p_1, p_2, \ldots, p_n$$

are all logically equivalent, that is, that each implies the other.

The most efficient way to do this is to prove a *circle of implications*

$$p_1 \rightarrow p_2 \rightarrow \cdots \rightarrow p_n \rightarrow p_1$$

Notice that this proves that all the sentences p_k are logically equivalent in pairs. For example, since

$$p_1 \rightarrow p_2 \rightarrow p_3$$

it follows that $p_1 \rightarrow p_3$, and since

$$p_3 \rightarrow \cdots \rightarrow p_n \rightarrow p_1$$

it follows that $p_3 \rightarrow p_1$.

To prove that n statements are logically equivalent using a circle of implications, it is necessary to prove n implications. To prove that each of the n statements is logically equivalent in pairs, it is necessary to prove

$$2 \cdot \binom{n}{2} = n \cdot (n-1) = n^2 - n$$

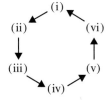

Figure 5.1
Six Characterizations
of a Tree

implications. For example, to prove that six statements are logically equivalent by using a circle of implications, we must prove six implications. If we try to prove that the statements are equivalent in pairs, we must prove $6^2 - 6 = 30$ implications, which is far more work.

In Section 3 of Chapter 4, we have given a significant example of a theorem which is proved by a circle of implications, namely the six different characterizations of a tree, as in Figure 5.1.

APPLICATION   *The Liar Problem*

We began this section with a statement of the Liar Paradox. This paradox, studied since classical times, arises through a combination of self-reference and negation, and is one of the paradoxes of self-reference.

Arguments which are similar to the argument in the Liar Paradox can give legitimate proofs. The argument in Cantor's Theorem, which we have just proved, is similar to the argument in Russell's Paradox.

Now we shall describe the Liar Problem, whose solution is given by an argument which, while it is correct, is closely related to the argument in a paradox of self-reference.

The Liar Problem   In a certain country, all the people belong to one of two tribes: the Truth-Tellers, who always tell the truth, and the Liars, who always lie. A traveller in this country comes to a fork in the road, and wants to know whether the left or the right fork will lead him to a certain city. At the fork of the road is a man who belongs to one of the two tribes, but the traveller does not know which one.

The man will answer one question and only one. How can the traveller phrase the question so as to find the right way?

Solution   The traveller points to the right-hand road and asks, "If you belonged to the other tribe, would you say that this was the road to the city?"

There are four cases: the man may be a Truth-Teller or a Liar, and the right-hand road may or may not be the road to the city.

Suppose that the right-hand road is the road to the city. If the man is a Truth-Teller, he must truthfully say that if he were a Liar he would answer "no." If the man is a Liar, he must say, falsely, that if he were a Truth-Teller he would answer "no."

Similarly, if the right-hand road is not the road to the city, both the Truth-Teller and the Liar must answer "yes."

Therefore the traveller may find the right road by asking the question that we have just given, and doing the opposite of what the man suggests.

EXERCISES

1. Take a piece of paper. On one side write "The sentence on the other side of this paper is true." On the other side write "The sentence on the other side of this paper is false." Can either sentence be true or false?

2. A barber in Cambridge, England, advertised that he

> "...shaves those, and only those, who
> do not shave themselves."

Did the barber shave himself?

3. What number, if any, is denoted by the phrase "the least integer not nameable in fewer than nineteen syllables"? *Hint:* Count the number of syllables in this phrase.

Let p, q, and r be the following sentences:

> p: *"John is at the office."*
> q: *"Joan is at the office."*
> r: *" Laura is at the office."*

Write the following English sentences using logical connectives.

4. "Laura is not at the office."

5. "If John and Joan are at the office then Laura is at the office."

6. "If John is at the office then Joan is not at the office."

7. "If John or Laura is at the office then Joan is at the office."

8. "John and Joan are at the office, and Laura is not at the office."

9. "If John is at the office or Joan is not at the office, then Laura is not at the office."

10. "If John is at the office and if, whenever John is at the office Laura is also at the office, then Joan is at the office."

With the same interpretation of the variables p, q, and r, write the following logical notation sentences as English sentences.

11. $(\sim q)$.   12. $(p \to r)$.

13. $((p \lor r) \to q)$.   14. $((p \land (p \to r)) \to q)$.

15. $(((\sim q) \to r) \to p)$.   16. $((q \land \sim r) \to (\sim p))$.

17. $((\sim p) \land (q \lor r))$.

Prove the following statements using the contrapositive.

18. If m is an integer and m^2 is divisible by 3, then m is divisible by 3. *Hint:* Every integer is of the form $3k$, $3k + 1$, or $3k + 2$. Assume that m is of the form $3k + 1$ or $3k + 2$, and consider m^2.

19. If m is an integer and m^2 is divisible by 5, then m is divisible by 5. *Hint:* Every integer is of the form $5k$, $5k + 1$, $5k + 2$, $5k + 3$, or $5k + 4$. Assume that m is not of the form $5k$, and consider m^2.

Prove the following statements by contradiction.

20. $\sqrt{6}$ is irrational. *Hint:* Assume that $\sqrt{6} = \frac{m}{n}$ is a ratio of integers with no common factor, and prove that in fact 2 must divide them both.

21. $\sqrt{10}$ is irrational.

22. $\sqrt{3}$ is irrational. *Hint:* Use the contrapositive, [*contradiction*] and Exercise 18.

23. $\sqrt{5}$ is irrational. *Hint:* Use the contrapositive, [*contradiction*] and Exercise 19.

24. If A is a finite set with n elements, $n = |A|$, we have proved that the number of subsets of A, $|P(A)| = 2^n$. Use induction to prove that for all $n \geq 0$, $n < 2^n$, and hence that there can never be a 1–1 correspondence between A and $P(A)$ if A is finite.

25. Prove that if m is a positive integer, then the following statements are logically equivalent:

   (i) m is divisible by 3.
   (ii) m^2 is divisible by 3.

26. Prove that if m is a positive integer, then the following statements are logically equivalent:

   (i) m is divisible by 5.
   (ii) m^2 is divisible by 5.

27. Use a circle of implications to prove that if m is a positive integer, then the following statements are logically equivalent:

   (i) m is divisible by 3.
   (ii) m^2 is divisible by 3.
   (iii) m^3 is divisible by 3.

28. Use a circle of implications to prove that if m is a positive integer, then the following statements are logically equivalent:

   (i) m is divisible by 5.
   (ii) m^2 is divisible by 5.
   (iii) m^3 is divisible by 5.
   (iv) m^4 is divisible by 5.

ADVANCED EXERCISES

Here is a variant of the Liar Problem. Once again, a traveller is passing through a country where all the people are either Truth-Tellers, who always tell the truth, or Liars, who always lie. The traveller comes to a fork in the road, and at the fork of the road there is a man who is either a Truth-Teller or a Liar, but the traveller does not know which.

The traveller wants to go to a certain city. This time there are three forks in the road, a left fork, a middle fork, and a right fork, and only one of these leads to the city.

1. The traveller must ask the man questions to determine which fork in the road leads to the city. What is the smallest number of questions that are needed? Is one enough? Are two enough? What should the questions be?

COMPUTER EXERCISES

1. Use the factorization command in your computer algebra program to prove that

$$n^2 - n + 41$$

is prime for $0 \leq n \leq 40$, but not for $n = 41$.

2. Prove that

$$n^2 - 79n + 1601$$

is prime for $0 \leq n \leq 79$, but not for $n = 80$.

3. Write a program to evaluate the functions $\pi_1(x)$ and $\pi_3(x)$ (for integer values of x). Verify the statement in Example 4 that

$$\pi_3(26,861) = 1,472 < \pi_1(26,861) = 1,473$$

Is this the smallest value of x for which $\pi_3(x) < \pi_1(x)$?

5.2   LOGIC

"Its habit of getting up late you'll agree

That it carries too far, when I say

That it frequently breakfasts at five-o'clock tea

And dines on the following day.

"The third is its slowness in taking a jest.

Should you happen to venture on one,

It will sigh like a thing that is deeply distressed:

And it always looks grave at a pun."

Logic began and developed in three rather brief periods: a first period at the time of Aristotle and his immediate followers; a second period during the Middle Ages; and a third period beginning in the middle of the nineteenth century and continuing to the present day. This third period saw the development of *symbolic logic*.

Aristotle of Stagira was born in 384 B.C. and died in 322 B.C.

Aristotle's logical writings are the *Categories*, *On interpretation*, the *Prior analytics*, the *Posterior analytics*, the *Topics*, and *On sophistical refutations*. Collectively called the *Organon*, they total a few hundred pages.

The *Peripatetics*, among them Theophrastus (about 372–288 B.C.), continued Aristotle's logical work. It was also developed by the *Stoics*—the followers of Zeno (about 336–264 B.C.), and Chrysippus (about 280–205 B.C.).

The great philosophers of the Middle Ages continued the development of logic. Among these were Peter Abelard (1079–1142), Thomas Aquinas (about 1225–1274), John Duns Scotus (about 1266–1308), and William of Ockham (about 1285–1349).

Aristotle's point of view dominated the study of logic until late in the nineteenth century.

Aristotle's logic begins with the *classification of sentences* which are used in logical arguments into four types.

Universal Affirmative: "All *A* is *B*".
Interpretation Using Sets: $A \subset B$.
Code letter: *A*

Particular Affirmative: "Some *A* is *B*".
Interpretation Using Sets: $A \cap B \neq \emptyset$.
Code letter: *I*

Universal Negative: "No *A* is *B*".
Interpretation Using Sets: $A \cap B = \emptyset$.
Code letter: *E*

Particular Negative: "Some *A* is not *B*".
Interpretation Using Sets: $A \cap B^c \neq \emptyset$.
Code letter: *O*

[handwritten: $A' = 0$ $O' = A$ $E' = I$ $I' = E$]

The codeletters *A* and *I* for the affirmative sentences are the first two vowels of the Latin word "affirmo," "I assert that." The codeletters *E* and *O* for the negative sentences are the two vowels of the Latin word "nego," "I deny that."

The interpretation of these sentences using sets allows us to see what is essential in the sentence from a logical point of view. Sentences which are the same, from a logical point of view, can be expressed in various ways in ordinary language.

In this section, we shall give a brief description of Aristotelian logic. We shall describe a basic kind of argument called a *syllogism*. Arguments in Aristotelian logic are arguments which can be expressed as a sequence of syllogisms. Aristotle classified the valid types, or *moods* of syllogisms, and we shall describe this classification. We shall describe a four-letter code for the arguments of Aristotelian logic, and show how it is analogous to the genetic code.

EXAMPLE 1   Give the type and the codeletter of each of the following three sentences. Find an interpretation using sets for each sentence.

"All men are mortal."
"Socrates is a man."
"Socrates is mortal."

Solution   Let A be the set of men, and B the set of beings who are mortal. Then the first sentence can be interpreted as the assertion that $A \subset B$.

Let A be the set consisting just of Socrates, and B the set of men. Then the second sentence can be interpreted as the assertion that $A \subset B$.

Let A be the set consisting just of Socrates, and B the set of beings who are mortal. Then the third sentence can be interpreted as the assertion that $A \subset B$.

We see that each of the three sentences is of the universal affirmative form, and has the code letter A. ●

Notice how the interpretation using sets allows us to detect the same logical structure in sentences with different grammatical forms.

EXAMPLE 2   Give the type and codeletter of each of the following three sentences. Find an interpretation using sets for each sentence.

    E *"No junior faculty are members of the Committee on Committees."*
    A *"All wealthy faculty are members of the Committee on Committees."*
    E *"No junior faculty are wealthy."*

Solution   Let A be the set of junior faculty members, and B the set of members of the Committee on Committees. Then the first sentence can be interpreted as the assertion that $A \cap B = \emptyset$.

Let A be the set of wealthy faculty members, and B the set of members of the Committee on Committees. Then the second sentence can be interpreted as the assertion that $A \subset B$.

Let A be the set of junior faculty members, and B the set of wealthy faculty members. The third sentence can be interpreted as the assertion that $A \cap B = \emptyset$.

The first sentence has codeletter E, the second sentence has codeletter A, and the third sentence has codeletter E. ●

There are relations between the four basic types of sentences. Suppose that S is a sentence of the universal affirmative type. S can be interpreted by the assertion that for certain sets A and B, $A \subset B$. Consider the sentence

    "not S" .

This sentence can be interpreted as meaning that A is not a subset of B. This means that some element of A is not an element of B; in other words that "Some A is not B". This is a sentence of the particular negative type. This and similar obervations give us the rules:

I.   If S is a sentence with code letter A, then "not S" is a sentence with codeletter O.

II. If S is a sentence with code letter O, then "not S" is a sentence with codeletter A.

Figure 5.2
The Square
of Opposition

III. If S is a sentence with code letter E then "not S" is a sentence with codeletter I.

IV. If S is a sentence with code letter I then "not S" is a sentence with codeletter E.

These relations among the sentence types can be summarized in a diagram called the *Square of Opposition* (Figure 5.2).

After the description of the basic sentence types and their logical relations, the next step in Aristotelian logic is to consider the simplest type of argument, the *syllogism*. Aristotle originally defined a syllogism to be any logically correct argument, but the term soon came to have the more special meaning of the next definition.

DEFINITION 5.2.1   A *syllogism* consists of three sentences of type A, E, I, or O in order. Each of these sentences contains two terms, and these terms must occur in pairs in the three sentences. The *mood* of the syllogism is the ordered triplet of types A, E, I, O of the three sentences.

Let us call the three terms which occur in a syllogism A, B, and C. One of these terms, say B, must occur in each of the first two sentences. The three terms can be arranged in the three sentences in any of the four following ways. These four possible arrangements are called *figures* (the first figure, the second figure, the third figure, and the fourth figure):

1		2		3		4	
B	A	A	B	B	A	A	B
C	B	C	B	B	C	B	C
C	A	C	A	C	A	C	A

We shall call the mood of a syllogism its *AEIO code*, or simply its code. The *form* of a syllogism is given by its code and figure. A syllogism is *valid* if its third sentence follows correctly from the first two, otherwise it is *invalid*. The validity of a syllogism is entirely determined by its form.

How many forms of a syllogism are possible? There are four code letters, and the code letter of each of the three sentences can be chosen independently. Then there are four figures. By the Product Rule, the total number of possible forms is

$$4 \cdot 4 \cdot 4 \cdot 4 = 256$$

EXAMPLE 3   Prove that the following syllogism is valid:

   (i)   Every man is mortal.

   (ii)   Socrates is a man.

   (iii)   Socrates is mortal.

Proof   Each of the three sentences is of the universal affirmative type. Let A be the set of beings who are mortal, let B be the set of men, and C the set consisting just of Socrates. Then the first sentence says that $B \subset A$, and the second sentence says that $C \subset B$. From this we conclude that $C \subset A$, which is the interpretation of the third sentence, in terms of sets.

The code for this syllogism is AAA and the figure is the first figure. We have actually proved that any syllogism with this form is valid.   ●

EXAMPLE 4   Prove that the following syllogism is valid:

   (i)   No junior faculty are members of the Committee on Committees.

   (ii)   All wealthy faculty are members of the Committee on Committees.

   (iii)   No junior faculty are wealthy.

Proof   Let A be the set of junior faculty members, let B be the set of members of the Committee on Committees, and C the set of wealthy faculty members. The first sentence says that $A \cap B = \emptyset$, and the second sentence says that $C \subset B$. From this we deduce that $A \cap C = \emptyset$, which is the interpretation of the third sentence, in terms of sets.

The code for this syllogism is EAE and the figure is the second figure. We have actually proved that any syllogism of this form is valid.   ●

A long series of arguments of this kind gives the following classification of the valid forms of the syllogism. We will denote the three terms in a syllogism by A, B, C in the same way as before.

THEOREM 5.2.1   The following fifteen forms of the syllogism are always valid, and no others are always valid:

Figure 1	Figure 2	Figure 3	Figure 4
AAA	AEE	AII	AEE
AII	AOO	EIO	EIO
EAE	EAE	IAI	IAI
EIO	EIO	OAO	

If assumptions are made about existence of objects of type A, B, or C then there are nine additional valid forms:

Figure 1	Figure 2	Figure 3	Figure 4	Assume
AAI EAO	AEO EAO		AEO	Some C exist
		AAI EAO	EAO	Some B exist
			AAI	Some A exist

The English logician William of Sherwood (about 1200–1270 A.D.) introduced mnemonic words for the valid forms of the syllogism:

> Barbara (code AAA in the first figure),
> Celarent (code EAE in the first figure),
> Darii (code AII in the first figure),
> Ferio (code EIO in the first figure),
> and so on.

William of Sherwood also wrote *De insolubilibus* (On insoluble propositions) about the paradoxes of self-reference.

Now we shall give examples of two forms of the syllogism, one valid and the other invalid.

EXAMPLE 5   Prove that a syllogism in the mood AEE and the second figure is valid.

Solution   This means that an argument of the following form must be correct:

(i)   All A is B.
(ii)   No C is B.
(iii)   Therefore, no C is A.

We can see that this argument is correct by using the interpretation with sets. The first sentence means that $A \subset B$. The second sentence means that $C \cap B = \emptyset$. From these, we deduce that $C \cap A = \emptyset$, which is the interpretation of the third sentence, with sets.   ●

EXAMPLE 6   Prove that a syllogism in the mood AAA and the second figure is invalid.

Solution   For such a syllogism to be valid, the following argument must be correct:

(i)   All A is B.
(ii)   All C is B.
(iii)   Therefore, all C is A.

We can show that this argument is false by finding a counterexample. Let $A = \{1\}$, $B = \{1, 2\}$, and $C = \{2\}$. Then $A \subset B$ and $C \subset B$, but it is false that $C \subset A$.   ●

If an argument can be expressed as a sequence of syllogisms we shall say that it can be *factored into syllogisms*.

> **DEFINITION 5.2.2** An *argument in Aristotelian logic* is one which can be factored into syllogisms.
>
> The *AEIO code*, or the *triplet code*, for such an argument is the ordered sequence of *AEIO* codes for the syllogisms into which it can be factored.

Late in his career, Lewis Carroll wrote a book on logic, *Symbolic Logic*, the first part of which was published in 1896. In spite of its title, it is for the most part a description of Aristotelian logic. Here is an example from Lewis Carroll's book:

EXAMPLE 7 Factor the following argument into syllogisms, and find its *AEIO* code.

  (i)   No ducks waltz.
  (ii) No officers ever refuse to waltz.
  (iii) All of my poultry are ducks.
  (iv) None of my poultry are officers.

Solution This argument can be factored into the following two syllogisms:

  (i)   No ducks waltz.
  (ii) No officers ever refuse to waltz.
  (iii) No ducks are officers.

and

  (i)   No ducks are officers.
  (ii) All of my poultry are ducks.
  (iii) None of my poultry are officers.

The *AEIO* code for each syllogism is *EAE*, so the *AEIO* code for the argument is *EAEEAE*. ●

EXAMPLE 8 Factor the following argument into syllogisms, and find its *AEIO* code.

  (i)   No trustworthy faculty are members of the Committee on Committees.
  (ii) All wealthy faculty are members of the Committee on Committees.
  (iii) All junior faculty are trustworthy.
  (iv) No junior faculty are wealthy.

Solution   This argument can be factored into the following two syllogisms:

    (i)   No trustworthy faculty are members of the Committee on Committees.

    (ii) All wealthy faculty are members of the Committee on Committees.

    (iii) No trustworthy faculty are wealthy.

and

    (i)   No trustworthy faculty are wealthy.

    (ii) All junior faculty are trustworthy.

    (iii) No junior faculty are wealthy.

The *AEIO* code for each syllogism is *EAE*, so the *AEIO* code for the argument is *EAEEAE*.   ●

APPLICATION   *Logic and the Genetic Code*

There is an analogy between the arguments of Aristotelian logic and sequences which occur in the genetic code. We shall describe it here.

The genetic material of plants and animals is stored in *DNA sequences*. DNA is arranged in two linked strands, called the *double helix*. Each strand consists of a sequence of four molecules called *nucleotides*: T = Thymine, A = Adenine, G = Guanine, and C = Cytosine.

In 1949 the chemist Erwin Chargaff observed that in the DNA from a given plant or animal, the proportions of thymine and adenine were very nearly equal, and the proportions of guanine and cytosine were very nearly equal. Later it was realized that in the double helix, thymine is always paired with adenine, and guanine is always paired with cytosine. These facts are referred to as *Chargaff's Rules*.

The DNA of a plant or animal serves to store the genetic material of that plant or animal. Plants and animals are composed entirely of *proteins*. In reproduction, these proteins must be constructed, using the DNA sequences of the plant or animal as a model.

The *Fundamental Dogma* of molecular biology (formulated by Francis Crick in 1956) describes how this is done:

$$DNA \rightarrow RNA \rightarrow Protein$$

In reproduction, the two DNA strands separate; the DNA sequences which code proteins are copied by messenger RNA; messenger RNA acts as a template for the construction of a protein. In RNA, T = Thymine is replaced by U = Uracil.

A protein is a sequence of *amino acids*. The number of amino acids in a protein ranges from a few dozen to a few thousand. There are twenty-two amino acids, and each amino acid is coded for by one or more triplets of nucleotides T, A, G, C in the DNA sequence. When the DNA sequence is converted to an RNA sequence, each amino acid is coded by a triplet of letters U, A, G, C.

Here is a portion of the genetic code (amino acids and the UAGC triplets which code them):

Alanine by GCU,GCC,GCA,GCG
Arginine by CGU,CGC,CGA,CGG

Asparagine by AAU, AAC
Aspartic acid by GAU, GAC
Cysteine by UGU, UGC
Glutamine by CAA, CAG
and so on.

Here are three aspects of the analogy between the arguments of Aristotelian logic and the DNA or RNA sequences which code for proteins:

I.   Chargaff's Rules, involving the pairings $T \leftrightarrow A$ or $U \leftrightarrow A$ and $G \leftrightarrow C$, are similar to the pairings $A \leftrightarrow O$ and $E \leftrightarrow I$ of the Square of Opposition.

II.   The amino acids which code for proteins are coded by triplets in the TAGC code (in DNA), or by triplets in the UAGC code (in RNA). The syllogisms which make up an argument in Aristotelian logic are coded by triplets in the AEIO code.

III.   There are twenty-two amino acids. The number of valid forms of the syllogism is close to this (twenty-four, if we count the nine conditionally valid forms).

EXERCISES

Give the type of each of the following sentences (universal affirmative, particular affirmative, universal negative, or particular negative), and its codeletter (A, I, E, or O). Give an interpretation in terms of sets for each sentence.

1. Every Cretan is a liar.

2. There's n'er a villian in all Denmark, but he's an arrant knave.

3. Some members of the Committee on Bylaws are members of the Committee on Committees.

4. Some men are mortal.

5. No members of the Rank and Tenure Committee are members of the Committee on Bylaws.

6. I never was, nor never will be false.

7. Some members of the Committee on Committees are not members of the Committee on Bylaws.

8. And all the clouds that loured upon our house, (are) in the deep bosom of the ocean buried.

Give the type of the sentences in each of the following syllogisms, and give its *AEIO* code. Prove that the argument of each syllogism is correct, using an argument with sets.

9.   (i) Every Cretan is a liar.
         (ii) Epimenides lived on the island of Crete.
         (iii) Epimenides was a liar.

10.   (i) No faculty member who is sincere is a member of the Committee on Committees.
          (ii) Some members of the Committee on Bylaws are members of the Committee on Committees.
         (iii) Some members of the Committee on Bylaws are not sincere.

11.   (i) Every Danish villain is an arrant knave.
          (ii) Claudius was a Danish villain.
         (iii) Claudius was an arrant knave.

12.   (i) Some members of the Committee on Bylaws are members of the Committee on Committees.
          (ii) No senior faculty are members of the Committee on Bylaws.
         (iii) Some members of the Committee on Committees are not senior faculty.

13. Here is a syllogism from Lewis Carroll's book *Symbolic Logic* (we have altered the grammar).

           (i) Your story about how you once met the sea serpent always makes me yawn.
          (ii) I never yawn at a story unless it is totally devoid of interest.
         (iii) Therefore your story about how you met the sea serpent is totally devoid of interest.

Factor each of the following arguments into syllogisms, and give its *AEIO* code.

14.

    (i) No faculty member who is sincere is a member of the Committee on Committees.

    (ii) Some members of the Committee on Bylaws are members of the Committee on Committees.

    (iii) A faculty member who is not sincere does not deserve praise.

    (iv) Some members of the Committee on Bylaws do not deserve praise.

15.

    (i) Some members of the faculty are wealthy.

    (ii) All wealthy faculty are members of the Committee on Committees.

    (iii) All wealthy faculty who are members of the Committee on Committees are also members of the Secret Subcommittee of the Committee on Committees.

    (iv) No one can be elected to the Secret Subcommittee of the Committee on Committees unless he or she is a senior faculty member.

    (v) Some senior faculty are wealthy.

ADVANCED EXERCISES

1. The following example is from Lewis Carroll's book *Symbolic Logic*. He called it the *Winds and Windows Problem*. Factor it into syllogisms, and give its *AEIO* code.

    (i) There is always sunshine when the wind is in the East.

    (ii) When it is cold and foggy, my neighbor practices the flute.

    (iii) When my fire smokes, I set the door open.

    (iv) When it is cold and I feel rheumatic, I light my fire.

    (v) When the wind is in the East and comes in gusts, my fire smokes.

    (vi) When I keep the door open, I am free from headache.

    (vii) Even when the sun is shining and it is not cold, I keep my window shut if it is foggy.

    (viii) When the wind does not come in gusts, and when I have a fire and keep the door shut, I do not feel rheumatic.

    (ix) Sunshine always brings on fog.

    (x) When my neighbor practices the flute, I shut the door, even if I have no headache.

    (xi) When there is a fog and the wind is in the East, I feel rheumatic.
Conclude that:

    (xii) When the wind is in the East, I keep my window shut.

5.3   THE PROPOSITIONAL CALCULUS

"The fourth is its fondness for bathing-machines,

Which it constantly carries about,

And believes that they add to the beauty of scenes—

A sentiment open to doubt.

"The fifth is ambition. It next will be right

To describe each particular batch:

Distinguishing those which have feathers, and bite,

From those which have whiskers, and scratch."

The idea of expressing logical arguments in some kind of calculus is a very old one. Gottfried Wilhelm Leibniz (1646–1716) created a logical calculus which might be viewed as the first symbolic logic. His ideas were taken up by George Boole and developed further in the *Laws of Thought*.

John Venn developed Boole's ideas. Venn's book *Symbolic Logic* was a catalyst for the steady development of the subject, although there was a continuing controversy at the time about the value of symbolic logic. In the preface to the second edition (1894) of *Symbolic Logic* (referring to the first edition, published in 1881) Venn wrote:

> *At the time when the first edition of this book was composed it would scarcely be too much to say that the conception of a Symbolic Logic was either novel or repugnant to every professional logician.*

Symbolic logic might have continued to develop slowly, amid controversies, were it not for a crisis which occurred in 1902.

The German logician Friedrich Ludwig Gottlob Frege (1848–1925) developed symbolic logic with the goal of giving a logical foundation for arithmetic. He described this theory in a book *Grundgesetze der Arithmetik* (Foundations of Arithmetic). He sent a copy of this book to Bertrand Arthur William Russell (1872–1970) in 1902. Russell wrote a letter to Frege explaining that the logical system which Frege was using contained an essential flaw, the paradox now known as Russell's Paradox, which we described earlier.

After Russell's Paradox was discovered, it was feared that it might be impossible to carry out Frege's program, that is, to find a logical foundation for mathematics which was free of paradoxes. Since mathematics can be founded on the theory of sets, the more specific problem was to find a theory of sets which was rich enough to develop mathematics, but which was free of paradoxes.

Several years later, Bertrand Russell and Alfred North Whitehead (1861–1947) proved that an adequate set theory could be founded on logic, without paradoxes, in *Principia Mathematica*, published in three volumes (in 1910, 1912, and 1913). Their work made the importance of symbolic logic clear, and it has been the dominant type of logic ever since. Here is a quote from the Introduction to *Principia Mathematica*:

> *The adaptation of the rules of the symbolism to the processes of deduction aids the intuition in regions too abstract for the imagination readily to present to the mind the true relation between the ideas employed.*

The Motivation for Symbolic Logic

I.   There are logically correct arguments which cannot be expressed as sequences of syllogisms—that is, which cannot be expressed as arguments in Aristotelian logic. Thus even from a purely logical point of view, we cannot regard Aristotelian logic as a completely adequate logical system.

II. There is a need to give a logical foundation for mathematics which avoids the paradoxes. We have just discussed the problems which arose after the discovery of Russell's Paradox. These problems were solved in *Principia Mathematica*, and symbolic logic was essential to their solution.

III. There is a need to prove the correctness of results which are obtained from computer programs. Computer programs are used both to prove theorems and to reach important practical conclusions. How can we be sure that the results given by these programs are correct? There is a need to *prove program correctness* using symbolic logic.

In this section, we shall describe the simplest logical language, the *propositional calculus*. At the end of the section, we shall discuss syllogisms using the propositional calculus.

The Propositional Calculus   The *propositional calculus* is a logical language with the following properties:

(i) There is an infinite list of symbols $p, q, r, s, \ldots$, called *variables*, which we may also denote with subscripts:

$$p_1, p_2, p_3, \ldots$$

(ii) There are *logical connectives* "$\sim$," "$\vee$," "$\wedge$," "$\rightarrow$," and "$\leftrightarrow$." "$\sim p$" means "not p," "$p \vee q$" means "p or q", "$p \wedge q$" means "p and q," "$p \rightarrow q$" means "if p, then q," and "$p \leftrightarrow q$" means "p if and only if q."

(iii) We shall give a recursive definition of a *sentence*, or *well-formed formula*, in the propositional calculus.

1. Any variable as in (i) is a sentence in the propositional calculus.

2. If P and Q are sentences in the propositional calculus then

$$(\sim P), (P \vee Q), (P \wedge Q), (P \rightarrow Q),$$
$$and\ (P \leftrightarrow Q)$$

are also sentences in the propositional calculus.

(iv) Each sentence S in the propositional calculus has a *truth value* T or F ("true" or "false"), when a truth value T or F is assigned to each variable. The truth value of S is determined recursively using a *truth table*. One does this by using the truth tables for the basic connectives given below.

(v) The *axioms* of the propositional calculus are obtained by substituting any sentences of the propositional calculus for P, Q, and R in the following expressions:

(A1) $(P \rightarrow (Q \rightarrow P))$.

(A2) $((P \rightarrow (Q \rightarrow R)) \rightarrow ((P \rightarrow Q) \rightarrow (P \rightarrow R)))$.

(A3) $(((\sim P) \rightarrow (\sim Q)) \rightarrow (Q \rightarrow P))$.

(vi)   A sentence Q of the propositional calculus is obtained from another sentence P, by *substitution* if Q arises from P, by substituting sentences of the propositional calculus for the variables of P. Substitution is a *rule of inference* in the propositional calculus. There is one other rule of inference, *modus ponens*, which says that a sentence Q follows from $(P \rightarrow Q)$ and P. A sentence in the propositional calculus has a *proof* if there is a sequence of sentences, beginning with an axiom and ending with the given sentence, such that each sentence in the sequence follows from earlier ones by using substitution or modus ponens.

It can be proved that any sentence which has a proof is a sentence in the sense of part (iii) of this definition.

Notice that the axioms for the propositional calculus involve only the connectives $\sim$ and $\rightarrow$. If we want to consider sentences which involve $\vee$, $\wedge$, or $\leftrightarrow$, we must define them in terms of $\sim$ and $\rightarrow$. The definitions are as follows. $(p \vee q)$ is $((\sim p) \rightarrow q)$; $(p \wedge q)$ is $(\sim (p \rightarrow (\sim q)))$; and $(p \leftrightarrow q)$ is $((p \rightarrow q) \wedge (q \rightarrow p))$.

"$\sim$" is called *negation*; "$\vee$" is called *disjunction*; "$\wedge$" is called *conjunction*; and "$\rightarrow$" is called *implication*.

EXAMPLE 1   Which of the following expressions are well-formed formulas in the propositional calculus?

$$((p \rightarrow q) \rightarrow (\sim q))$$
$$(((p \rightarrow q) \rightarrow p)$$
$$(p \rightarrow q) \rightarrow (r \rightarrow s)$$

Solution   The first expression is a well-formed formula, because it is constructed by our recursive rules. The second expression is not a well-formed formula. One way to see this is to observe that in a well-formed formula the number of left brackets is always equal to the number of right brackets. The expression has three left brackets and two right brackets, and so is not a well-formed formula. The third expression is not a well-formed formula although it has the same number of left and right parentheses, because the outside parentheses have been omitted. If we add them, we get $((p \rightarrow q) \rightarrow (r \rightarrow s))$, which is a well-formed formula.   ●

Now we shall describe how to construct a truth table for any well-formed formula in the propositional calculus. We start with the truth tables for the basic connectives:

Negation

p	$\sim p$
T	F
F	T

Disjunction

p	q	$p \vee q$
T	T	T
T	F	T
F	T	T
F	F	F

Conjunction

p	q	$p \wedge q$
T	T	T
T	F	F
F	T	F
F	F	F

Implication

p	q	$p \rightarrow q$
T	T	T
T	F	F
F	T	T
F	F	T

A truth table for a sentence which contains n variables has 2^n rows. We assign truth values T and F to the variables in columns. In the first column, there are half Ts, then half Fs. In the next column, there are one-fourth Ts, one-fourth Fs, one-fourth Ts, and one-fourth Fs. This procedure makes every possible assignment of truth values to the variables.

We construct the truth table of a well-formed formula by following the recursive construction of the formula, giving the truth values for each, until we have a column of truth values for the whole formula.

DEFINITION 5.3.1   A sentence of the propositional calculus is a *true sentence* or a *tautology* if its truth value is T for every assignment of truth values to the variables which occur in it.

We will show that each of the axioms of the propositional calculus is a tautology by constructing a truth table for each axiom. It is enough to do this for variables $p = P$, $q = Q$, and $r = R$.

EXAMPLE 2　Using a truth table, prove that $(p \rightarrow (q \rightarrow p))$ is a tautology.

Solution

p	q	$(q \rightarrow p)$	$(p \rightarrow (q \rightarrow p))$
T	T	T	T
T	F	T	T
F	T	F	T
F	F	T	T

The axiom is a tautology because every entry in the last column is a T.　●

EXAMPLE 3　Prove that $S = (((\sim p) \rightarrow (\sim q)) \rightarrow (q \rightarrow p))$ is a tautology.

Solution　Let $S_1 = ((\sim p) \rightarrow (\sim q))$.

p	q	$\sim p$	$\sim q$	$q \rightarrow p$	S_1	S
T	T	F	F	T	T	T
T	F	F	T	T	T	T
F	T	T	F	F	F	T
F	F	T	T	T	T	T

Again, the axiom is a tautology because every entry in the last column is a T.　●

EXAMPLE 4　Prove that $S = ((p \rightarrow (q \rightarrow r)) \rightarrow ((p \rightarrow q) \rightarrow (p \rightarrow r)))$ is a tautology.

Solution　Let $S_1 = (p \rightarrow q)$, $S_2 = (p \rightarrow r)$, $S_3 = (q \rightarrow r)$, $S_4 = (p \rightarrow (q \rightarrow r))$, and $S_5 = ((p \rightarrow q) \rightarrow (p \rightarrow r))$.

p	q	r	S_1	S_2	S_3	S_4	S_5	S
T	T	T	T	T	T	T	T	T
T	T	F	T	F	F	F	F	T
T	F	T	F	T	T	T	T	T
T	F	F	F	F	T	T	T	T
F	T	T	T	T	T	T	T	T
F	T	F	T	T	F	T	T	T
F	F	T	T	T	T	T	T	T
F	F	F	T	T	T	T	T	T

Once again, the axiom is a tautology because every entry in the last column is a T. ●

EXAMPLE 5   Prove that $S = (p \rightarrow (p \wedge q))$ is not a tautology.

Solution   The truth table is

p	q	$(p \wedge q)$	$(p \rightarrow (p \wedge q))$
T	T	T	T
T	F	F	F
F	T	F	T
F	F	F	T

S is not a tautology because there is an F in the last column of the truth table. ●

Remember that a Boolean function $f(x_1, x_2, \ldots, x_n)$ is a function such that the values of x_i and the values of F are 0 or 1. More generally, a Boolean function is a function such that the values of x_i and the values of f come from any two-letter alphabet.

In particular, we could choose the two-letter alphabet $\{T, F\}$. For any sentence S in the propositional calculus, the truth table computes the truth value of S, given the truth values of the variables p, q, r, etc. This function is called a *truth function*. We see that

> Any truth function is a Boolean function.

and

> Any truth function can be given an algorithmic description by a truth table.

We can ask which Boolean functions are truth functions. Every Boolean function is a truth function, as we shall prove in the next chapter.

Next we will illustrate the concept of proof by giving a proof of a sentence from the axioms.

EXAMPLE 6   Give a proof of the sentence $(p \rightarrow p)$ from the axioms.

wrong

Solution   We will list the sentences in the proof, with the justification for introducing each sentence on the right.

1. $((p \rightarrow ((p \rightarrow p) \rightarrow p)) \rightarrow ((p \rightarrow (p \rightarrow p)) \rightarrow (p \rightarrow p)))$ Axiom 2, with $(p \rightarrow p)$ for Q and p for R.

2. $(p \rightarrow ((p \rightarrow p) \rightarrow p))$ Axiom 1, with p for P and $(p \rightarrow p)$ for Q.

3. $((p \rightarrow (p \rightarrow p)) \rightarrow (p \rightarrow p))$ Modus ponens: (1) and (2).

4. $(p \rightarrow (p \rightarrow p))$ Axiom 1, with p for P and p for Q.

5. $(p \rightarrow p)$ Modus ponens: (3) and (4).   prove→true
                                                      ←…continues
●

A sentence is *true* if it is a true sentence, or a tautology. A sentence is *provable* if it has a proof from the axioms. The concept of truth, which depends on a truth table, and the concept of provability, which depends on the existence of a proof, do not at first sight have anything to do with each other.

The relation between these two concepts is nevertheless of fundamental importance in logic. A logical language in which every provable sentence is true is called *consistent*. A logical language in which every true sentence is provable is called *complete*.

If a logical language is not consistent, a sentence can be proved which is not a tautology. It then can be shown that a contradiction can be proved, that is, a sentence of the form

$$(p \wedge (\sim p))$$

If it is possible to prove a contradiction $(p \wedge (\sim p))$ in a logical system, then the sentence

$$((p \wedge (\sim p)) \rightarrow q)$$

is true for all sentences q, and by modus ponens, all sentences q can be proved. Thus in an inconsistent logical system, *any* sentence can be proved, and therefore such a system is useless.

The propositional calculus is both consistent and complete. The proofs of these two facts are too difficult to be given here.

In general, it can be far from obvious whether a logical language is consistent. One well-known logical language was used for years before it was discovered that it was inconsistent.

A *logical identity* is a sentence of the form

$$(p \leftrightarrow q)$$

Logical identities are important because they represent patterns of proof. How can we tell whether a logical identity is correct?

Remember that $(p \leftrightarrow q)$ is defined to mean $(p \rightarrow q)$ and $(q \rightarrow p)$. If any entry in the last column of the truth tables for $(p \rightarrow q)$ and $(q \rightarrow p)$ are different, then one of these implications must have the form $T \rightarrow F$, and therefore it is not a tautology. This means that

LOGICAL IDENTITY

> A logical identity $(p \leftrightarrow q)$ holds if and only if the last columns of the truth table for $(p \rightarrow q)$ and $(q \rightarrow p)$ are the same.

The following is a list of ten logical identities (the last two are tautologies rather than identities, and involve $\rightarrow$ rather than $\leftrightarrow$).

Logical Identities

The following sentences are logical identities for any sentences P and Q in the propositional calculus:

  I. <u>Double Negation:</u>

$$((\sim (\sim P)) \leftrightarrow P)$$

  II. <u>Commutativity:</u>

$$((P \vee Q) \leftrightarrow (Q \vee P))$$
$$((P \wedge Q) \leftrightarrow (Q \wedge P))$$

  III. <u>Associativity:</u>

$$((P \vee (Q \vee R)) \leftrightarrow ((P \vee Q) \vee R))$$
$$((P \wedge (Q \wedge R)) \leftrightarrow ((P \wedge Q) \wedge R))$$

  IV. <u>Distributivity:</u>

$$((P \wedge (Q \vee R)) \leftrightarrow ((P \wedge Q) \vee (P \wedge R)))$$

  V. <u>Definition of $\vee$:</u>

$$((P \vee Q) \leftrightarrow ((\sim P) \rightarrow Q))$$

  VI. <u>Definition of $\wedge$:</u>

$$((P \wedge Q) \leftrightarrow (\sim (P \rightarrow (\sim Q))))$$

  VII. <u>DeMorgan's Laws:</u>

$$((\sim (P \vee Q)) \leftrightarrow ((\sim P) \wedge (\sim Q)))$$
$$((\sim (P \wedge Q)) \leftrightarrow ((\sim P) \vee (\sim Q)))$$

  VIII. <u>Contrapositive:</u>

$$((P \rightarrow Q) \leftrightarrow ((\sim Q) \rightarrow (\sim P)))$$

  IX. <u>Modus Tollens:</u>

$$(((\sim Q) \wedge (P \rightarrow Q)) \rightarrow (\sim P))$$

  X. <u>Modus Ponens:</u>

$$((P \wedge (P \rightarrow Q)) \rightarrow Q)$$

EXAMPLE 7   Prove that the contrapositive $((P \rightarrow Q) \leftrightarrow ((\sim Q) \rightarrow (\sim P)))$ is a logical identity.

Solution   We must compare the last columns of the truth tables of $(p \rightarrow q)$ and $((\sim q) \rightarrow (\sim p))$. The first of these is

p	q	$(p \rightarrow q)$
T	T	T
T	F	F
F	T	T
F	F	T

and the second is

p	q	$(\sim p)$	$(\sim q)$	$((\sim q) \rightarrow (\sim p))$
T	T	F	F	T
T	F	F	T	F
F	T	T	F	T
F	F	T	T	T

Since the last columns of the truth tables are the same, the identity holds.   ●

As we try to express arguments in a logical language, it is natural to consider the propositional calculus first. Other logical languages have more structure, and allow more complicated arguments to be carried out. In the next chapter, we shall describe the most important of these, the *predicate calculus*.

It is important to understand what can and can not be said in each logical language that we consider.

What Can and Can Not Be Said in the Propositional Calculus   The propositional calculus is a logical language which allows us to express sentences broken down into their component parts, connected by the logical connectives "not," "or," "and," and "if-then."

The propositional calculus can be used to express many correct arguments (represented by tautologies), and many correct patterns of proof (represented by logical identities).

In the propositional calculus, we can not express the concepts "for all", or "there exists".

We can not define the notion of a prime number in the propositional calculus, because this involves the notion of divisibility, which involves equations. The propositional calculus does not contain an "=" sign. By the same token, we can not discuss properties of the integers in the propositional calculus, such as the Euclidean Algorithm or unique factorization into primes.

APPLICATION   *Syllogisms*

Recall that in the previous section we gave a discription of syllogisms, a basic type of logical argument. The identity

$$((P \rightarrow Q) \rightarrow ((Q \rightarrow R) \rightarrow (P \rightarrow R)))$$

is called *The Principle of the Syllogism*. It gives the general form of the argument in a syllogism, where P, Q, and R are the three terms in the three sentences involved. However, we can not use it to distinguish the different moods or figures of the syllogism.

> *The propositional calculus is not adequate to describe the different forms of the syllogism.*

EXERCISES

Which of the following expressions is a well-formed formula in the propositional calculus?

1. $((p \vee q) \rightarrow q)$.   **2.** $(\sim p \rightarrow q)$.

3. $(p \rightarrow (q \rightarrow p)))$.   **4.** $((p \wedge q) \rightarrow (p \vee q))$.

5. $(((p \vee q) \vee (p \vee r)) \leftrightarrow ((p \vee r) \vee q)$.

Use truth tables to determine which of the following sentences in the propositional calculus are tautologies.

6. $(p \rightarrow (\sim (\sim p)))$.   **7.** $((p \vee q) \rightarrow (p \wedge q))$.

8. $(p \rightarrow (p \vee q))$.

9. $((p \rightarrow q) \leftrightarrow (p \vee q))$.

10. $(((\sim q) \wedge (p \rightarrow q)) \rightarrow (\sim p))$.

11. $((\sim (p \vee q)) \leftrightarrow ((\sim p) \wedge (\sim q)))$.

12. $((p \rightarrow q) \rightarrow ((q \rightarrow r) \rightarrow (p \rightarrow r)))$.

Use truth tables to prove that each of the following identities, from the list of logical identities in this section,

is in fact a logical identity. You may substitute $P = p$, $Q = q$, and $R = r$.

13. Double negation   **14.** Commutativity

15. Associativity   **16.** Distributivity

17. Definition of $\vee$ **18.** Definition of $\wedge$

19. DeMorgan's Laws

20. Is $((((p \rightarrow (q \wedge r)) \leftrightarrow ((p \rightarrow q) \wedge (p \rightarrow r)))$ a logical identity?

21. Is $(((p \wedge q) \rightarrow r) \leftrightarrow ((p \rightarrow q) \wedge (q \rightarrow r)))$ a logical identity?

22. Prove that the Principle of the Syllogism, which we stated at the end of this section, is a tautology.

23. Is $((p \rightarrow q) \vee (q \rightarrow p))$ a tautology?   Answer is wrong

24. Is $((p \rightarrow q) \rightarrow (q \rightarrow p))$ a tautology?

ADVANCED EXERCISES

Give a proof of each of the following sentences in the propositional calculus. Recall that if a sentence to be proved from the axioms involves $\vee$ or $\wedge$, then you must use the definitions that we have given for these connectives in terms of $\sim$ and $\rightarrow$.

You will find that it is difficult to prove any of these sentences directly from the axioms. You are allowed to use the *Deduction Theorem*, which makes the proofs a great deal easier. The Deduction Theorem states that if a sentence Q of the propositional calculus follows from a sentence P and the axioms, then $(P \rightarrow Q)$ follows from the axioms. The theorem is often stated "If $P \vdash Q$, then $\vdash (P \rightarrow Q)$."

1. $(\sim (\sim (p)) \rightarrow p)$.   **2.** $(p \rightarrow (\sim (\sim p)))$.

3. $((p \wedge q) \rightarrow p)$.   **4.** $(p \rightarrow (p \vee q))$.

5. $(((\sim p) \wedge (p \vee q)) \rightarrow q)$.

In this section, we saw that whether an expression in the propositional calculus is a well-formed formula is related to the way in which parentheses appear in the expression. There is a closely related combinatorial problem.

Catalan's Problem   Recall that we have discussed the Catalan numbers $C_n = \frac{1}{n+1} \binom{2n}{n}$ several times. Here is another problem, due to Catalan and named after him, which can be solved with Catalan numbers.

Suppose that we have a sequence of letters $a, b, c, d, \ldots$ which we wish to parenthesize. This can be done in one way for two letters ab; in two ways for three letters $(ab)c$ and $a(bc)$; in five ways for four letters $a(b(cd))$, $a((bc)d)$, $(ab)(cd)$, $((ab)c)d$, and $(a(bc))d$.

In general, *the number of ways in which $n+1$ letters in order can be parenthesized is the Catalan number C_n.*

6. List all ways to parenthesize five letters a, b, c, d, e, and show that the number of ways is given by the Catalan number C_4.

7. List all ways to paranthesize six letters a, b, c, d, e, f, and show that the number of ways is given by the Catalan number C_5.

COMPUTER EXERCISES

Write a computer program whose output is a list of all ways to parenthesize n letters in order, as in the previous exercises. Print your lists in the following cases, and verify Catalan's formula in each case.

1. For seven letters

2. For eight letters

3. For nine letters

5.4   CHAPTER SUMMARY AND SUPPLEMENTARY EXERCISES

For although common Snarks do no manner of harm,

Yet, I feel it my duty to say,

Some are Boojums—" The Bellman broke off in alarm,

For the Baker had fainted away.

Chapter Summary

Section 5.1

In this section, we discussed the basic proof techniques. One theme of the section was the paradoxes of self-reference.

Section 5.2

In this section, we discussed classic Aristotelian logic. We described Aristotle's classification of the moods and figures of the syllogism. Together these are called the forms of the syllogism.

Section 5.3

In this section, we described the propositional calculus, the simplest logical language. We described tautologies and logical identities. We described what can and can not be said in the propositional calculus.

EXERCISES

Using an argument with sets, prove that each of the following is a valid form of the syllogism:

1. Mood *AII* in the first figure.
2. Mood *EAE* in the first figure.
3. Mood *EIO* in the first figure.
4. Mood *AOO* in the second figure.
5. Mood *EAE* in the second figure.
6. Mood *EIO* in the second figure.
7. Mood *EIO* in the third figure.
8. Mood *IAI* in the third figure.
9. Mood *OAO* in the third figure.
10. Mood *AEE* in the fourth figure.
11. Mood *EIO* in the fourth figure.
12. Mood *IAI* in the fourth figure.

Use a counterexample with sets to prove that each of the following is not a valid form of the syllogism.

13. Mood *AOO* in the first figure.
14. Mood *AEE* in the first figure.
15. Mood *AII* in the second figure.
16. Mood *IAI* in the second figure.
17. Mood *AAA* in the third figure.
18. Mood *AEE* in the third figure.
19. Mood *AOO* in the fourth figure.
20. Mood *EAE* in the fourth figure.

Boolean Algebras, Boolean Functions, and Logic

6.1 BOOLEAN ALGEBRAS AND FUNCTIONS

They roused him with muffins—they roused him with ice—

They roused him with mustard and cress—

They roused him with jam and judicious advice—

They set him conundrums to guess.

There was silence supreme! Not a shriek, not a scream,

Scarcely even a howl or a groan,

As the man they called "Ho!" told his story of woe

In an antidiluvian tone.

In the first chapter of *The Laws of Thought*, Boole wrote

There is not only a close analogy between the operations of the mind in general reasoning and its operations in the particular science of Algebra, but there is to a considerable extent an exact agreement in the laws by which the two classes of operations are conducted.

The Laws of Thought contains Boole's unified theory of logic, combinatorics, and probabilities. Central to Boole's theory is the structure that we now call a *Boolean algebra*.

In this section, we shall describe Boolean algebras and their applications. We gave a preliminary description of Boolean algebras in Section 1.4 as algebras of sets. Here we shall begin with an abstract description of a Boolean algebra, explain why it is useful, and then explain why every Boolean algebra can be considered to be an algebra of sets. We shall describe how Boolean algebras can be applied to problems in logic and how elements in a Boolean algebra can be written in two normal forms, *disjunctive normal form* and *conjunctive normal form*.

The elements of a Boolean algebra are called *Boolean expressions*. Each Boolean expression gives rise to a Boolean function.

In the next section, we shall describe one of the basic applications of Boolean functions. Boolean expressions can be used to construct circuits, and Boolean identities can be used to simplify circuits. *Karnaugh maps* are also very useful.

In the final section of this chapter, we shall continue our discussion of logic and describe the *predicate calculus*, a logical language with more structure than the propositional calculus which we described in the previous chapter.

A *binary operation* on a set X, denoted "o," is a function $o : X \times X \to X$: that is, for each pair of elements x and y in X, xoy is also in X.

In Section 1.4, we gave a preliminary definition of a Boolean algebra. Here we shall give a more general definition, which is due to E. V. Huntington (a professor of mathematics and mechanics at Harvard), in the paper *Postulates for the Algebra of Logic*, published in 1904.

DEFINITION 6.1.1   A *Boolean algebra* is a quadruple $(X, +, \cdot, \bar{x})$ where X is a set; where for each $x \in X$, $\bar{x} \in X$; and where $+$ and $\cdot$ are binary operations on X such that the following axioms are satisfied:

A_1: The operations $+$ and $\cdot$ are commutative:

$$x + y = y + x \text{ and } x \cdot y = y \cdot x$$

for all $x \in X$ and all $y \in X$.

A_2: In X there are two elements $0 \in X$ and $1 \in X$ such that

$$0 + x = x \text{ and } 1 \cdot x = x$$

for all $x \in X$.

A_3: The operations $+$ and $\cdot$ are each distributive over the other:

$$x \cdot (y + z) = x \cdot y + x \cdot z$$

and

$$x + (y \cdot z) = (x + y) \cdot (x + z)$$

A_4: For every $x \in X$ there exists an $\bar{x} \in X$ such that

$$x + \bar{x} = 1 \text{ and } x \cdot \bar{x} = 0$$

We shall use the notation "xy" for "$x \cdot y$" as Boole did in *The Laws of Thought*.

Next we shall prove three theorems which show how general properties of Boolean algebras follow from the axioms.

THEOREM 6.1.1   For every element x in a Boolean algebra X,

$$x + x = x \text{ and } xx = x$$

Proof

$$x = x + 0 \text{ by Axiom 1}$$
$$= x + x\bar{x} \text{ by Axiom 4}$$
$$= (x + x)(x + \bar{x}) \text{ by Axiom 3}$$
$$= (x + x)(1) \text{ by Axiom 4}$$
$$= x + x \text{ by Axiom 2}$$

and

$$x = x(1) \text{ by Axiom 1}$$
$$= x(x + \bar{x}) \text{ by Axiom 4}$$
$$= xx + x\bar{x} \text{ by Axiom 3}$$
$$= xx + 0 \text{ by Axiom 4}$$
$$= xx \text{ by Axiom 2}$$ ●

The work involved in proving theorems about Boolean algebras is cut in half by the Principle of Duality.

Principle of Duality   Every theorem which follows from the axioms for a Boolean algebra remains valid if the operations $+$ and $\cdot$, and the elements 0 and 1, are exchanged throughout.

Proof   This follows from the symmetry of the axioms with respect to the two operations and the two elements.   ●

THEOREM 6.1.2   For each element x in a Boolean algebra X,

$$x + 1 = 1 \text{ and } x0 = 0$$

Proof

$$1 = x + 1 \text{ by Axiom 4}$$
$$= x + \bar{x}(1) \text{ by Axiom 2}$$
$$= (x + \bar{x})(x + 1) \text{ by Axiom 3}$$
$$= 1(x + 1) \text{ by Axiom 4}$$
$$= x + 1 \text{ by Axiom 2}$$

The other assertion follows by applying the Principle of Duality. ●

THEOREM 6.1.3   For each pair of elements x and y in a Boolean algebra,

$$x + xy = x \text{ and } x(x + y) = x$$

Proof

$$x = 1x \text{ by Axiom 2}$$
$$= (1 + y)x \text{ by the previous theorem}$$
$$= 1x + yx \text{ by Axiom 3}$$
$$= x + yx \text{ by Axiom 2}$$
$$= x + xy \text{ by Axiom 1}$$

The other assertion follows by applying the Principle of Duality. ●

Other properties of Boolean algebras can be proved in a similar way. Here is a brief list of such properties:

I.   The operations $+$ and $\cdot$ are associative. For any three elements x, y, and z of a Boolean algebra X,

$$x + (y + z) = (x + y) + z \text{ and } x(yz) = (xy)z$$

II.   For each element x of a Boolean algebra X, the element $\bar{x}$ is unique.

III.   For each element x of a Boolean algebra X, $\bar{\bar{x}} = x$.

IV.   In any Boolean algebra, $\bar{0} = 1$ and $\bar{1} = 0$.

V.   DeMorgan's Laws:

$$\overline{xy} = \bar{x} + \bar{y} \text{ and } \overline{x + y} = \bar{x}\bar{y}$$

Why have we discussed Boolean algebras in such generality? By doing so, we can discover facts about any particular Boolean algebra, and we can relate properties of different Boolean algebras. Algebras of sets are important examples of Boolean algebras.

THEOREM 6.1.4   Let S be any set. Then the set of all subsets of S, $X = P(S)$, is a Boolean algebra if $+$ is $\cup$, $\cdot$ is $\cap$, $\bar{x}$ is the complement, $0 = \emptyset$, and $1 = S$.

Proof   We will verify that the four axioms are satisfied, replacing elements $x, y, z, \ldots$ of the abstract Boolean algebra by subsets $A, B, C, \ldots$ of S.

Axiom 1 is satisfied because $A \cup B = B \cup A$ and $A \cap B = B \cap A$. Axiom 2 is satisfied because $\emptyset \cup A = A$ and $S \cap A = A$. Axiom 3 is satisfied. We can check that $A \cap (B \cup C) = (A \cap B) \cup (A \cap C)$ and $A \cup (B \cap C) = (A \cup B) \cap (A \cup C)$. Axiom 4 is satisfied because $A \cup A^c = S$ and $A \cap A^c = \emptyset$. ●

EXAMPLE 1   If $S = \{a\}$, the set of subsets $P(S) = \{\emptyset, \{a\}\}$ is a Boolean algebra with two elements. ●

EXAMPLE 2   If $S = \{a, b\}$, the set of subsets $P(S) = \{\emptyset, \{a\}, \{b\}, \{a, b\}\}$ is a Boolean algebra with four elements.

More generally, a Boolean algebra X is an *algebra of sets* if there is a set S such that the elements of X are subsets of S. We should stress that an algebra of sets need not consist of all subsets of S.   ●

EXAMPLE 3   Let S be any set. Then the set $X = \{\emptyset, S\}$, with the operations of union, intersection, and complement, is a Boolean algebra with two elements.   ●

EXAMPLE 4   Let $S = \{a, b, c, d\}$. Then the set $X = \{\emptyset, \{a, b\}, \{c, d\}, S\}$, with the operations of union, intersection, and complement, is a Boolean algebra with four elements.   ●

It is clear that there is a close relationship between an arbitrary Boolean algebra and an algebra of sets. The exact nature of this relationship was described by M. H. Stone in the paper *The Theory of Representations of Boolean Algebras*, which was published in 1936.

To describe Stone's Theorem, we must explain the notion of *isomorphism* of two algebraic structures. Roughly speaking, if two algebraic structures are isomorphic they are "essentially the same".

Every algebraic structure consists of a set with laws of operation defined on it, for example binary operations, and perhaps other items of structure. An isomorphism of two algebraic structures is a one–one correspondence of the underlying sets, which preserves all the laws of operation and the other items of structure. Recall that an isomorphism of graphs is a one–one correspondence of the sets of nodes, which preserves the edge sets. Here is a precise definition of isomorphism of Boolean algebras.

DEFINITION 6.1.2   Let X and Y be Boolean algebras. An *isomorphism*

$$f : X \rightarrow Y$$

is a one–one correspondence f between the sets X and Y such that

$$f(x + y) = f(x) + f(y) \text{ for all } x, y \in X$$
$$f(xy) = f(x)f(y) \text{ for all } x, y \in X$$
$$f(\overline{x}) = \overline{f(x)} \text{ for all } x \in X$$
$$f(0) = 0 \text{ and } f(1) = 1$$

EXAMPLE 5   Let $X = \{0, 1\}$ be a Boolean algebra with two elements, and let $Y = \{\emptyset, S\}$ be a Boolean algebra of sets with two elements. Then the function $f : X \rightarrow Y$ defined by $f(0) = \emptyset$ and $f(1) = S$ is an isomorphism of Boolean algebras.   ●

The following is Stone's Theorem.

THEOREM 6.1.5   Let X be any Boolean algebra. Then there is a set S, a Boolean algebra Y of subsets of S, and an isomorphism of Boolean algebras $f : X \rightarrow Y$.

The main difficulty in proving Stone's Theorem is in finding the set S, given the Boolean algebra X. It would take too long to explain the algebraic ideas which are used to resolve this difficulty. The technique of the proof gives more detail, as in the following corollary.

COROLLARY 6.1.1   Let X be any finite Boolean algebra. Then there is a finite set S and an isomorphism of Boolean algebras $f : X \to P(S)$.

In other words, any finite Boolean algebra is isomorphic to the Boolean algebra of *all* subsets of some finite set. Let the set S have n elements.

COROLLARY 6.1.2   If X is any finite Boolean algebra, then $|X| = 2^n$ for some positive integer n, and up to isomorphism there is only one Boolean algebra of order 2^n.

This Boolean algebra can be identified with the Boolean algebra of subsets of $S = \{1, 2, \ldots, n\}$. Its elements can be identified with the nodes of the graph of the n-cube, drawn in Chapter 4.

The structure of infinite Boolean algebras is more complicated. For example, the Boolean algebra $PF(S)$ in the following theorem is not isomorphic to the Boolean algebra of all subsets of *any* set.

THEOREM 6.1.6   Let $S = \{1, 2, 3, \ldots\}$ be the set of natural numbers. Let $PF(S)$ be the set of all subsets of S which are either finite, or have finite complement, with the operations of union, intersection, and complement. Then $PF(S)$ is a Boolean algebra, and $PF(S)$ is *not* isomorphic to $P(S)$.

Proof   It is easy to verify that Axioms 1 through 4 for Boolean algebras hold for $PF(S)$. It can be proved that there is a one–one correspondence $f : S \to PF(S)$. (The proof is long, and we shall omit it.) Cantor's Theorem, proved in Section 5.1, states that there can not be a one–one correspondence between S and $P(S)$, so there can not be a one–one correspondence between $PF(S)$ and $P(S)$. Thus they are not isomorphic.   ●

The propositional calculus has the structure of a Boolean algebra.

THEOREM 6.1.7   The propositional calculus is a Boolean algebra if $+$ is $\vee$, $\cdot$ is $\wedge$, $\overline{x}$ is negation, 1 is any tautology, 0 is the negation of any tautology, and $=$ is $\leftrightarrow$.

Proof   We shall verify that the four axioms are satisfied, replacing elements $x, y, z, \ldots$ of the abstract Boolean algebra by variables $p, q, r, \ldots$ of the propositional calculus.

Axiom 1 is satisfied because of the identities $((p \vee q) \leftrightarrow (q \vee p))$ and $((p \wedge q) \leftrightarrow (q \wedge p))$. Axiom 2 is satisfied because, if P is any tautology, we have the identities $(((\sim P) \vee p) \leftrightarrow p)$ and $((P \wedge p) \leftrightarrow p)$. Axiom 3 is satisfied because of the identities $(((p \wedge (q \vee r)) \leftrightarrow ((p \wedge q) \vee (p \wedge r)))$ and $(((p \vee (q \wedge r)) \leftrightarrow (p \vee q) \wedge (p \vee r)))$. Axiom 4 is satisfied because, if P is any tautology, we have identities $(((p \vee (\sim p)) \leftrightarrow P)$ and $((p \wedge (\sim p)) \leftrightarrow (\sim P))$.   ●

By Stone's Theorem, the propositional calculus is isomorphic as a Boolean algebra to an algebra of sets. Under this isomorphism, the logical identity

$$(((p \wedge (p \vee q)) \leftrightarrow p)$$

becomes the set identity

$$A \cap (A \cup B) = A$$

Under the isomorphism, implication $\rightarrow$ is replaced by inclusion $\subset$. The tautology

$$(p \rightarrow (p \vee q))$$

becomes the following statement about sets

$$A \subset (A \cup B)$$

The isomorphism of Boolean algebras between the propositional calculus and a Boolean algebra of sets, whose existence follows from Stone's Theorem, makes precise the comment in the quote from Boole, at the beginning of this section.

DEFINITION 6.1.3   Let X be a Boolean algebra. A *Boolean expression* is any element of X. Let S be a Boolean algebra with two elements. A *Boolean function of n variables* is a function

$$f : S^n \rightarrow S$$

Up to isomorphism there is only one Boolean algebra with two elements, and we shall choose $S = \{0, 1\}$, where $0 + 0 = 0, 0 + 1 = 1 + 0 = 1 + 1 = 1, \bar{0} = 1, \bar{1} = 0$, $00 = 10 = 01 = 0$, and $11 = 1$.

By substituting 0 or 1 for the terms in a Boolean expression, we can define a Boolean function:

Each Boolean expression gives rise to a Boolean function.

EXAMPLE 6   Explain how the Boolean expression $xyz + \bar{x}\bar{y} + 1$ gives rise to a Boolean function.

Solution   Define a Boolean function $f(x, y, z)$ by letting x, y, and z take the values 0 and 1, and substituting them in the expression above. For example, if $x = y = z = 0$, we have

$$f(0, 0, 0) = 000 + \bar{0}\bar{0} + 1 = 0 + 1 + 1 = 1$$

and if $x = 1, y = 0, z = 0$, we have

$$f(1, 0, 0) = 100 + \bar{1}\bar{0} + 1 = 0 + 0 + 1 = 1. \qquad \bullet$$

The operations in S give rise to operations on Boolean functions: sum, product, and complement:

$$f(x_1, \ldots, x_n) + g(x_1, \ldots, x_n)$$
$$f(x_1, \ldots, x_n)g(x_1, \ldots, x_n)$$
$$\overline{f(x_1, \ldots, x_n)}$$

There are two normal forms for Boolean expressions, *disjunctive normal form* and *conjunctive normal form*.

DEFINITION 6.1.4

i. A Boolean expression $f(x_1, x_2, \ldots, x_n)$ is in *disjunctive normal form*, or *sum-of-products form* if it is a sum of *minterms*

$$a_1 a_2 \cdot \ldots \cdot a_n$$

where $a_i = x_i$ or $\overline{x_i}$ and all the terms are different.

ii. A Boolean expression $f(x_1, x_2, \ldots, x_n)$ is in *conjunctive normal form* if it is a product of terms

$$a_1 + a_2 + \ldots + a_n$$

where each $a_i = x_i$ or $\overline{x_i}$ and all the terms are different.

For example,

$$f(x, y, z) = xyz + \bar{x}yz + xy\bar{z} + \bar{x}\bar{y}\bar{z} \qquad \text{5ণ}$$

is a Boolean expression of three variables in disjunctive normal form, and

$$f(x, y, z) = (x + y + z)(\bar{x} + \bar{y} + z)(\bar{x} + \bar{y} + \bar{z}) \qquad \text{PS}$$

is a Boolean expression of three variables in conjunctive normal form.

We shall use the same symbol $f(x_1, x_2, \ldots, x_n)$ for a Boolean expression in the variables x_i, and for the Boolean function determined by this expression.

THEOREM 6.1.8

    (i)   Each Boolean expression is equal to a Boolean expression in disjunctive normal form.

    (ii)   Each Boolean expression is equal to a Boolean expression in conjunctive normal form.

Proof of (i)   Let $f(x_1, x_2, \ldots, x_n)$ be a Boolean expression in n variables. By applying DeMorgan's Law

$$\overline{a + b} = \bar{a}\bar{b} \text{ and } \overline{ab} = \bar{a} + \bar{b}$$

repeatedly, we can arrange it such that bars are applied only to the variables x_i. By applying the distributive law

$$a(b + c) = ab + ac$$

we can make f a sum of products of variables x_i and $\overline{x_i}$.

If any term does not contain all n variables, we can multiply it by terms of the form

$$x_i + \overline{x_i} = 1$$

without changing it, until all variables appear. Finally, multiplying out and eliminating repeated terms with the identity

$$a + a = a$$

we obtain the desired normal form.

Proof of (ii)   Suppose that $f(x_1, x_2, \ldots, x_n)$ is to be put in conjunctive normal form. By part (i), $\overline{f}$ may be put in disjunctive normal form. Applying DeMorgan's Laws to $f = \overline{\overline{f}}$, we see that f is in conjunctive normal form.   ●

EXAMPLE 7   Find disjunctive and conjunctive normal forms for $f = xy + \overline{z}$.

Solution

$$f = xy + \overline{z} = xy(z + \overline{z}) + (x + \overline{x})(y + \overline{y})\overline{z}$$
$$= xyz + xy\overline{z} + xy\overline{z} + x\overline{y}\,\overline{z} + \overline{x}y\overline{z} + \overline{x}\,\overline{y}\,\overline{z}$$
$$= xyz + xy\overline{z} + x\overline{y}\,\overline{z} + \overline{x}y\overline{z} + \overline{x}\,\overline{y}\,\overline{z}$$

is the disjunctive normal form of f. The disjunctive normal form of $\overline{f}$ is

$$\overline{f} = x\overline{y}z + \overline{x}yz + \overline{x}\,\overline{y}z$$

Notice that the terms which occur in the disjunctive normal form for $\overline{f}$ are exactly those which do *not* occur in the disjunctive normal form for f. This always happens. So, applying DeMorgan's Laws, the conjunctive normal form of $f = \overline{\overline{f}}$ is

$$f = (\overline{x} + y + \overline{z})(x + \overline{y} + \overline{z})(x + y + \overline{z})$$ ●

EXAMPLE 8   Find the disjunctive normal form and the conjunctive normal form of $f = xz + xy$.

Solution

$$f = xz + xy$$
$$= x(y + \overline{y})z + xy(z + \overline{z}) = xyz + x\overline{y}z + xyz + xy\overline{z}$$
$$= xyz + x\overline{y}z + xy\overline{z}$$

is the disjunctive normal form of f. The disjunctive normal form of $\overline{f}$ is

$$\overline{f} = \overline{x}yz + \overline{x}\,\overline{y}z + \overline{x}y\overline{z} + x\overline{y}\,\overline{z} + \overline{x}\,\overline{y}\,\overline{z}$$

The conjunctive normal form of $f = \overline{\overline{f}}$ is

$$f = (x + \overline{y} + \overline{z})(x + y + \overline{z})(x + \overline{y} + z)(\overline{x} + y + z)(x + y + z)$$ ●

By the Product Rule, there are 2^n minterms in n variables. Minterms have a simple but important property:

> *Each minterm is equal to 1 for exactly one assignment of values 0 and 1 to the variables.*

EXAMPLE 9   For which values of the variables is $x\bar{y}z\bar{w}$ equal to 1?

Solution   Clearly this is only possible if $x = 1$, $y = 0$, $z = 1$, $w = 0$.   ●

This simple observation about minterms gives another very useful principle:

> *A Boolean function f with any desired output can be constructed by taking the sum of minterms for values of the variables for which $f = 1$.*

The same observation proves the following theorem.

THEOREM 6.1.9

> **(i)** The Boolean functions determined by two Boolean expressions are equal if and only if the disjunctive normal forms of these two expressions are equal.
>
> **(ii)** The Boolean functions determined by two Boolean expressions are equal if and only if the conjunctive normal forms of these two expressions are equal.

APPLICATION   *Normal Forms in the Propositional Calculus*

Disjunctive and conjunctive normal forms can be found in any Boolean algebra, for example in the propositional calculus. We will take the Boolean expression in Example 7 and translate it into the propositional calculus.

EXAMPLE 10   Find the disjunctive and conjunctive normal forms of

$$f = ((p \wedge q) \vee (\sim r))$$

Solution   We have only to translate the expressions in Example 7 into the propositional calculus:

$$f \leftrightarrow ((p \wedge q \wedge r) \vee (p \wedge q \wedge (\sim r)) \vee (p \wedge (\sim q) \wedge (\sim r))$$
$$\vee ((\sim p) \wedge q \wedge (\sim r)) \vee ((\sim p) \wedge (\sim q) \wedge (\sim r)))$$

is the disjunctive normal form of f, and

$$f \leftrightarrow (((\sim p) \vee q \vee (\sim r)) \wedge (p \vee (\sim q) \vee (\sim r)) \wedge (p \vee q \vee (\sim r)))$$

is the conjunctive normal form of f.   ●

EXERCISES

Use Venn diagrams to verify the following identities in a Boolean algebra. Let A and B be subsets of a set S, let $x = A$ and $y = B$, $+ = \cup$, and $\cdot = \cap$.

1. $x + x = x$.   **2.** $xx = x$.

3. $x + 1 = 1$.   **4.** $x0 = 0$.

5. $x + xy = x$.   **6.** $x(x + y) = x$.

Let $S = \{a, b, c\}$. Which of the following sets of subsets of S are Boolean algebras if $+ = \cup$, $\cdot = \cap$, $0 = \emptyset$, $1 = S$, and $\bar{x} =$ complement?

7. $X = \{\emptyset, \{a\}, \{b, c\}\}$.

8. $X = \{\{a\}, \{b, c\}, S\}$.

9. $X = \{\emptyset, \{a\}, S\}$.

10. $X = \{\emptyset, \{a\}, \{b, c\}, S\}$.

11. $X = \{\emptyset, \{a\}, \{b\}, \{a, b\}, S\}$.

12. $X = \{\emptyset, \{a\}, \{b\}, \{c\}, \{a, b\}, \{a, c\}, \{b, c\}, \{a, b, c\}\}$.

Give a table of values $0, 1$ for the Boolean functions determined by the following Boolean expressions.

13. $x + y$.   **14.** $x + \bar{y}$.

15. xy.   **16.** $x\bar{y}$

17. $x + y + z$.   **18.** $xy + xz + yz$.

19. $\bar{x} + yz$.   **20.** $x + y + z + w$.

Put the following expressions into disjunctive and conjunctive normal form.

21. $x + y$.   **22.** $x + yz$.

23. $\bar{x}y + z$.   **24.** $x + y + z$.

25. $xy + xz + yz$.   **26.** $\bar{x} + \bar{y}z + z\bar{w}$.

Translate the following expressions into the propositional calculus.

27. $x + y$.   **28.** xy.

29. $x + yz + x\bar{z}$.   **30.** $(x + y)(z + \bar{w})$.

ADVANCED EXERCISES

Give proofs, from the four axioms for a Boolean algebra, of the following identities.

1. In any Boolean algebra, the operation $+$ is associative: $x + (y + z) = (x + y) + z$.

2. For each element x of a Boolean algebra X, the element $\bar{x}$ is unique.

3. For each element x of a Boolean algebra X, $\bar{\bar{x}} = x$.

4. In any Boolean algebra, $\bar{0} = 1$ and $\bar{1} = 0$.

5. One of DeMorgan's Laws: $\overline{xy} = \bar{x} + \bar{y}$.

COMPUTER EXERCISES

Put the following Boolean expressions in disjunctive normal form.

1. $x + y + z$.   **2.** $xy + xz + yz$.

3. $x + y + z + w$.

4. $xy + xz + xw + yz + yw + zw$.

5. $u + v + x + y + z + w$.

Simplify the following expressions, which are in disjunctive normal form.

6. $xyz + xy\bar{z} + x\bar{y}z + x\bar{y}\bar{z} + \bar{x}\bar{y}z$.

7. $xyz + x\bar{y}z + xy\bar{z} + x\bar{y}\bar{z} + \bar{x}yz + \bar{x}\bar{y}z$.

8. $xyzw + xy\bar{z}w + xyz\bar{w} + xy\bar{z}\bar{w} + x\bar{y}zw + \bar{x}yzw + \bar{x}\bar{y}zw$.

6.2   BOOLEAN FUNCTIONS AND CIRCUITS

"My father and mother were honest, though poor—"

"Skip all that!" cried the Bellman in haste.

"If it once becomes dark there's no chance of a Snark—

We have hardly a minute to waste!"

"A dear uncle of mine (after whom I was named)

Remarked, when I bade him farewell—"

"Oh, skip your dear uncle!" the Bellman exclaimed,

As he angrily tingled his bell.

In this section, we shall show how Boolean functions arising from Boolean expressions can be used to describe electrical circuits. In the paper *A Symbolic Analysis of Relay and Switching Circuits*, which was published in 1939, Claude E. Shannon described how Boolean algebra can be applied to the design and analysis of electrical circuits. This kind of analysis is used in the design of circuits for a wide variety of machines and their components, for example computers and computer chips.

One basic problem in the study of Boolean functions and their applications is that the number of Boolean functions grows extremely rapidly with the number of variables. For example, there are

$$2^{2^6} = 18,446,744,073,709,551,616$$

Boolean functions of six variables.

For this reason, it is very important to find ways of simplifying Boolean functions, and of describing Boolean functions which are not completely arbitrary. The problem of simplification is connected with the problem of *minimizing a circuit*. It can be studied by manipulating Boolean expressions, as in the previous section. It can also be solved by using *Karnaugh maps*. At the end of the section, we shall show how Boolean functions can be used to describe regions in logic diagrams.

The first problem in dealing with Boolean functions is to find ways of representing them. We shall describe four ways of doing this:

    I.   By using tables

    II.   By using a Gray code

    III. By using the full binary tree

    IV. By using the n-cube

In discussing Boolean functions, we shall assume that the values of the variables are in the Boolean algebra with two elements $S = \{0, 1\}$, discussed in the previous section. Recall that in S, $1 + 1 = 1$.

By abuse of language, we shall identify Boolean expressions with the Boolean functions which they define. For example, we shall write

$$f(x, y, z) = x + y + z$$

for the Boolean function defined by substituting 0 or 1 for x, y, and z, and so on. The values of this Boolean function can be represented in a *table*:

x	y	z	$f(x, y, z) = x + y + z$
0	0	0	$0 + 0 + 0 = 0$
0	0	1	$0 + 0 + 1 = 1$
0	1	0	$0 + 1 + 0 = 1$
0	1	1	$0 + 1 + 1 = 1$
1	0	0	$1 + 0 + 0 = 1$
1	0	1	$1 + 0 + 1 = 1$
1	1	0	$1 + 1 + 0 = 1$
1	1	1	$1 + 1 + 1 = 1$

Notice that the entries in the rows of values for x, y, and z are arranged just like the values of variables in a truth table. In the first column, the first half of the entries are 0, the rest 1. In the next column, the first quarter of the entries are 0, the next quarter 1, the next quarter 0, and the last quarter 1. In the next column, the first eighth of the entries are 0, ..., and so on. This arrangement gives every possible assignment of values to the variables exactly once. If the Boolean function has n variables, the corresponding table will have 2^n rows.

EXAMPLE 1   Construct a table for the Boolean function $f(x, y, z) = xy + xz + yz$

Solution

x	y	z	$f(x, y, z) = xy + xz + yz$
0	0	0	$00 + 00 + 00 = 0$
0	0	1	$00 + 01 + 01 = 0$
0	1	0	$01 + 00 + 10 = 0$
0	1	1	$01 + 01 + 11 = 1$
1	0	0	$10 + 10 + 00 = 0$
1	0	1	$10 + 11 + 01 = 1$
1	1	0	$11 + 10 + 10 = 1$
1	1	1	$11 + 11 + 11 = 1$

If we are to represent a Boolean function of n variables, we must somehow list all binary vectors of length n. The tables we have just constructed utilize one way of doing this. Notice that the list of binary vectors in these tables is not a Gray Code (since successive vectors may differ in more than one place). Any Gray Code may be used instead, for example the Standard Gray Code, with the value of the Boolean function given after each vector.

EXAMPLE 2   Use the Standard Gray Code to represent $f(x, y, z) = x + y + z$.

Solution

(x, y, z)	$f(x, y, z) = x + y + z$
$(0, 0, 0)$	0
$(0, 0, 1)$	1
$(0, 1, 1)$	1
$(0, 1, 0)$	1
$(1, 1, 0)$	1
$(1, 1, 1)$	1
$(1, 0, 1)$	1
$(1, 0, 0)$	1

Recall that binary vectors of length n can be represented by paths in the full binary tree, where 1 corresponds to an edge to the right, and 0 corresponds to an edge to the left, and also that there is a unique path in this tree from the root to any terminal node. We can represent any Boolean function of n variables by using the full binary tree, with the terminal nodes labelled with the values of the function.

EXAMPLE 3   Use the full binary tree with three levels to represent the Boolean function $f(x, y, z) = x + y + z$.

Solution   This is shown in Figure 6.1, where the values of the function are given in the circles at the terminal nodes. ●

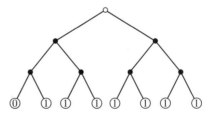

Figure 6.1   Representing a Boolean Function with the Full Binary Tree

In Chapter 4, we drew the graph of the n-cube. We labelled the nodes of this graph with binary vectors of length n. In particular, the graph of the n-cube is a way of representing all binary vectors of length n. This in turn is a way to represent a Boolean function of n variables. We label each node of the graph of the n-cube with a binary vector and the corresponding value of the Boolean function.

EXAMPLE 4   Use the graph of the 3-cube to represent the Boolean function $f(x, y, z) = x + y + z$.

Solution   The graph of the 3-cube, labelled with the binary vectors of length 3 and the corresponding values of $f(x, y, z)$ is shown in Figure 6.2.   ●

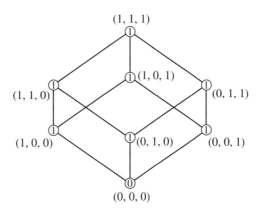

Figure 6.2   Representing a Boolean Function with the 3-Cube

At the end of the previous section, we made the point that any Boolean function
f can be put in disjunctive normal form, or sum-of-products form, by setting f equal
to the sum of minterms for which $f = 1$. This representation is extremely useful in
computer science, for example in constructing the circuits which we shall discuss later
in this section.

There is a serious practical problem associated with the representation of a Boolean
function in disjunctive normal form. Such a Boolean function can often be represented
by a far simpler Boolean expression. This is the problem of *simplification*:

> *Given a Boolean function f in sum-of-products form, how can we
> represent f by a simpler Boolean expression?*

For example, consider the Boolean function $f(x, y, z) = 1$. Clearly all eight
minterms in x, y, z will occur in the representation of f:

$$1 = xyz + \bar{x}yz + x\bar{y}z + xy\bar{z} + \bar{x}\bar{y}z + \bar{x}y\bar{z} + x\bar{y}\bar{z} + \bar{x}\bar{y}\bar{z}$$

so the representation of f in sum-of-products form can be greatly simplified.

Boolean exressions and the corresponding Boolean functions can be represented and
simplified using a rectangular diagram called a *Karnaugh map*. These diagrams were
introduced by Maurice Karnaugh, following earlier work of E. W. Veitch, in the paper
The map method for synthesis of combinational logic circuits, which was published in
1953.

A Karnaugh map for a Boolean function $f(x_1, \ldots, x_n)$ is a rectangular array of 2^n
squares, which are labelled to represent the 2^n possible minterms of f. The labelling
of the squares is such that the minterms which label adjacent squares differ in exactly
one entry. "Adjacent" is used here in a cyclic fashion. Two squares are adjacent if they

are next to each other, or if they occur at opposite ends of any row or column of the Karnaugh map.

Karnaugh maps are constructed by a recursive procedure. Karnaugh maps for Boolean functions of $n + 1$ variables are obtained from Karnaugh maps for Boolean functions of n variables. This is done by taking two copies of the map for n variables, reversing the order of one, pasting the two together, and relabelling the squares. Karnaugh maps of one, two, three, or four variables are shown in Figure 6.3.

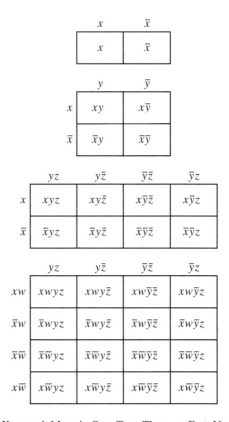

Figure 6.3   Karnaugh Maps in One, Two, Three, or Four Variables

The minterms which label each cell of a Karnaugh map are obtained as products of the minterms at the beginning of each row and column.

It is convenient to use Karnaugh maps for Boolean functions of only six or fewer variables. Our examples shall only use Boolean functions of four or fewer variables, so you can simply copy the diagrams in Figure 6.3.

The problem of simplifying a Boolean function can be approached in several ways, particularly if the number of variables is large. (We give several references for this in Appendix C.)

Following are the rules for using a Karnaugh map to simplify a Boolean function.

The Technique of Karnaugh Maps   Given a Boolean function $f(x_1, \ldots, x_n)$ in sum-of-products form, construct a Karnaugh map for f and use it to simplify f by following these steps:

I. Draw the Karnaugh map for the appropriate number of variables (you can copy the diagram in Figure 6.3).

II. For each minterm in f, put a 1 in the corresponding square of the diagram.

III. Decompose the 1s into sets of rectangles of adjacent 1s, and try to satisfy the following two conditions, in the following order of importance:

    A. The number of elements in each row of the rectangle, or the number of elements in each column, or both if possible, should be even.

    B. The rectangles should be as large as possible.

    Remember that the squares at the ends of each row and column are considered to be adjacent. Circle each rectangle. The rectangles may overlap.

IV. Simplify the Boolean expression corresponding to the sum of minterms in each rectangle.

V. Add the simplified Boolean expressions obtained in IV to obtain the simplified expression for f.

Evenness is emphasized in the first part of rule III so that identities like $x + \bar{x} = 1$ can be exploited. The fact that the rectangles are allowed to overlap is related to the identity $x = x + x$.

Perhaps the hardest thing to understand about Karnaugh maps is the proper use of rule III. The difficulty is to understand that condition A should take precedence over condition B. The following examples illustrate the use of all the rules.

EXAMPLE 5   Use a Karnaugh map to simplify the sum-of-products form of $f = x + y$.

Solution   The sum-of-products expression for f is $f(x, y) = xy + \bar{x}y + x\bar{y}$. The Karnaugh map for f is given in Figure 6.4.

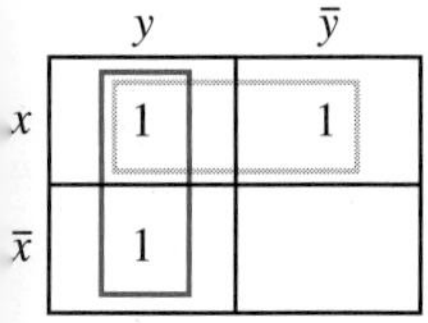

In the Karnaugh map, there are two rectangles, corresponding to the decomposition

$$f(x, y) = (xy + x\bar{y}) + (xy + \bar{x}y)$$
$$= x(y + \bar{y}) + (x + \bar{x})y$$
$$= x(1) + (1)y$$
$$= x + y$$

Figure 6.4

The Karnaugh Map for $f(x, y) = x + y$

$X(Y + \bar{Y})(Z + \bar{Z})$

EXAMPLE 6   Use a Karnaugh map to simplify the sum-of-products form of $f = x + y + z$.

Solution   The sum-of-products expression for f is

$$f = xyz + \bar{x}yz + x\bar{y}z + xy\bar{z} + \bar{x}\bar{y}z + \bar{x}y\bar{z} + x\bar{y}\bar{z}.$$

The Karnaugh map for f is given in Figure 6.5.

	yz	$y\bar{z}$	$\bar{y}\bar{z}$	$\bar{y}z$
x	1	1	1	1
$\bar{x}$	1	1		1

Figure 6.5   The Karnaugh Map for $f(x, y, z) = x + y + z$

There are three rectangles in the Karnaugh map: two squares, which are chosen first (by rule III A), and the first row of the diagram, which corresponds to the decomposition

$$f = (xyz + xy\bar{z} + \bar{x}yz + \bar{x}y\bar{z}) + (x\bar{y}z + \bar{x}\bar{y}z + xyz + \bar{x}yz)$$
$$+ (xyz + xy\bar{z} + x\bar{y}z + x\bar{y}\bar{z})$$
$$= y + z + x = x + y + z$$

  ●

EXAMPLE 7   Use a Karnaugh map to simplify the sum-of-products form of $f = xy + xz + yz$.

Solution   The sum-of-products form of f is

$$f = \bar{x}yz + xy\bar{z} + xyz + x\bar{y}z$$

The Karnaugh map for f is given in Figure 6.6. There are three 2-by-1 rectangles in the Karnaugh map, corresponding to the decomposition

$$f = (xyz + x\bar{y}z) + (xyz + xy\bar{z}) + (xyz + \bar{x}yz)$$
$$= xz + xy + yz = xy + xz + yz.$$

  ●

	yz	$y\bar{z}$	$\bar{y}\bar{z}$	$\bar{y}z$
x	1	1		1
$\bar{x}$	1			

Figure 6.6   The Karnaugh Map for $f = xy + xz + yz$

EXAMPLE 8   Use a Karnaugh map to simplify $f = xwyz + xwy\bar{z} + xw\bar{y}\bar{z} + \bar{x}wyz + \bar{x}wy\bar{z} + \bar{x}w\bar{y}\bar{z} + \bar{x}\bar{w}\bar{y}\bar{z} + x\bar{w}\bar{y}z$.

Solution   The Karnaugh map for f is given in Figure 6.7. Notice that it consists of a 2-by-3 rectangle, and two other points. We decompose the 2-by-3 rectangle into two squares (by rule III A), put the 1 adjacent to the 2-by-3 rectangle in a 1-by-2 rectangle (again by rule III A), and put the isolated 1 in a square by itself. This gives the decomposition

$$f = (xwyz + \bar{x}wyz + xwy\bar{z} + \bar{x}wy\bar{z}) + (xwy\bar{z} + xw\bar{y}\bar{z} + \bar{x}wy\bar{z} + \bar{x}w\bar{y}\bar{z})$$
$$+ (\bar{x}w\bar{y}\bar{z} + \bar{x}\bar{w}\bar{y}z) + x\bar{w}\bar{y}z$$
$$= wy + w\bar{z} + \bar{x}\bar{y}z + x\bar{w}\bar{y}z.$$

	yz	$y\bar{z}$	$\bar{y}\bar{z}$	$\bar{y}z$
xw	1	1	1	
$\bar{x}w$	1	1	1	
$\bar{x}\bar{w}$			1	
$x\bar{w}$				1

Figure 6.7   A Karnaugh Map for a Boolean Function of Four
Variables

One basic application of Boolean functions is the construction of *logic circuits*.
Logic circuits are electrical circuits which are composed of elements called *logic gates*.
Boolean functions in sum-of-products form can be used to construct logic circuits with a
desired output. Some solution of the simplification problem, for example using Karnaugh
maps, will then give a simpler circuit with the same output.

Logic circuits can be constructed using three logic gates, an *OR gate*, an *AND gate*,
and a *NOT gate*. More precisely, a logic circuit is a circuit with multiple inputs, composed
of OR, AND, and NOT gates, with a single output. OR, AND, and NOT gates are shown
in Figure 6.8.

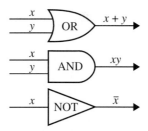

Figure 6.8   OR, AND, and NOT Gates

The outputs of these three gates, for a given input, are shown in the following tables:
An *OR gate*:

x	y	$x + y$
1	1	1
1	0	1
0	1	1
0	0	0

An *AND gate*:

x	y	xy
1	1	1
1	0	0
0	1	0
0	0	0

A *NOT gate*:

x	$\bar{x}$
1	0
0	1

An OR gate corresponds to the sum in a Boolean algebra, an AND gate corresponds to the product, and a NOT gate corresponds to the complement. They are defined in precise analogy with the operations Or, And, and Not in the propositional calculus.

By using the logic gates which we have just defined, each of which has one or two inputs, we can define similar logic gates with any number of inputs. For example, we can define an OR gate with three inputs, as in Figure 6.9.

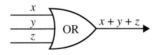

Figure 6.9   An OR Gate with Three Inputs

Similarly, we can define an OR gate with any finite number of inputs, or an And gate with any finite number of inputs.

Now the operations Or, And, and Not are the operations of a Boolean algebra, and therefore

Logic circuits form a Boolean algebra.

The input and output of logic circuits consists of binary strings. This is illustrated in Figure 6.10.

Since all logic circuits form a Boolean algebra, each single logic circuit corresponds to a Boolean expression, and hence to a Boolean function. Conversely, every Boolean function can be represented by a logic circuit.

THEOREM 6.2.1   Every logic circuit corresponds to a Boolean expression, and so to a Boolean function. Any Boolean function f in sum-of-products form can be obtained from a logic circuit by the following two steps:

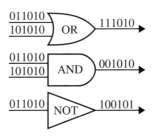

Figure 6.10   Logic Gates with Input and Output

I.   Construct a logic circuit for each minterm of f by using NOT gates and a multiple AND gate.

II.   Combine the gates in I by a multiple OR gate.

EXAMPLE 9   Construct a logic circuit for the sum-of-products form of $f = xy + xz + yz$, and simplify this circuit.

Solution   We have seen that the sum-of-products form of f is $f = xy + xz + yz = \bar{x}yz + xy\bar{z} + xyz + x\bar{y}z$. The logic circuit for the sum-of-products form of f is shown in Figure 6.11, and the simplified form, for $f = xy + xz + yz$, in Figure 6.12.   ●

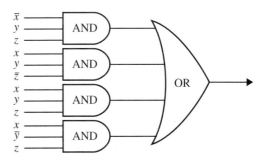

Figure 6.11   A Circuit for Example 9

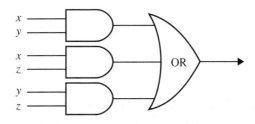

Figure 6.12   A Circuit for $f = xy + xz + yz$

EXAMPLE 10   Construct a logic circuit for the sum-of-products form of $f = x + y + z$, and simplify this circuit.

Solution   The sum-of-products form of $f = x + y + z$ is $f = xyz + \bar{x}yz + x\bar{y}z + xy\bar{z} + \bar{x}\bar{y}z + \bar{x}y\bar{z} + x\bar{y}\bar{z}$, and the corresponding circuit is shown in Figure 6.13. The simplified circuit was already given in Figure 6.9. ●

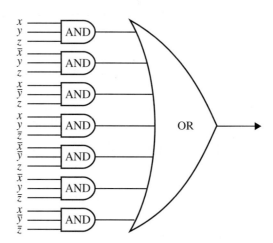

Figure 6.13   A Circuit for Example 10

AND and OR are *binary connectives*, each determined by a Boolean function in two variables. Since there are

$$2^{2^2} = 16$$

Boolean functions in two variables, there are 16 binary connectives in all. We can describe all of them with the following table:

		1	2	3	4	5	6	7	8	9	10	11	12	13	14	15	16
1	1	0	1	0	0	0	1	1	1	0	0	0	1	1	1	0	1
1	0	0	0	1	0	0	1	0	0	1	1	0	1	1	0	1	1
0	1	0	0	0	1	0	0	1	0	1	0	1	1	0	1	1	1
0	0	0	0	0	0	1	0	0	1	0	1	1	0	1	1	1	1

Two of these binary connectives are closely related to AND and OR. They are called NAND ("NOT-AND") and NOR ("NOT-OR"), the negations of AND and OR. NAND is described in column 15 of the table, and NOR in column 5.

Diagrams for NAND gates and NOR gates are shown in Figure 6.14.

A logical connective is called *complete* if every other connective can be expressed in terms of it.

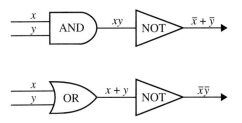

Figure 6.14 A NAND Gate and a NOR Gate

THEOREM 6.2.2   NAND and NOR are complete.

Proof   It is convenient to prove this by using relations from the propositional calculus. NAND can be exressed as $\sim (p \wedge q)$ and NOR can be expressed as $\sim (p \vee q)$.

To prove that any connective is complete, it is sufficient to express AND, OR, and NOT in terms of this connective.

To prove that NAND is complete, first express NOT in terms of NAND by setting $p = q$:

$$\sim (p \wedge p) \leftrightarrow (\sim p \vee \sim p) \leftrightarrow \sim p$$

Once NOT has been expressed in terms of NAND, we can clearly express AND in terms of NAND. Using DeMorgan's Laws, we can also express OR in terms of NAND:

$$\sim (\sim p \wedge \sim q) \leftrightarrow (p \vee q)$$

The proof that NOR is complete is exactly similar.   ●

APPLICATION   *Regions in Logic Diagrams*

We have seen how Venn diagrams can be used to model set operations and logical arguments. In the Advanced Exercises of Section 1.4, we showed how to draw a Venn diagram for any finite number of sets. Most of the diagrams that we drew, however, involved three or fewer sets. In fact, it becomes awkward to use Venn diagrams when more sets are involved. How can we describe regions in logic diagrams which involve any finite number of sets in an efficient way?

For example, consider the region in Figure 6.15. How can we describe this region efficiently, as well as all similar ones? We shall describe a method for doing this which is closely related to the Principle of Inclusion-Exclusion.

We have described several ways in which the n-cube can be used as a model. The n-cube can also be used as a model for all the intersections of n sets. Suppose that

$$A_1, A_2, \ldots, A_n$$

are all subsets of some set X. We shall describe the intersections of these sets with binary vectors of length n, where a set A_k occurs in the intersection if there is a 1 in the kth

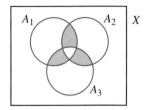

Figure 6.15   A Region in a Venn Diagram

entry of the vector, and does not occur if there is a 0 in the kth entry. By convention, X corresponds to the 0-vector.

Suppose that $n = 3$. The vectors and the corresponding intersections are

$$(0, 0, 0) \leftrightarrow X$$
$$(1, 0, 0) \leftrightarrow A_1$$
$$(0, 1, 0) \leftrightarrow A_2$$
$$(0, 0, 1) \leftrightarrow A_3$$
$$(1, 1, 0) \leftrightarrow A_1 \cap A_2$$
$$(1, 0, 1) \leftrightarrow A_1 \cap A_3$$
$$(0, 1, 1) \leftrightarrow A_2 \cap A_3$$
$$(1, 1, 1) \leftrightarrow A_1 \cap A_2 \cap A_3$$

A *logic diagram with n sets* is a collection of sets

$$D = \{X, A_1, A_2, \ldots, A_n\}$$

where the sets A_i are subsets of X.

A *primitive region* in a logic diagram is a difference of sets

$$(A_{i_1} \cap A_{i_2} \cap \cdots \cap A_{i_r}) - S$$

where S is the union of all the r-fold intersections of the A_i except the given one. A *region* in a logic diagram as above is a Boolean function of n variables.

If we have a Venn diagram of the sets in a logic diagram, the Boolean function describes regions in the diagram by describing which primitive regions are included (if the value of the function is 1), and which primitive regions are excluded (if the value of the function is 0). For example, the region in Figure 6.15 is described by the Boolean function

$$f(0, 0, 0) = 0$$
$$f(1, 0, 0) = 0$$
$$f(0, 1, 0) = 0$$
$$f(0, 0, 1) = 0$$
$$f(1, 1, 0) = 1$$

$$f(1, 0, 1) = 1$$
$$f(0, 1, 1) = 1$$
$$f(1, 1, 1) = 0$$

From the definitions we obtain the following theorem.

THEOREM 6.2.3   In a logic diagram with n sets there are 2^n primitive regions, and 2^{2^n} regions altogether.

EXERCISES

Represent each of the following Boolean functions in the four ways described at the beginning of this section: by using a table, by using a Gray Code, by using the full binary tree, and by using the n-cube.

1. $f(x, y) = x + y.$ **2.** $f(x, y) = xy.$

3. $f(x, y) = x\bar{y}.$ **4.** $f(x, y) = x + \bar{x}y.$

5. $f(x, y, z) = x + y + \bar{z}.$

6. $f(x, y, z) = x\bar{y} + \bar{x}z + yz.$

7. $f(x, y, z) = \bar{x} + yz + xyz.$

8. $f(x, y, z, w) = x + y + z + w.$

9. $f(x, y, z, w) = xy + xz + xw + yz + yw + zw.$

10. $f(x, y, z, w) = x + \bar{y}z + xyz + x\bar{y}z\bar{w}.$

Represent each of these Boolean functions in sum-of-products form.

11. The function in (1).

12. The function in (2).

13. The function in (3).

14. The function in (4).

15. The function in (5).

16. The function in (6).

17. The function in (7).

18. The function in (8).

19. The function in (9).

20. The function in (10).

Construct a Karnaugh map for each of the functions in (11)–(20), and use it to simplify the function in sum-of-products form.

21. The function in (11).

22. The function in (12).

23. The function in (13).

24. The function in (14).

25. The function in (15).

26. The function in (16).

27. The function in (17).

28. The function in (18).

29. The function in (19).

30. The function in (20).

For each of the functions in (1)–(10), give a corresponding logic circuit, and a simplified circuit, obtained by use of a Karnaugh map — this will usually correspond to the Boolean expression in (1)–(10).

31. The function in (1).

32. The function in (2).

33. The function in (3).

34. The function in (4).

35. The function in (5).

36. The function in (6).

37. The function in (7).

38. The function in (8).

39. The function in (9).

40. The function in (10).

ADVANCED EXERCISES

Express each of the three following logic circuits entirely in terms of NAND, and then in terms of NOR.

For each of the following regions in a logic diagram with three sets, give a Boolean function describing the region.

4.

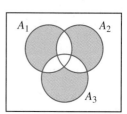

1.

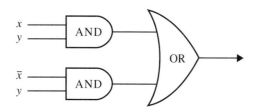

5.

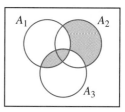

2.

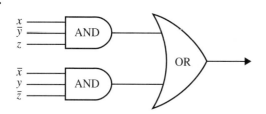

6.

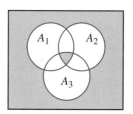

3.

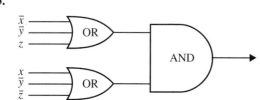

6.3 THE PREDICATE CALCULUS

"You may seek it with thimbles—and seek it with care;

You may hunt it with forks and hope;

You may threaten its life with a railway-share;

You may charm it with smiles and soap—"

"But oh, beamish nephew, beware of the day,

If your Snark be a Boojum, for then

You will softly and silently vanish away,

And never be met with again!"

The logical system which we call the *predicate calculus* was first described in detail in *Principia Mathematica* (1912) by Bertrand Russell and Alfred North Whitehead.

A *predicate* is a property which the objects of a theory may or may not have. The notation "$A(x)$" means that "x has the property A."

The basic difference between the predicate calculus and the propositional calculus, described in the previous chapter, is that the predicate calculus allows expression of the phrases

> *"for all x"*

symbolized by

> $\forall(x)$

and

> *"there exists an x"*

symbolized by

> $\exists(x)$

The symbols $\forall$ and $\exists$ are called *quantifiers*. $\forall$ is the *universal quantifier* and $\exists$ is the *existential quantifier*. Typical expressions in the predicate calculus are

> $\forall(x)A(x)$

meaning "for all x, x has the property A," and

> $\exists(x)A(x)$

meaning "there exists an x such that x has the property A."

In this section, we shall show how the predicate calculus gives better expression to arguments than the propositional calculus. In particular, we shall explain how Aristotle's classification of syllogisms, which could not be carried out in the propositional calculus, can be carried out in the predicate calculus.

We might suggest this as an instance of the following quote from the preface to *Principia Mathematica*:

> *Hence the scope of mathematics is enlarged by the addition of new subjects and by a backward extension into provinces hitherto abandoned to philosophy.*

In this section, we develop a working acquaintance with the predicate calculus. We shall describe well-formed formulas, identities, and proofs, just as we did for the propositional calculus.

We shall clarify certain technical problems which often cause difficulties, such as the problem of negating a sequence of universal and existential quantifiers.

At the end of this section is a brief description of the application of the predicate calculus to one of the fundamental problems of computer science, that of *proving program correctness*.

There is a notion of well-formed formula (abbreviated *wff*) for sentences in the predicate calculus, just as there was for sentences in the propositional calculus. The definition is recursive, as it was for the propositional calculus.

DEFINITION 6.3.1 A *well-formed formula* in the predicate calculus is defined as follows:

    i. If $A(x_1, x_2, \ldots, x_n)$ is a predicate involving the indicated n variables, then $A(x_1, x_2, \ldots, x_n)$ is a well-formed formula in the predicate calculus.

    ii. If A and B are well-formed formulas, then so are $(\sim A)$, $(A \rightarrow B)$, and $(\forall x)A(x)$.

We also make the following definitions:

    iii. $(\exists x)A(x)$ is an abbreviation for $(\sim (\forall x)(\sim A(x)))$.

    iv. $(A \wedge B)$ is an abbreviation for $(\sim (A \rightarrow (\sim B)))$.

    v. $(A \vee B)$ is an abbreviation for $((\sim A) \rightarrow B)$.

EXAMPLE 1 Which of the following three formulas are well-formed formulas in the predicate calculus?

    I. $((\forall x)A(x)$.

    II. $(\forall x)(\exists y)(A(x) \rightarrow B(y))$.

    III. $(\forall x)(\forall y)A(x, y)$.

Solution   The formula in I is not a well-formed formula, because the number of left parentheses is not equal to the number of right parentheses. The formulas in II and III are generated by the rules of the definition, and therefore are well-formed formulas. We will use the word "sentence" to mean the same thing as "well-formed formula."   ●

The predicate calculus described here is called the *first-order predicate calculus*, to distinguish it from more complicated logical systems. The basic difference is that in the first-order predicate calculus, quantifiers may be applied to variables but not to predicates. For example, the expression

$$(\forall A)A(x)$$

is not a well-formed formula in the first-order predicate calculus.

Recall that truth tables can be used to determine whether a sentence in the propositional calculus is true or not. There is also a notion of truth in the propositional calculus.

An *interpretation* of the predicate calculus is a set S, where the values of the variables are understood to lie. The sentence

$$(\forall x)A(x)$$

means that x has the property A for all elements of the set S, and

$$(\exists x)A(x)$$

means that at least one element of the set S has the property A. When we work with a sentence in the predicate calculus, we always understand that some interpretation has been chosen.

A sentence is *true* in a given interpretation S, if it is true however elements of S are substituted for the variables it contains. A sentence in the predicate calculus is *true* if it is true in every interpretation. A true sentence will also be called a tautology.

There is a basic difference between the notions of truth for the propositional calculus and the predicate calculus. Truth for a sentence in the propositional calculus can be checked by a finite algorithm, namely constructing a truth table. Because the predicate calculus has infinitely many models, there is no finite algorithm which will determine whether an arbitrary sentence in the predicate calculus is true or not.

EXAMPLE 2   Give an example of a sentence which is true in one interpretation and not in another. Give an example of a true sentence.

Solution   Let $(\forall x)A(x)$ be the sentence "all subsets of x have two or fewer elements." This sentence is true for the interpretation where S is chosen to be the set of subsets of $\{1, 2\}$, and x is understood to be one of these subsets, but false if we choose S to be the set of subsets of $\{1, 2, 3\}$.

For an example of a true sentence, choose $(\forall x)(A(x) \rightarrow A(x))$. This sentence is true for any choice of predicate $A(x)$ whatever, for no matter what interpretation we choose, if x has the property A then x certainly has the property A.   ●

In the definition of the existential quantifier given above, we introduced the identity

$$(\exists x)A(x) \leftrightarrow (\sim (\forall x)(\sim A(x)))$$

There is a corresponding identity

$$(\forall x)A(x) \leftrightarrow (\sim (\exists x)(\sim A(x)))$$

These identities are very useful. We will illustrate this by using Aristotle's classification of sentences, discussed in Section 5.2. Recall that there are four types of sentences. We shall express each one in the predicate calculus.

Universal Affirmative: "All A is B."

$$(\forall x)(A(x) \rightarrow B(x))$$

Particular Affirmative: "Some A is B."

$$(\exists x)(A(x) \wedge B(x))$$

Universal Negative: "No A is B."

$$(\forall x)(A(x) \rightarrow (\sim (B(x))))$$

Particular Negative: "Some A is not B."

$$(\exists x)(A(x) \wedge (\sim (B(x))))$$

Because these are the sentences which occur in Aristotle's classification of syllogisms, we see that

> *Aristotle's classification of syllogisms can be carried out in the predicate calculus.*

Recall the Square of Opposition, discussed in Section 5.2, in which the negation of a universal affirmative is a particular negative, and the negation of a particular affirmative is a universal negative. The first of these identities can be expressed in the predicate calculus (we omit some parentheses) as:

$$(\sim (\forall x)(A(x) \rightarrow B(x)) \leftrightarrow (\sim (\forall x)((\sim A(x)) \wedge B(x))$$

$$\leftrightarrow (\exists x)(\sim ((\sim A(x)) \vee B(x)))$$

$$\leftrightarrow (\exists x)(A(x) \wedge (\sim (B(x))))$$

The second of these identities can be expressed as

$$(\sim (\exists x)(A(x) \wedge B(x))) \leftrightarrow (\forall x)(\sim (A(x) \wedge B(x))$$

$$\leftrightarrow (\forall x)(\sim A(x) \vee (\sim B(x)))$$

$$\leftrightarrow (\forall x)(A(x) \rightarrow (\sim B(x)))$$

EXAMPLE 3   Express a syllogism in the predicate calculus.

Solution   We will choose the syllogism "Every man is mortal"; "Socrates is a man"; therefore "Socrates is mortal." Let $B(x)$ mean that "x is a man"; let $A(x)$ mean that "x is mortal"; and let $C(x)$ mean that "x is Socrates." Then the syllogism can be expressed as follows. The two premises are

$$(i) \quad (\forall x)(B(x) \rightarrow A(x))$$

$$(ii) \quad (\forall x)(C(x) \rightarrow B(x))$$

and from these we derive the conclusion

$$(iii) \quad (\forall x)(C(x) \rightarrow A(x)) \qquad \bullet$$

Next we shall express a sentence involving various kinds of predicates.

EXAMPLE 4   Express the sentence "Every woman is the niece of her father's brother" in the predicate calculus.

Solution   Let $W(x)$ mean "x is a woman"; let $F(x, y)$ mean "y is the father of x"; let $B(y, z)$ mean "y and z are brothers"; and let $N(x, z)$ mean "x is the niece of z." Then our sentence can be expressed as follows (we omit some of the parentheses associated with the quantifiers).

$$(\forall x)(\forall y)(\forall z)(W(x) \rightarrow (F(x, y) \wedge B(y, z) \rightarrow N(x, z))) \qquad \bullet$$

The next example involves negating a sequence of quantifiers.

EXAMPLE 5   Negate the sentence

$$(\forall x)(\exists y)(\forall z)(A(x) \rightarrow B(y, z))$$

Solution

$$\sim (\forall x)(\exists y)(\forall z)(A(x) \rightarrow B(y, z)) \leftrightarrow (\exists x)(\sim (\exists y)(\forall z)(A(x) \rightarrow B(y, z)))$$
$$\leftrightarrow (\exists x)(\forall y)(\sim (\forall z)(A(x) \rightarrow B(y, z)))$$
$$\leftrightarrow (\exists x)(\forall y)(\exists z)(\sim (\sim A(x) \vee B(y, z)))$$
$$\leftrightarrow (\exists x)(\forall y)(\exists z)(A(x) \wedge \sim B(y, z))) \qquad \bullet$$

What is the importance of the predicate calculus for computer science? One application of the predicate calculus is to *proving program correctness*, which we shall discuss in the Application for this section. Another basic reason that the predicate calculus is important in computer science has to do with two theorems which were proved by the logician Kurt Gödel (1906–1978).

The Gödel Completeness Theorem   The first-order predicate calculus is *complete*. That is, every true sentence in the first-order predicate calculus can be proved in the first-order predicate calculus.

Gödel also studied the *augmented first-order predicate calculus*. Notice that in the predicate calculus we cannot describe the properties of numbers, except indirectly, using predicates. In the augmented first-order predicate calculus, symbols are added for the integers, $0, 1, -1, 2, -2, \ldots$, for equality, $=$, and for addition and multiplication $+$ and $\cdot$. In the resulting system, we can discuss *formal number theory*. We can describe the properties of the integers, prove unique factorization, and so on.

The Gödel Incompleteness Theorem   The augmented first-order predicate calculus is *incomplete*. That is, there is a true sentence in the augmented first-order predicate calculus which can not be proved in the augmented first-order predicate calculus.

The Gödel Incompleteness Theorem sets limits to computation in ways which are still not well understood. It can be reformulated in terms of ideal computers (*Turing machines*). We shall do this in the final section of this book.

APPLICATION   *Proving Program Correctness*

Computer programs have been used to prove important theorems and to reach important practical conclusions. How can we be sure that the output of a program is correct? Anyone who has written a long computer program knows how easy it is for errors to creep in. For this reason, computer scientists are concerned with the problem of proving that the output of a program is correct.

In 1969 C. A. R. Hoare published the paper *An Axiomatic Basis for Computer Programming*, which gives an approach to the problem of proving program correctness. The paper begins

*Computer programming is an exact science in that all the properties
of a program and all the consequences of executing it in any given
environment can, in principle, be found out from the text of the*

> *program itself by purely deductive reasoning. Deductive reasoning involves the application of valid rules of inference to sets of valid axioms. It is therefore desirable and interesting to elucidate the axioms and rules of inference which underlie our reasoning about computer programs.*

Hoare's analysis is based on the notion of *partial correctness*, for which he uses the symbol

$$P\{Q\}R$$

Here Q is a program, P is an initial assertion, and R is a final assertion. Partial correctness means that if P is true for the input values of Q, and if Q terminates, then R is true for the output values of Q. Partial correctness does *not* assume that the program Q terminates. If the program is partially correct and if it also terminates, then it is *correct*.

Hoare gave several rules of composition for his symbol, for example that

$$\text{if } P\{Q\}R \text{ and } R \to S, \text{ then } P\{Q\}S;$$

$$\text{and if } P\{Q\}R \text{ and } S \to P \text{ then } S\{Q\}R;$$

$$\text{and if } P\{Q_1\}R_1 \text{ and } R_1\{Q_2\}R, \text{ then } P\{Q_1; Q_2\}R$$

where $\{Q_1; Q_2\}$ denotes the program Q_1 followed by the program Q_2.

EXAMPLE 6   Show that the program

$$y := 2$$

$$z := xy$$

is correct with respect to the initial assertion $P : x = 2$ and the final assertion $R : z = 4$.

Solution   If P is true, so that $x = 2$ as the program begins, then since y is assigned the value 2, and z has the value $z = 2 \cdot 2 = 4$, the final assertion R is true after the program runs.

EXERCISES

State which of the following are well-formed formulas in the predicate calculus, and give a reason.

1. $\sim A(x)$

2. $(\sim (A(x))$

3. $(\sim A(x))$

4. $A(x) \to B(x)$

5. $(A(x) \to B(x))$

6. $\forall x(A(x) \to B(x))$

7. $(\forall x)(A(x) \to B(x))$

8. $A(x) \to \exists x\, B(x)$

9. $(A(x) \to \exists x\, B(x))$

10. $(A(x) \to (\exists x)B(x))$

11. Give an example of a sentence in the predicate calculus which is true in one interpretation and not in another.

12. Give an example of a true sentence in the predicate calculus.

Refer to Example 3 of this section and the classification of syllogisms in Section 5.2. Give the general form of the syllogism, and then give a specific example of such a syllogism, as in Example 3.

13. Code AII in the First Figure.

14. Code EAE in the First Figure.

15. Code AOO in the Second Figure.

16. Code EIO in the Second Figure.

17. Code IAI in the Third Figure.

18. Code OAO in the Third Figure.

19. Code AEE in the Fourth Figure.

20. Code IAI in the Fourth Figure.

Express the following sentences in the predicate calculus.

21. Every man is the nephew of his father's brother.

22. For every integer x, there is an integer y which is larger than x.

23. If x is larger than 1, there is an integer between x and $2x$.

24. If y is nearer to x than z is, and if w is nearer to x than y is, then w is nearer to x than z is.

Simplify the negation of each of the following sentences, as in Example 5 of this section.

25. $(\forall x)(A(x) \rightarrow B(x))$

26. $(\exists x)(\forall y)(A(x) \vee B(x, y))$

27. $(\exists x)(\forall y)(\exists z)(A(z) \rightarrow B(x, y))$

6.4   CHAPTER SUMMARY AND SUPPLEMENTARY EXERCISES

> *"It is this, it is this that oppresses my soul,*
>
> *When I think of my uncle's last words:*
>
> *And my heart is like nothing so much as a bowl*
>
> *Brimming over with quivering curds!*
>
> *"It is this, it is this—" "We have heard that before!"*
>
> *The Bellman indignantly said.*
>
> *And the Baker replied " Let me say it once more.*
>
> *It is this, it is this that I dread!"*

Section 6.1

In this section we described Boolean algebras and their basic properties, and gave several examples. We described Boolean expressions and how they give rise to Boolean functions. We described the conjunctive and disjunctive normal forms for a Boolean expression.

Section 6.2

We began this section by describing several ways of representing Boolean functions or Boolean expressions. We pointed out that it is important to have a method for simplifying Boolean expressions, and described the technique of Karnaugh maps for doing this. We described the relation between Boolean expressions and circuits. We showed how Boolean functions can be used to describe regions in logic diagrams.

Section 6.3

In this section we described the predicate calculus, which in a sense is the next logical language after the propositional calculus. We showed how Aristotle's classification of syllogisms can be carried out in the predicate calculus. We briefly discussed Gödel's Completeness Theorem and Incompleteness Theorem. We discussed the role of logic in proving computer program correctness.

EXERCISES

Use a Gray Code to give the values of the following Boolean functions.

1. $f = \bar{x} + y$.

2. $f = \bar{x}y$.

3. $f = x + \bar{y} + \bar{z}$.

4. $f = x + yz$.

5. $f = xz + yz$.

6. $f = x + yz + w$.

7. $f = xy + xz + wz$.

8. $f = x\bar{y} + yzw$.

9. $f = x + yzw$.

10. $f = xyz + \bar{x}\bar{y}\bar{w}$.

11–20. Put each of the functions in 1–10 into sum-of-products form.

21–30. Draw a Karnaugh map for each of the functions in 11–20, and use it to simplify the given function.

Graphs and Relations

7.1 GRAPH COLORING AND MATCHING

The Bellman looked uffish, and wrinkled his brow,

"If only you'd spoken before!

It's excessively awkward to mention it now,

With the Snark, so to speak, at the door!"

In this chapter, we shall explore several closely related themes in the theory of graphs. One of these is *graph coloring:* a graph G has a *node coloring* with k colors if the nodes can be colored with k colors in such a way that no two nodes with the same color are joined by an edge.

The simplest result about graph coloring is that a graph G has a two-coloring if and only if G is bipartite. This suggests the importance of bipartite graphs in discrete mathematics, and in fact bipartite graphs occur in the solution of many important problems. In this section, we will discuss and prove a basic result about bipartite graphs, the *marriage theorem* of Philip Hall, and a closely related result, *Menger's theorem*. Graph coloring has a significant application in computer science, to *register allocation*.

In the second section of this chapter, we shall discuss partially ordered sets, and a special kind of partially ordered set called a *lattice*. We shall discuss and prove a result about general partially ordered sets, *Dilworth's theorem*. A Boolean algebra is a special type of lattice.

Recall that a graph G is bipartite if the vertex set $V(G)$ is a disjoint union $V(G) = V_1 \cup V_2$ of two subsets such that no edge of G joins a node in V_1 to another node in V_1 or a node in V_2 to another node in V_2.

In this chapter, we will discuss only node colorings of a graph.

225

THEOREM 7.1.1 The following conditions are equivalent for any graph G:

> **(i)** G is bipartite.
>
> **(ii)** G has a two-coloring.
>
> **(iii)** G has no cycle of odd length.

Proof If G is bipartite, so that $V(G) = V_1 \cup V_2$, we can color the nodes in V_1 with the color A and the nodes in V_2 with the color B. If G has a two-coloring, it cannot contain a cycle of odd length since a cycle of odd length has a three-coloring but not a two-coloring. Now assume that G has no cycle of odd length. Passing to a connected component of G, we can assume that G is connected. Now fix a node v of G, and let V_1 be the set of nodes of G joined to v by a path of even length, and V_2 the set of nodes of G joined to v by a path of odd length. If any node w of G is in both these sets, then w lies on a closed walk of G of odd length which contains a cycle of odd length, so $V_1 \cap V_2 = \emptyset$. No node of V_1 can be joined to a node of V_1 and no node of V_2 can be joined to a node of V_2 because again G would contain a cycle of odd length, so G is bipartite. ●

The n-cube is a notable example of a bipartite graph.

EXAMPLE 1 Prove that the graph of the n-cube is bipartite.

Solution Remember that the nodes of the graph of the n-cube can be labelled with the subsets of $\{1, 2, \ldots, n\}$, and that two nodes are joined if their labels differ in exactly one element. Let V_1 be the set of nodes labelled with subsets with an even number of elements, and V_2 the set of nodes labelled with subsets with an odd number of elements. This represents the graph of the n-cube as a bipartite graph. ●

Figure 7.1 shows the graph of the 3-cube as a bipartite graph. The elements of the two subsets are labelled A and B. We can also view this as a two-coloring with colors A and B. In general, we will denote colors by letters $A, B, C, D, \ldots$.

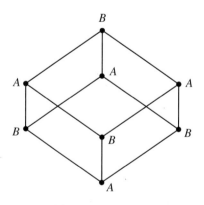

Figure 7.1 The 3-Cube as a Bipartite Graph

Figure 7.2 shows the graph of the octahedron.

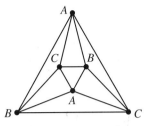

Figure 7.2   A Coloring of the Graph of the Octahedron

↳ LABELING, NO TWO EDGES THAT ARE
CONNECTED HAVE SAME
COLOR

EXAMPLE 2   Is the graph of the octahedron bipartite?

Solution   The graph of the octahedron contains triangle-cycles of length three, so it is not bipartite and does not have a two-coloring. It does have a three-coloring, as shown in Figure 7.2.

●

EXAMPLE 3   How can we color the complete graph K_n?

Solution   The graph K_n has n nodes, each connected to every other node by an edge. Thus we can color K_n by coloring each vertex with a different color. K_n has an n-coloring, but no k-coloring if k is less than n.

●

This leads to a simple but useful observation:

If a graph G contains K_n as a subgraph, then at least n colors are required to color G.

For example, the graph in Figure 7.3 contains the complete graph K_4, and so cannot be colored with fewer than four colors. A four-coloring is shown in Figure 7.3.

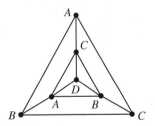

Figure 7.3   A Four-Coloring of a Graph

Now we shall discuss the concept of a *matching,* which can be expressed in terms of bipartite graphs.

Let X be a finite set and

$$A_1, A_2, \ldots, A_n$$

a set of non-empty subsets of X. A *system of distinct representatives* for this set of subsets is a set of n different elements

$$a_1 \in A_1, a_2 \in A_2, \ldots, a_n \in A_n$$

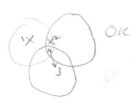

We can think of the element $a_i \in A_i$ as representing the set A_i.

The following theorem of Philip Hall was published in 1935 in the paper *On representatives of subsets.* It describes when a system of distinct representatives exists. The proof that we give here was published by P. A. Halmos and H. E. Vaughan in 1950.

Hall's Theorem   If X is a finite set and $A_1, A_2, \ldots, A_n$ are non-empty subsets of X, then the following conditions are equivalent:

(i) $A_1, A_2, \ldots, A_n$ has a system of distinct representatives.

(ii)   For each integer k with $1 \leq k \leq n$, the union of any k of the subsets A_i contains at least k elements.

makes sure each class has own rep even though there's overlap

Proof   Condition (i) clearly implies condition (ii). Our problem is to prove that if condition (ii) holds, then condition (i) must hold.

Condition (ii) clearly implies condition (i) if $n = 1$, and this starts a proof by induction. Assume first that the union of any k subsets contains $k + 1$ elements. Let $a_1 \in A_1$ be any element, and remove a_1 from X. Then condition (ii) holds for the remaining $n - 1$ sets, so by induction we can complete the system of distinct representatives.

Now assume that there is some set of k subsets whose union contains exactly k elements. We can assume that $k < n$, for otherwise we can simply take the n elements as a system of distinct representatives. Begin choosing a system of distinct representatives by taking $a_1, a_2, \ldots, a_k$ to be these k elements. Now any h of the remaining $n - k$ subsets must have a union containing at least h elements, for otherwise the union of these h subsets together with the k subsets above would contain fewer than $h + k$ elements. Therefore, we can complete the choice of a system of distinct representatives by induction.   ●

A *matching* is a bipartite graph G in which each node of one component is joined to one or more nodes of the other component. The matching is *complete* if each node of the first component is joined by an edge to a different node of the other component. Hall's theorem, translated into the language of bipartite graphs, gives a condition for a complete matching to exist.

Hall's theorem is often stated as *Hall's marriage theorem.* If n boys each know a certain number of girls, when can each boy be married to a girl that he knows? The answer is that this can be done if and only if each set of k boys collectively knows at least k girls.

EXAMPLE 4   Suppose that we have four boys and four girls. Suppose that the first boy knows all four girls, the second boy knows the first and third girls, the third boy knows only the second girl, and the fourth boy knows the first girl and the fourth girl. Can each boy be married to a girl that he knows?

Solution   The answer is yes. The first boy can marry the first girl, the second boy can marry the third girl, the third boy can marry the second girl, and the fourth boy can marry the fourth girl. This solution corresponds to a complete matching, which is shown in Figure 7.4.   ●

EXAMPLE 5   Suppose again that we have four boys and four girls. The first boy knows all four girls, the second boy knows the second and third girls, the third boy knows the second girl, and the fourth boy knows the third girl. Can each boy be married to a girl that he knows?

Boys   Girls

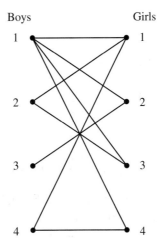

Figure 7.4   A Matching

Solution   The answer is no. The problem is that the sets of girls which the last three boys know are {2, 3}, {2}, and {3}, and the union of these three sets has only two elements, so the condition of Hall's theorem is violated.   ●

If A and B are two vertices of a graph G, an AB-*cutset* is a set of edges of G whose removal separates A from B. Two paths from A to B are *edge-disjoint* if they have no edges in common. The following theorem is attributed to K. Menger, who stated a closely related form of it (for node cutsets and node-disjoint paths) in a paper about topology which was published in 1927. The theorem in the following form was first proved by Ford and Fulkerson in 1955. We shall omit the proof.

Menger's Theorem   The maximum number of edge-disjoint paths connecting two vertices A and B of a connected graph G is equal to the minimum size of an AB-cutset.

Figure 7.5 is an example of a graph which has a minimal AB-cutset consisting of three edges (circled in blue), and such that A and B are connected by three edge-disjoint paths (circled in gray).

Menger's theorem and Hall's theorem are closely related. It can be proved that Menger's theorem and Hall's theorem are logically equivalent.

THEOREM 7.1.2   If the largest degree of a node of a graph G is d, then G has a $d + 1$-coloring.

Proof   The proof is by induction on the number n of nodes of G. The statement is clearly true if $n = 1$, and this starts the induction. If n is larger than one, remove a node v from G and the adjacent edges, leaving a graph with $n - 1$ nodes. This graph has a $d + 1$-coloring, by induction. Now introduce the node v and its adjacent edges again. There are d nodes adjacent to v, which are colored with at most d colors. Use the $d + 1$st color to color v.   ●

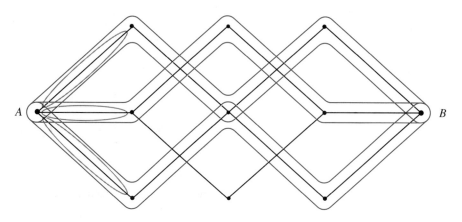

Figure 7.5   An Example of Menger's Theorem

A stronger form of the theorem that we have just proved was given by R. L. Brooks in 1941 in the paper *On coloring the nodes of a network*. The proof is more difficult, and we shall omit it.

Brooks' Theorem   If G is a simple graph which is not a complete graph or an odd-length cycle, and if the largest degree of a vertex of G is d, then G has a d-coloring.

Every node of the graph of the dodecahedron has degree three, so Brooks' theorem says that this graph must have a three-coloring. One such three-coloring is shown in Figure 7.6.

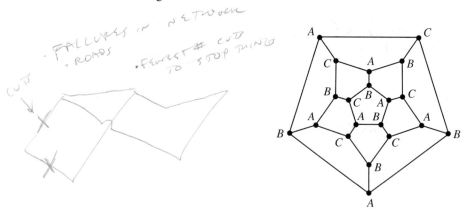

Figure 7.6   A Three-Coloring of the Graph of the Dodecahedron

In 1879, Alfred Bray Kempe (1849–1922) announced a proof that every planar graph has a four-coloring. In 1890, Percy John Heawood (1861–1955) showed that Kempe's proof was incomplete. However, Kempe's argument can be used to prove that every planar graph has a five-coloring. Before giving the proof, we will need a lemma.

Lemma   Every simple planar graph G contains a node of degree no more than five.

Proof   First, we can assume that G is connected.

Suppose that G has n nodes, m edges and f faces (the faces are the regions into which the edges of G divide the plane, and we include the "infinite face" outside G). By Euler's theorem, $n - m + f = 2$. Since every face is bounded by at least three edges, $3f \leq 2m$, so since $3n - 3m + 3f = 6$, $3n - 3m + 2m \geq 6$, or $3n - m \geq 6$, or $m \leq 3n - 6$. If every node of G has degree at least 6, then $6n \leq 2m$, or $3n \leq m$, so $m \leq m - 6$, and this is a contradiction.   ●

The Five-Color Theorem   Any planar graph G has a five-coloring.

Proof   Clearly we can assume that G is connected.

The proof is by induction on the number of nodes n of G. The theorem is obviously true if $n = 1$, and this starts the induction. By the lemma, G has a node v of degree no more than five. If the degree of v is less than five, the theorem is proved by deleting v, coloring the resulting graph by induction, and then replacing v and giving v a color different from any of the adjacent nodes.

Now suppose that the degree of v is five, and consider v and the five adjacent nodes. If the nodes are all joined by edges, then G contains K_5 which is not planar, a contradiction. Therefore at least two of the nodes adjacent to v, say a and b, must not be joined by an edge. Contract the edges joining nodes a and b to v. By induction, the resulting graph has a five-coloring. Now replace nodes a and b and the edges joining a and b to v, and color a and b with the color used for v. Then color v with one of the five colors which has not been used for a, b, or the other three nodes adjacent to v.   ●

It seems that the first person to ask whether every planar graph has a four-coloring was Francis Guthrie (1831–1899). He mentioned the question to his brother Frederick Guthrie, who asked Augustus DeMorgan about it. DeMorgan referred to the problem in a letter to Hamilton in 1852, and this appears to be the first written reference to the four-color problem.

Mathematicians worked on the four-color problem for more than a century. Many incomplete proofs were given, some of which nevertheless involved useful ideas. Finally, in 1976, a proof was given by Kenneth Appel and Wolfgang Haken. The idea of the proof is to reduce the problem to the problem of coloring a large but finite number of graphs called *unavoidable configurations*, and then using a computer program to show that each of these has a four-coloring. The proof is extremely long and we cannot give it here.

Why is the proof of the four-color theorem so much harder than the proof of the five-color theorem? Mathematicians make a distinction between *local problems*, which involve the properties of a space in the neighborhood of a point, and *global problems*, which involve the properties of the whole space. For graphs, a local problem involves the properties of the graph near a single node, while a global problem involves the properties of the whole graph. Global problems are almost always harder than local problems. If you examine the proof of the five-color theorem that we have given, you will see that it is local in nature. By contrast, the four-color problem appears to be global in nature. That is, choices which are made to color a few nodes of a graph appear to affect the choices which must be made to color all the other nodes of the graph.

The Four-Color Theorem   Any planar graph has a four-coloring.

Every node of the graph of the icosahedron has degree five, so Brooks' theorem says that this graph has a five-coloring. The four-color theorem says that it must have a four-coloring, as in Figure 7.7 (if you examine this figure, you will see that the graph of the icosahedron cannot have a three-coloring).

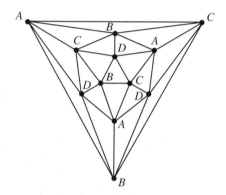

Figure 7.7   A Four-Coloring of the Graph of the Icosahedron

Here is another way to think about graph coloring. Let G be a simple graph, and let

$$P_G(k)$$

be the number of ways of coloring the vertices of G with k or fewer colors. $P_G(k)$ is called the *chromatic function* of G. We shall state the following theorem without proof.

THEOREM 7.1.3   The chromatic function $P_G(k)$ of a simple graph G is a polynomial. If G has n nodes, the degree of $P_G(k)$ is n, and the constant term of $P_G(k)$ is always zero.

Because of this theorem, the chromatic function $P_G(k)$ of G is called the *chromatic polynomial* of G. The chromatic polynomial can be used to give a characterization of trees.

THEOREM 7.1.4   If T is a simple graph with n nodes, then the following conditions are equivalent:

    (i) T is a tree.

    (ii) $P_T(k) = k(k-1)^{n-1}$.

EXAMPLE 6   Find the number of three-colorings of a tree T with three nodes.

Solution   By the formula above, there are $P_T(3) = 3(3-1)^{3-1} = 12$ of these colorings, as listed in Figure 7.8.   ●

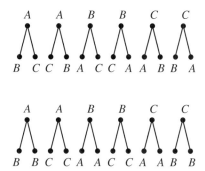

Figure 7.8   Twelve Colorings of a Tree

APPLICATION   *Register Allocation*

Register allocation is an important step in compiling a program, that is, in translating it from a high-level language like C, C⁺⁺, or Pascal into instructions which can be executed by a computer. Although many approaches to good register allocation have been developed, most recent techniques involve algorithms for graph coloring.

When a program is executed by a computer, data may be entered by the user or from files (for example, files on a hard drive). Data manipulations, however, use the computer's internal memory because access to files is very slow. There are often several layers of internal memory including *main memory* (for example, RAM in a PC), and *registers* (fast memory within the central processing unit itself). A register holds a single word of memory. The number of registers usable at any one time is quite limited in most computer architectures.

Computers cannot process source language statements as such. This means, for example, that an expression such as

$$a \star (b + c - 2) + d/(a \star c)$$

must be broken down into binary operations (like $a + b$) and unary operations (like $-c$) before a computer can evaluate it. In each case, the result is stored in a *temporary variable* and we will denote these variables by $t1, t2, t3, \ldots$.

Thus

$$x := a \star (b + c - 2) + d/(a \star c)$$

becomes

$$t1 := b + c$$
$$t2 := t1 - 2$$
$$t3 := a \star t2$$
$$t4 := a \star c$$
$$t5 := d/t4$$
$$t6 := t3 + t5$$
$$x := t6$$

In addition, we need to get the values of a, b, c, d into memory (LOAD), and the value of x back into memory (STORE).

$$\text{LOAD } b \text{ into } t7$$
$$\text{LOAD } c \text{ into } t8$$
$$t1 := t7 + t8$$
$$t2 := t1 - 2$$
$$\text{LOAD } a \text{ into } t9$$
$$t3 := t9 \star t2$$
$$\text{LOAD } d \text{ into } t10$$
$$t4 := t9 \star t8$$
$$t5 := t10/t4$$
$$t6 := t3 + t5$$
$$\text{STORE } t6 \text{ into } x$$

In register allocation, we begin with a set of registers, say four registers $r0, r1, r2, r3$, and we need to store each temporary variable ti in a register rj. The same register can be used for different temporary variables if the times that they are used (their *live ranges*) do not overlap. The live range of a temporary variable extends from the line after it is first defined until the last time it is used. For the code above, the temporary variables are defined (D) and used (U) as follows:

$t1$	$t2$	$t3$	$t4$	$t5$	$t6$	$t7$	$t8$	$t9$	$t10$
						D			
							D		
							U		
D						U	D		
U	D								
	U	D							
								D	
									D
			D					U	U
			U	D				U	
		U		U	D				
					U				

This gives an *interference graph* in which each temporary variable is a node, and there is an edge between two nodes if the live ranges of the corresponding temporary variables overlap. This graph is pictured in Figure 7.9. If we think of the registers as colors, one coloring of this graph is as follows:

$$r0 : t1, t2, t4, t5, t6, t7, t10$$
$$r1 : t8$$
$$r2 : t3$$
$$r3 : t9$$

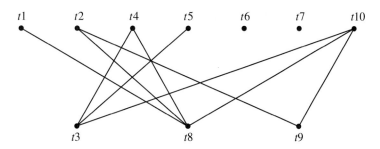

Figure 7.9 A Coloring of an Interference Graph

The code now becomes:

$$\text{LOAD } b \text{ into } r0$$
$$\text{LOAD } c \text{ into } r1$$
$$r0 := r0 + r1$$
$$r0 := r0 - 2$$
$$\text{LOAD } a \text{ into } r3$$
$$r2 := r3 \star r0$$
$$\text{LOAD } d \text{ into } r0$$
$$r0 := r0/r3$$
$$r0 := r0 \star r1$$
$$r0 := r2 + r0$$
$$\text{STORE } r0 \text{ into } x$$

EXERCISES

In these exercises, we give several examples of Hall's marriage problem. If the problem can be solved, give a system of distinct representatives and draw the corresponding matching. If the problem cannot be solved, state why Hall's condition fails.

1. Four boys know four girls. The first boy knows the fourth girl, the second boy knows the first and second girls, the third boy knows the second and third girls, and the fourth boy knows the third girl.

2. Four boys know four girls. The first boy knows the third girl, the second boy knows the first girl and the third girl, the third boy knows the third girl and the fourth girl, and the fourth boy knows the third girl.

3. Four boys know five girls. The first boy knows the first girl and the second girl, the second boy knows the second girl and the third girl, the third boy knows the third girl and the fourth girl, and the fourth boy knows the fourth girl and the fifth girl.

4. Four boys know five girls. The first boy knows the second, third, and fifth girls, the second boy knows the first, second, and fourth girls, the third boy knows the third girl, and the fourth boy knows the third girl.

In the following exercises, you are asked to work out the statement of Menger's theorem. Find an AB-cutset of minimal size, and find the same number of edge-disjoint paths from A to B.

5.

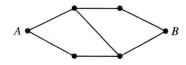

6.

7.

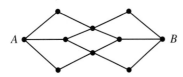

8.

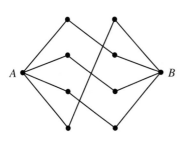

Each of the following graphs is *regular*. That is, each vertex has the same degree *d*. By Brooks' theorem, each of the graphs must have a *d*-coloring. Find such a coloring.

9.

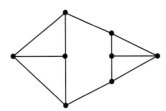

10.

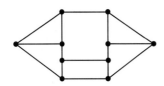

11.

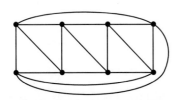

12.

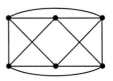

Find a four-coloring of each of the following graphs.

13.

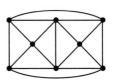

14.

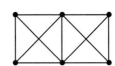

15.

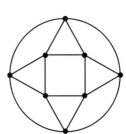

16.

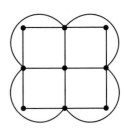

Find ten three-colorings of each of the following trees.

17.

18.

19.

20.

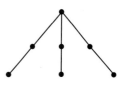

ADVANCED EXERCISES

The graphs which we used in the application about register allocation are called *interval graphs*. One type of interval graph is constructed as follows. Take a finite number of closed intervals of real numbers, and let them correspond to the nodes of a graph G. Join two nodes by an edge if the corresponding intervals have a non-empty intersection.

For each of the following sets of intervals, give the corresponding interval graph, and color it with the minimum possible number of colors.

1. $[1, 3], [2, 4], [2, 3], [0, 5]$
2. $[0, 2], [3, 4], [5, 7], [1, 6]$
3. $[1, 3], [2, 4], [4, 5], [2, 5]$
4. $[0, 1], [2, 3], [4, 5], [6, 7], [0, 7]$
5. $[0, 2], [1, 3], [2, 4], [-1, 0], [4, 5]$

Which of the following graphs is an interval graph?

6.

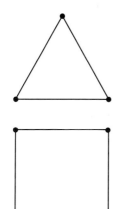

7.

8.

9.

10.

7.2   POSETS AND LATTICES

We should all of us grieve, as you may well believe,

If you never were met with again—

But surely my man, when the voyage began,

You might have suggested it then?

Partially ordered sets give models which are useful in discrete mathematics. We have already seen several examples of partially ordered sets: the *n*-cube is one. Many of the properties of partially ordered sets (*posets*) can be expressed in terms of *chains* and *antichains*. Chains and antichains are related by *Dilworth's theorem*, which is the fundamental result about the structure of posets. The statement of Dilworth's theorem is simple and beautiful, and it is a very pleasant exercise to interpret the statement of this theorem for a given poset.

If we think of the *n*-cube as the set of subsets of $\{1, 2, \ldots, n\}$, it is natural to use the operations of union and intersection, and when working with the set $\text{Div}(n)$ of all divisors of n, it is natural to consider the operations of least common multiple and greatest common divisor. There is a type of partially ordered set called a *lattice* which includes these examples and many others. The *n*-cube is a lattice, and so is $\text{Div}(n)$.

Every Boolean algebra is a lattice, and a *complemented distributive* lattice is a Boolean algebra. Every Boolean algebra can be viewed as an algebra of sets, and every finite Boolean algebra can be identified with the *n*-cube for some *n*. This is one reason why we have mentioned the *n*-cube throughout this book.

We shall discuss the *subspaces* of the *n*-cube, which also form a lattice. It is very likely that this lattice of subspaces will be useful in working out the problems associated with *parallelism* in computer science. We shall show how subspaces of the *n*-cube can be used to give a winning strategy for the game of *Nim*, and how they occur in a solution of the *Kirkman schoolgirl problem*.

If S is a partially ordered set and x and y are elements of S, we shall use the notation $x \le y$ to denote the partial ordering.

Recall that a partial ordering is reflexive, antisymmetric, and transitive. That is, for all $x, y, z \in S$:

  (i) $x \le x$.

 (ii) $x \le y$ and $y \le x$ imply $x = y$.

(iii) $x \le y$ and $y \le z$ imply $x \le z$.

Two elements x and y of a partially ordered set S are *comparable* if $x \le y$ or $y \le x$.

DEFINITION 7.2.1   Let S be a partially ordered set. A subset $T \subset S$ is a *chain* if each pair of elements of T are comparable, and an *antichain* if no pair of elements of T are comparable.

Many properties of a poset can be expressed in terms of chains and antichains. A poset is always a disjoint union of chains. Two examples are the 3-cube (Figure 7.10) and the 4-cube (Figure 7.11).

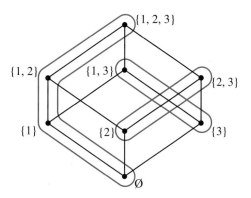

Figure 7.10   The 3-Cube as a Union of Three Chains

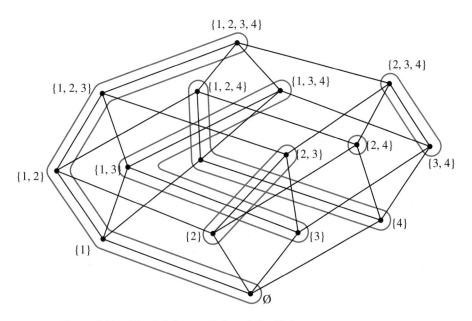

Figure 7.11   The 4-Cube as a Union of Six Chains

The maximal length of an antichain in the 3-cube is 3, and the maximal length of an antichain in the 4-cube is 6. These figures show that, in these two cases, a poset is a disjoint union of as many chains as the maximal length of an antichain in the poset.

If we think of the n-cube as the set of subsets of the set $\{1, 2, \ldots, n\}$, what is the maximal length of an antichain? Figures like 7.10 and 7.11 suggest that this maximal length cannot be greater than the middle binomial coefficient

$$\binom{n}{\lfloor n/2 \rfloor}$$

Sperner's lemma asserts that this is exactly the maximal length of an antichain in the *n*-cube. This result was published by the German mathematician E. Sperner in 1928, and has since become one of the most fundamental results in combinatorics. The elegant proof that we give here was published by D. Lubell in 1966.

Sperner's Lemma   The maximal length of an antichain in the *n*-cube is the middle binomial coefficient

$$\binom{n}{\lfloor n/2 \rfloor}$$

$\frac{6!}{3!3!}$

Proof   Let Γ be an antichain in the *n*-cube, that is, a set of subsets of $\{1, 2, \ldots, n\}$, none of which contains another. Since the middle binomial coefficient is maximal among the binomial coefficients in the *n*th row of the Pascal triangle, it is enough to prove the slightly stronger statement

$$\sum_{A \in \Gamma} \binom{n}{|A|}^{-1} \leq 1$$

For each subset $A \subset \{1, 2, \ldots, n\}$, exactly $|A|!(n - |A|)!$ maximal chains of $\{1, 2, \ldots, n\}$ contain A. Since none of the $n!$ maximal chains of $\{1, 2, \ldots, n\}$ meet Γ more than once, we have

$$\sum_{A \in \Gamma} |A|!(n \doteq |A|)! \leq n!$$

and this proves the lemma.   ●

Notice that the length of a maximal antichain may well be less than the maximal length of an antichain. The n-cube contains many antichains whose length is less than the bound given by Sperner's lemma.

EXAMPLE 1   Make a list of all antichains of maximal length in the 3-cube.

Solution   We view the 3-cube as subsets of $\{1, 2, 3\}$. Notice that $\emptyset$ and $\{1, 2, 3\}$ are antichains which cannot be extended. The list of all maximal antichains in the 3-cube is as follows:

$$\emptyset$$
$$\{1, 2, 3\}$$
$$\{1\}, \{2, 3\}$$
$$\{2\}, \{1, 3\}$$
$$\{3\}, \{1, 2\}$$
$$\{1\}, \{2\}, \{3\}$$
$$\{1, 2\}, \{1, 3\}, \{2, 3\}$$

so there are seven maximal antichains in the 3-cube.   ●

Julius Wilhelm Richard Dedekind (1831–1916) asked: what is the number of antichains in the n-cube? A. Kisielewicz published a solution to this problem in 1988. There are more than 2^{70} antichains in the 8-cube.

In 1948, Robert Palmer Dilworth (1914–) published the paper *A decomposition theorem for partially ordered sets* which contains the following theorem, now known as Dilworth's theorem. We should stress that this is only one of several significant theorems that Dilworth proved.

Dilworth's theorem holds for all posets, but we shall prove it only for finite posets. In Figures 7.10 and 7.11 are two examples of the statement of Dilworth's theorem. The proof of the theorem that we give here was published in 1967 by the Norwegian mathematician H. Tverberg.

Dilworth's Theorem   Let S be a finite poset. Suppose that S contains an antichain with N elements, but no antichain with more than N elements. Then S is the union of N chains.

Proof   The proof is by induction on $|S|$. If $|S| = 0$, the theorem is vacuously true.

Now assume that $|S| > 0$ and let C be a maximal chain in S. That is, C is a subset of S which is a chain, and which is not a proper subset of any other chain contained in S.

We consider two cases:

Case I: No antichain in $S - C$ has N elements. Then, by the inductive assumption, $S - C$ is the union of $N - 1$ chains, and hence S is the union of N chains.

Case II: $S - C$ contains an antichain with N elements. Choose such an antichain

$$A = \{a_1, a_2, \ldots, a_N\}$$

Define

$$S^- = \{x \in S : x \leq a_i\}$$

for some i.

Define also

$$S^+ = \{x \in S : x \geq a_i\}$$

for some i.

Let c^+ be the maximal element of C. If $c^+ \in S^-$, then $c^+ \leq a_i$ for some i, and since c^+ is not in $S - C$, $c^+ < a_i$ for some i. But this contradicts the maximality of C. Therefore, $c^+ \notin S^-$ and by the inductive assumption,

$$S^- = S_1^- \cup S_2^- \cup \ldots \cup S_N^-$$

where $a_i \in S_i^-$ and S_i^- is a chain.

If $x \in S_i^-$, then $x \leq a_j$ for some j. Hence if $a_i < x$, we would have $a_i < a_j$, a contradiction. This means that a_i is the maximal element of S_i^-.

An exactly similar argument shows that

$$S^+ = S_1^+ \cup S_2^+ \cup \ldots \cup S_N^+$$

where S_i^+ is a chain with minimal element a_i. Then

$$S = S^+ \cup S^- = \left(S_1^- \cup S_1^+\right) \cup \ldots \cup \left(S_N^- \cup S_N^+\right)$$

is the union of N chains.   ●

The theory of lattices began in 1897 with the publication of Richard Dedekind's paper *Über Zerlegung von Zahlen durch ihre grössten gemeinsamen Teiler* (On the decomposition of numbers by their greatest common divisors).

Let a and b be two elements of a poset L. The *meet* $a \wedge b$ of a and b is a unique element of L which is less than a and b in the partial order, and is maximal among all elements of L with this property. Similarly, the *join* $a \vee b$ of a and b is a unique element of L which is greater than a and b in the partial order, and is minimal among all elements of L with this property.

In general, the meet or join of two elements in a partially ordered set may not exist.

DEFINITION 7.2.2   A *lattice L* is a partially ordered set in which every two elements have a meet and a join.

In a lattice $a \leq b$, if and only if $a \wedge b = a$ or $a \vee b = b$.
Here are some basic identities which hold in any lattice:

(i)   Idempotence: $a \wedge a = a$ and $a \vee a = a$.

(ii)   Commutativity: $a \wedge b = b \wedge a$ and $a \vee b = b \vee a$.

(iii) Associativity: $(a \wedge b) \wedge c = a \wedge (b \wedge c)$ and $(a \vee b) \vee c = a \vee (b \vee c)$.

(iv)   Absorption: $a \wedge (a \vee b) = a$ and $a \wedge (a \vee b) = a$.

How can a poset fail to be a lattice? In Figure 7.12 are two posets: the first is a lattice and the second is not. These figures are Hasse diagrams: $x \leq y$ if there is a rising path from x to y.

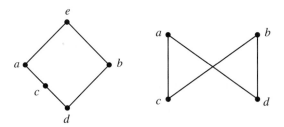

Figure 7.12   A Lattice, and a Poset Which Is Not a Lattice

The first poset in Figure 7.12 is a lattice because each pair of elements has a join and a meet. Notice that the joins and meets of different pairs of elements may be the same. For example, $a \wedge b = c \wedge b = d$, and $a \vee b = c \vee b = e$.

The second poset in Figure 7.12 fails to be a lattice for several reasons. Elements a and b have no join, and elements c and d have no meet. Also, elements c and d have no join, because elements a and b are minimal over both c and d, and the join of two elements in a lattice must be unique.

> **DEFINITION 7.2.3**   A lattice L is
>
> **i.** *Distributive* if $a \wedge (b \vee c) = (a \wedge b) \vee (a \wedge c)$ and $a \vee (b \wedge c) = (a \vee b) \wedge (a \vee c)$ for all elements a, b, and c of L.
>
> **ii.** *Bounded* if there are elements O and I in L such that $O \le a$ and $a \le I$ for all elements a of L.
>
> **iii.** *Complemented* if L is bounded and if for each $a \in L$ there is a $b \in L$ such that $a \wedge b = O$ and $a \vee b = I$.

EXAMPLE 2   The n-cube is a lattice.

Solution   If we think of the n-cube as subsets A and B of $\{1, 2, , \ldots, n\}$, the lattice operations are $A \wedge B = A \cap B$ and $A \vee B = A \cup B$.

The n-cube is a distributive lattice because of the set identities

$$A \cap (B \cup C) = (A \cap B) \cup (A \cap C)$$

$$A \cup (B \cap C) = (A \cup B) \cap (A \cup C)$$

The n-cube is bounded because we can set $O = \emptyset$ and $I = \{1, 2, \ldots, n\}$.

The n-cube is a complemented lattice because $A \cup A^c = \{1, 2, \ldots, n\}$ and $A \cap A^c = \emptyset$.   ●

EXAMPLE 3   If n is a positive integer, Div(n) is a lattice.

Solution   The set Div(n) consists of the divisors of n. The lattice operations are the greatest common divisor $a \wedge b = g.c.d.(a, b)$ and the least common multiple $a \vee b = l.c.m.(a, b)$.

The number-theoretic relations

$$g.c.d.(a, l.c.m.(b, c)) = l.c.m.(g.c.d.(a, b), g.c.d.(a, c))$$

and

$$l.c.m.(a, g.c.d.(b, c)) = g.c.d.(l.c.m.(a, b), l.c.m.(a, c))$$

show that Div(n) is a distributive lattice.   ●

Div(n) is not a complemented lattice in general. For example, in the lattice Div(12), the elements 2 and 6 have no complements, as shown in Figure 7.13.

When is Div(n) a complemented lattice? An integer is *squarefree* if it is not divisible by a square, or equivalently if it is a product of distinct primes. If m is squarefree and m is the product of n distinct primes, then the divisors of m are determined by the set of prime factors of m, and so Div(m) can be identified with the n-cube. Then, Div(m) will be a complemented lattice. If m is not squarefree, like $m = 12$, then Div(m) will always have elements without complements.

In a distributive lattice, complements are unique if they exist.

THEOREM 7.2.1   In a bounded distributive lattice L, complements are unique if they exist.

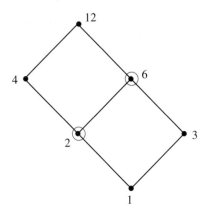

Figure 7.13   Div(12) Is Not a Complemented Lattice

Proof   Let $a \in L$ and let b and c be complements for a, that is, elements of L such that $a \wedge b = O$ and $a \vee b = I$, and also such that $a \wedge c = O$, and $a \vee c = I$.
Then

$$b = b \wedge I = b \wedge (a \vee c) = (b \wedge a) \vee (b \wedge c) = O \vee (b \wedge c) = b \wedge c$$

so $b \leq c$. By symmetry, $c \leq b$, so $b = c$.   ●

A Boolean algebra is a lattice with lattice operations $a \vee b = a + b$ and $a \wedge b = ab$. Complements exist by assumption, and since a Boolean algebra is distributive, these complements are unique. We shall state the following two theorems without proof.

THEOREM 7.2.2   A bounded complemented distributive lattice is a Boolean algebra.

If the lattice is finite, we can make a more precise statement. Notice that a finite lattice must be bounded, because the meet of all elements of the lattice is O and the join of all elements of the lattice is I. An *atom* in a bounded lattice is an immediate successor of O.

THEOREM 7.2.3   If L is a finite complemented distributive lattice with atoms $a_1, a_2, \ldots, a_n$, then each element $b \in L$ can be expressed uniquely in the form

$$b = a_{i_1} \vee a_{i_2} \vee \ldots \vee a_{i_k}$$

and the mapping

$$f(b) = \{i_1, i_2, \ldots, i_k\}$$

is a lattice isomorphism of L with the n-cube realized as subsets of $\{1, 2, \ldots, n\}$.
The *subspaces* of the n-cube form a lattice. Think of the n-cube as the set of binary vectors of length n, Z_2^n. A subspace $V \subset Z_2^n$ is a subset which contains the zero vector $(0, 0, \ldots, 0)$ and is such that if $v \in V$ and $w \in V$, then $v + w \in V$.

EXAMPLE 4   The set of subspaces of Z_2^n is a lattice. The lattice operations are

$$V \wedge W = V \cap W$$

and

$$V \vee W = V + W = \{v + w : v \in V \text{ and } w \in W\}$$ ●

We will illustrate this concept by describing the subspaces of the 3-cube. A subspace of the n-cube always contains 2^d elements, where d is called the *dimension* of the subspace.

There is only one zero-dimensional subspace, $\{(0, 0, 0)\}$, and only one three-dimensional subspace, the whole 3-cube.

The one-dimensional subspaces of the 3-cube consist of the zero vector and one non-zero vector. For example, $\{(0, 0, 0), (1, 0, 0)\}$ is a one-dimensional subspace. Since there are seven non-zero vectors in the 3-cube, there are seven one-dimensional subspaces.

The two-dimensional subspaces consist of the zero vector, two non-zero vectors, and the sum of these two vectors. For example,

$$\{(0, 0, 0), (1, 0, 0), (0, 1, 0), (1, 1, 0)\}$$

is a two-dimensional subspace.

A two-dimensional subspace is determined by the choice of one of the three pairs of non-zero vectors it contains, so there are $\binom{7}{2}/3 = 7$ two-dimensional subspaces.

The one- and two-dimensional subspaces of Z_2^3 can be arranged into a useful configuration. Let us call the one-dimensional subspaces *points*, and the two-dimensional subspaces *lines*. Clearly each line, defined in this way, contains three points. The resulting configuration of points and lines is called the *Fano plane*, after the Italian mathematician Gino Fano (1871–1952), and it is shown in Figure 7.14. The Fano plane is a *projective plane*.

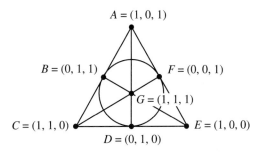

Figure 7.14   The Fano Plane

EXAMPLE 5   A man wants to invite seven guests to seven dinners. His table seats three persons besides himself. He wants to arrange the invitations in such a way that no two guests are ever invited at the same time. Can this be done?

Solution   This problem can be solved by using the design given by the points and lines of the projective plane. The man can invite three guests at a time, corresponding to the three points on each of the seven lines of the projective plane. The resulting solution is given in Figure 7.15.   ●

ABC
AEF
ADG
BDF
BEG
CDE
CFG

Figure 7.15

Using the Fano Plane
to Solve a Problem

What other practical use can be made of subspaces of the n-cube? The idea of regarding subspaces of a space as *states* of a system of some kind is a very powerful one. In quantum mechanics, the state of a physical system is defined in this way. The notion of a state, defined by a certain subspace, can be used to give a winning strategy for the game of Nim.

EXAMPLE 6   The game of Nim is played as follows. A certain number of counters are placed into k piles. Player B takes any number of counters out of any pile, but at least one. Player A then takes any number of counters out of any pile, but at least one. The player who removes the last counter or counters wins.

What strategy will win this game?

Solution   All sets of counters in any game of Nim are divided into two types: *safe states* and *unsafe states* for Player A. If the game begins with a safe state, then Player A will always win with correct play. If the game begins with an unsafe state, then Player B will always win with correct play.

For example, suppose that the game begins with two piles containing two counters each. Player B must remove one or two counters from one of the piles. If Player B removes one counter, then Player A removes one counter from the other pile, and Player B must lose on the next move. If Player B removes both counters from one of the piles, then Player A removes both counters from the other pile and wins. Thus, this is a safe state for Player A.

The safe states for Player A can be completely described as follows. Take the binary expansion of the number of counters in each pile, and then take the corresponding binary vector. Add these vectors. The state is safe for Player A if and only if the sum of these vectors is the zero vector.

For example, if there are two piles each with two counters, the binary expansions are $2 = 10$ and $2 = 10$, and the binary vectors are $(1, 0)$ and $(1, 0)$. Since $(1, 0) + (1, 0) = (0, 0)$, this is a safe state for Player A.

Since the vectors whose binary sums are zero lie in a certain subspace, this solution involves the states of a system.   ●

The lattice of subspaces of the n-cube is unlike the n-cube or $\text{Div}(n)$ in that it is not distributive.

EXAMPLE 7   Prove that the lattice of subspaces of the n-cube is not distributive.

Solution   We will prove this by considering the following three subspaces of the 2-cube:

$$V_1 = \{(0, 0), (1, 0)\}$$
$$V_2 = \{(0, 0), (0, 1)\}$$
$$V_3 = \{(0, 0), (1, 1)\}$$

Now $V_1 \cap V_2 = \{(0, 0)\}$ and $V_1 \cap V_3 = \{(0, 0)\}$, so $(V_1 \cap V_2) + (V_1 \cap V_3) = \{(0, 0)\}$. But $V_2 + V_3 = Z_2^2$, so $V_1 \cap (V_2 + V_3) = V_1$, and therefore the lattice of subspaces of Z_2^2 is not distributive.

Of course it follows that the lattice of subspaces of Z_2^n is not distributive for any $n \geq 2$. ●

The lattice of subspaces of Z_2^n satisfies a weaker condition than distributivity called *modularity*.

DEFINITION 7.2.4 A lattice L is *modular* if whenever $a \leq c$, $a \vee (b \wedge c) = (a \vee b) \wedge c$.

The lattice of subspaces of Z_2^n is modular. Any distributive lattice is modular.

Posets and lattices can be applied to logic, as we have seen. Classical logic is modelled by distributive lattices, but there is strong evidence that modular lattices are better models for logic in many situations. This is already well known to be the case in quantum mechanics.

One of the basic problems in computer science is the problem of *parallelism*. In order to speed up computation, operations are carried out simultaneously. Often, the n-cube is used as a model for the connection of the computers carrying out these operations or for the connection of chips in a single computer. Computer scientists have exploited part of an observation that physicists have recently made, which is that spaces of dimensions 10 and 26 give particularly useful models.

Several computers have already been designed and built which use the 10-cube as a model for the chips which they contain. It seems that the n-cube model will continue to be important in the development of parallel computer architecture and software. It is possible that parallel architecture is already being used in the human brain.

EXAMPLE 8 Describe a structure in the mammalian cortex and give a mathematical model for it.

Solution Recall that the *cortex* is a region in the front part of the brain which is concerned with reasoning. A microsurgical study has shown that in the brains of humans and many other mammals, the cortex contains about 100,000,000 vertical columns of cells, which are connected in a complicated way by neurons.

Perhaps these columns of cells function as chips, and perhaps they are connected in a cubical pattern. Notice that

$$2^{26} = 67,108,864$$

is very close to the observed number of columns of cells. Is it possible that the mammalian cortex contains a model of the n-cube, where n is an integer close to 26? ●

We will end this section with a question:

Can modular lattices, like the lattice of subspaces of the n-cube, be used to study the problems of parallelism?

| APPLICATION | *The Kirkman Schoolgirl Problem* |

We have shown how subspaces of the *n*-cube can be used to solve such problems as determining the safe states of the game of Nim. Now we shall show how these states arise in the solution of a more difficult problem.

The *Kirkman schoolgirl problem* was published by Thomas Penyngton Kirkman (1806–1895) in the *Cambridge and Dublin Mathematical Journal* in 1847. A schoolmistress takes her fifteen pupils for a walk each day, with the schoolgirls arranged in five rows of three. Can the girls be taken for walks on seven days in such a way that each girl has two different companions on each of the seven days?

The problem can be solved, and many methods have been introduced to do this. The problem is discussed in the tenth chapter of *Mathematical Recreations and Essays*, by W. W. Rouse Ball and H. S. M. Coxeter.

Here is one solution to the problem:

Day 1	*Day 2*	*Day 3*	*Day 4*	*Day 5*	*Day 6*	*Day 7*
1, 2, 3	1, 4, 5	1, 6, 7	1, 8, 9	1, 10, 11	1, 12, 13	1, 14, 15
4, 8, 12	2, 8, 10	2, 9, 11	2, 12, 14	2, 13, 15	2, 4, 6	2, 5, 7
5, 10, 15	3, 13, 14	3, 12, 15	3, 5, 6	3, 4, 7	3, 9, 10	3, 8, 11
6, 11, 13	6, 9, 15	4, 10, 14	4, 11, 15	5, 9, 12	5, 11, 14	4, 9, 13
7, 9, 14	7, 11, 12	5, 8, 13	7, 10, 13	6, 8, 14	7, 8, 15	6, 10, 12

Michael Goldberg made a remarkable observation about this solution:

> *Each triplet which occurs in this solution to the Kirkman schoolgirl problem is a safe state for a game of Nim.*

For example, take the triplet 6, 8, 14. In binary, $6 = 0110$, $8 = 1000$, and $14 = 1110$. Adding the corresponding binary vectors we have $(0, 1, 1, 0) + (1, 0, 0, 0) + (1, 1, 1, 0) = (0, 0, 0, 0)$, so this triplet is a safe state.

EXERCISES

For each of the following posets, find the maximal length of an antichain N and verify the assertion of Dilworth's theorem by representing the poset as a union of N chains.

2.

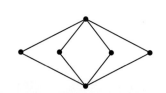

1.

3.

4.

5.

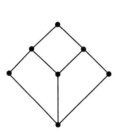

State whether each of the following posets is or is not a lattice, and give a reason. If the poset is a lattice, state whether it is distributive or complemented.

6.

7.

8.

9.

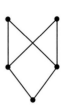

10.

11. Div(30) is a complemented lattice. Find a complement for each element.

12. Draw the Hasse diagram for Div(40). Is Div(40) a complemented lattice?

In each of the following lattices, find an element with two different complements.

13.

14.

15. Draw the Hasse diagram for the lattice of subspaces of the 3-cube.

16. Draw the Hasse diagram for the lattice of subspaces of the 4-cube.

In the following exercises, we give the number of counters in several piles for a game of Nim. Determine if each is a safe state for Player A.

17. 2, 2, 2

18. 2, 2, 2, 2

19. 2, 3, 2, 3

20. 1, 2, 3, 4, 5, 6, 7

ADVANCED EXERCISES

1. Prove the following more precise form of Sperner's lemma: There are exactly one or two antichains of maximal length $\binom{n}{\lfloor n/2 \rfloor}$ in the n-cube, according to whether n is even or odd.

2. Make a list of all the antichains in the 4-cube, and show in particular that, in this case, the assertion in Exercise 1 is correct.

3. Prove that the following two lattices are not distributive:

and

4. Prove that a finite lattice is not distributive if and only if it contains an isomorphic copy of one of the two lattices in Exercise 3.

5. Prove that the lattice of subspaces of the n-cube is modular. That is, if V_1, V_2, V_3 are any three subspaces of the n-cube, and $V_1 \subset V_3$, that $V_1 + (V_2 \cap V_3) = (V_1 + V_2) \cap V_3$.

7.3   CHAPTER SUMMARY AND SUPPLEMENTARY EXERCISES

"The rest of my speech" (he explained to his men)

"You shall hear when I've leisure to speak it.

But the Snark is at hand, let me tell you again!

'Tis your glorious duty to seek it!"

Section 7.1

In this section, we discussed matching and graph coloring. We proved Hall's theorem and stated Menger's theorem. We stated and proved several results about graph coloring. We stated the famous four-color theorem, and gave examples of four-colorings of several planar graphs. We described a significant application of graph coloring to computer science, in register allocation.

Section 7.2

In this section, we stated and proved two basic results about posets, Sperner's lemma and Dilworth's theorem. We described lattices, and proved several results about them. We described several examples of lattices: the n-cube, $\text{Div}(n)$, Boolean algebras, and the lattice of subspaces of the n-cube. We showed how lattice concepts can be used to describe the game of Nim, and a solution to the Kirkman schoolgirl problem.

EXERCISES

1. Seven boys know eleven girls. The first boy knows the first and third girls; the second boy knows the second girl; the third boy knows the second girl and the sixth girl; the fourth boy knows the third girl and the fourth girl; the fifth boy knows the sixth girl; the sixth boy knows the fifth, seventh, and eighth girls; and the seventh boy knows the ninth, tenth, and eleventh girls. Can each boy be married to a girl that he knows?

2. Find a 3-coloring of the following graph.

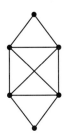

3. Find a 4-coloring of the following graph.

Find the maximum length of an antichain N in each of the following posets, and illustrate Dilworth's theorem by representing the poset as a union of N chains.

4.

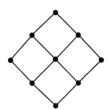

5.

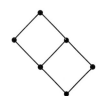

6.

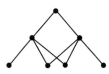

ADVANCED EXERCISES

If G is a finite graph, the chromatic function for G satisfies the following recursion:

$$P_G(k) = P_{G'}(k) - P_{G''}(k)$$

where G' and G'' are the graphs obtained from G by deleting and contracting a single edge e. It is this recursion which implies that $P_G(k)$ is a polynomial in k. Use this recursion to find the chromatic polynomials of the following graphs:

1.

2.

3.

4.

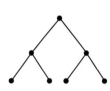

A rooted tree defines a poset. If T is a rooted tree and x and y are labels of nodes of T, then $x \leq y$ if both x and y lie on a path to the root of T, and if the length of the path from x to the root is at least as large as the length of the path from y to the root. For each of the following trees, label each node, and list the ordered pairs of nodes in the partial ordering.

5.

6.

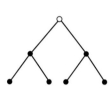

7.

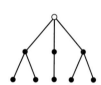

Algorithms 8

8.1 SORTING, SEARCHING, AND LISTING

> *"The thing can be done," said the Butcher, "I think*
>
> *The thing must be done, I am sure.*
>
> *The thing shall be done! Bring me paper and ink,*
>
> *The best there is time to procure."*

The Arab ruler Harun al-Rashid, who figures in the *Arabian Nights*, is believed to have founded the Dar al-Hikma or House of Wisdom, a learned society. A member of this society under Harun al-Rashid's successor caliph al-Ma'mun, who reigned from A.D. 813 to A.D. 833, was Abu Ja'far Muhammad Ibn Musa al-Khwarizmi (before A.D. 800 to sometime after A.D. 847). Al-Khwarizmi dedicated a book to al-Ma'mun with the title *al-Kitab al-muktasar fi hisab al-jabr wa'l-muqabala* (*The Compendious Book on Calculation by Completion and Balancing*).

The Arabic phrase "al-Khwarizmi" means "from Khwarizm," the region south of the Aral sea in Asia: this meant that he or his family came from there. "al-Khwarizmi" was translated into Latin as "algorismus," and this developed into the word *algorithm*. Thus the word *algorithm* is derived from a place name.

The phrase "al-jabr" can be translated as "completion," and refers to the elimination of negative quantities. This developed into the word *algebra*.

We shall give a more precise version of a definition which we gave earlier (in Section 1.3).

> **DEFINITION 8.1.1**   An *algorithm* is a finite set of rules which convert a well-defined *input* into a well-defined *output* in a finite number of steps.

Any algorithm can be analyzed as follows:

I.   First, the input and output of the algorithm should be described.

II.   Next, the finite list of rules of the algorithm should be given.

III.   Finally, a proof should be given that the algorithm will convert the input into the output in a finite number of steps.

EXAMPLE 1   Describe and analyze an algorithm which computes the sum of the first n positive integers.

Solution

I.   The input is the set $\{1, 2, 3, \ldots, n\}$ and the output is the sum of these n integers.

II.   The algorithm has one rule, the formula

$$1 + 2 + 3 + \ldots + n = n(n+1)/2$$

III.   We gave a proof by induction that this rule is correct in Section 1.2.   ●

EXAMPLE 2   Describe and analyze the Euclidean Algorithm.

Solution

I.   The input of the Euclidean Algorithm is a pair of positive integers $\{n, m\}$, with $n \leq m$. The output of the Euclidean Algorithm is the greatest common divisor of these two integers, g.c.d.(m, n).

II.   The Euclidean Algorithm has two rules:

(i)   Set $n = r_0$, and use the Division Algorithm to define a sequence of remainders

$$m = q_1 n + r_1$$
$$n = q_2 r_1 + r_2$$
$$\ldots$$
$$r_{N-2} = q_{N-1} r_{N-1} + r_N$$

(ii)   When $r_{N+1} = 0$ stop, and set $r_N = $ g.c.d.(m, n).

III.   Since the remainders form a strictly decreasing sequence of non-negative integers, the sequence of remainders must terminate. It cannot terminate with a positive integer because then another smaller remainder will be defined the next time (i) is applied. Therefore some remainder must be zero, so the algorithm must terminate after a finite number of steps.

Any integer d which divides both m and n must divide all the remainders, and in particular r_N, so g.c.d.(m, n) divides r_N. Since $r_{N-1} = q_N r_N$, r_N divides r_{N-1} and hence r_{N-2}. Continuing backwards through the equations above, we see that r_N divides all the remainders, and in particular r_N divides n and m. Therefore r_N divides g.c.d.(m, n), so $r_N =$ g.c.d.(m, n), and so the Euclidean Algorithm terminates with the proper output.   ●

A basic problem in computer science involves the processing of large amounts of data. Data is usually presented at first in a rather chaotic way. In order to process it, the data must first be *sorted* somehow, and then it will be necessary to *search* through the sorted list to find the data values which are of interest. Again, given a set of objects, we may want to list them in an efficient way. Thus we have three related problems:

The Sorting Problem: Given a set of data values, to sort them into a list.

The Searching Problem: Given a sorted list, to search through it quickly to find those entries in the list which are of interest.

The Listing Problem: Given a set of objects, to list them in an efficient way.

There are sorting algorithms and searching algorithms which solve the first two problems, and listing algorithms which solve the third. Because sorting must occur before searching, sorting algorithms are in a certain sense the most fundamental. We will give two examples of sorting algorithms, Mergesort and Quicksort; then a searching algorithm, Binary Search; and then several examples of listing algorithms.

We shall apply Mergesort and Quicksort to lists of numbers. There is very little loss of generality here, since data can usually be represented as numbers of some kind. For example, the problem of listing 10,000 students in a large university is the equivalent to the problem of listing their Social Security numbers.

In describing Mergesort it is convenient to assume that the input is a list of numbers whose length is a power of two. This is not a significant restriction, because we can add elements any list until the length of the list is a power of two, and it can be shown that this assumption does not significantly affect the rate at which the algorithm runs.

Mergesort   Given a list of 2^k positive integers, this algorithm sorts the list into a list of positive integers in increasing order.

(i)   Divide the list of integers into successive pairs, and order each pair of integers in increasing order.

(ii)   If the successive 2^r-tuples of integers have been ordered in increasing order, merge the first pair of 2^r-tuples, the next pair of 2^r-tuples, and so on, into 2^{r+1}-tuples and order each 2^{r+1}-tuple in increasing order.

(iii)   Repeat (ii) until the whole list is sorted in increasing order.

EXAMPLE 3   Sort the following list using Mergesort:

$$5, 2, 6, 1, 2, 9, 7, 4$$

Solution   We shall indicate the effect of each step of the algorithm with parentheses:

$$(2, 5)(1, 6)(2, 9)(4, 7)$$
$$(1, 2, 5, 6)(2, 4, 7, 9)$$
$$(1, 2, 2, 4, 5, 6, 7, 9)$$

Quicksort is another basic sorting algorithm. The implementation of Quicksort depends on a Partition Algorithm, which partitions the input list of numbers into two sublists. We shall call the input list an *array*, and the two sublists *subarrays*. Again we shall assume that the numbers to be listed are positive integers.   ●

PARTITION ALGORITHM   Given an array

$$a(1), a(2), \ldots, a(r)$$

of positive integers, this algorithm will partition the array into two subarrays, in such a way that the elements of the first subarray are all less than or equal to a_1, and the elements of the second subarray are all greater than or equal to a_1.

i. Introduce two pointers i and j, and start i to the left of the array and j to the right of the array.

ii. Set $x = a(1)$.

iii. Decrease j until $a(j) \leq x$.

iv. Increase i until $a(i) \geq x$.

v. If $i \lessgtr j$, exchange $a(i)$ and $a(j)$.

vi. If $i < j$, go to step (iii). *stop when* $i = j$

The result is a sorted array partitioned into two subarrays, in such a way that the first subarray terminates with the jth element, this being the final value of the j pointer.

Quicksort   This algorithm sorts an array of positive integers into an array with elements in increasing order.

(i)   If the array has only one element, do nothing.

(ii)   If the array has more than one element, use the Partition Algorithm, and apply Quicksort recursively to each of the subarrays given by this algorithm.

EXAMPLE 4   Apply the Partition Algorithm to the array:

$$5, 2, 6, 1, 2, 9, 7, 4$$

4 2 6 1 2 9 7 5
4 2 2 1 6 9 7 5   5 9 7 6
5 7 9 6
6 4

Solution   We shall give the results of iterating the Partition Algorithm, and indicate the two subarrays at the end with parentheses.

$$5, 2, 6, 1, 2, 9, 7, 4$$
$$5, 2, 6, 1, 2, 9, 7, 4$$
$$4, 2, 6, 1, 2, 9, 7, 5$$
$$4, 2, 6, 1, 2, 9, 7, 5$$
$$4, 2, 2, 1, 6, 9, 7, 5$$
$$(4, 2, 2, 1)(6, 9, 7, 5)$$ ●

Quicksort continues sorting by applying the Partition Algorithm to each of these subarrays. Several more steps are necessary for each subarray, although the first subarray 4, 2, 2, 1 is put in order 1, 2, 2, 4 after one more application of the Partition Algorithm.

Once a list has been put in order by a sorting algorithm, it will often be necessary to search through it for an entry in which one is interested. For example, if we have a list of numbers in increasing order, we may wish to search through it for a given number, which may or may not be on the list.

We shall describe one searching algorithm, *binary search*. In this algorithm we shall use the following notation: if x is a real number, the *floor* of x, $\lfloor x \rfloor$, is the largest integer less than or equal to x.

Binary Search   This algorithm searches through an ordered list

$$a(1) \leq a(2) \leq \ldots \leq a(r)$$

for a number a which may or may not be on the list.

(i) Define two counters i and j, and set $i = 1$ and $j = r$ as initial values.

(ii) If $j < i$, the algorithm terminates without finding a. If not, set $k = \lfloor (i + j)/2 \rfloor$.

(iii) If $a < a(k)$, go to (iv). If $a > a(k)$, go to (v). If $a = a(k)$ the algorithm finds a.

(iv) Set $j = k - 1$ and repeat (ii).

(v) Set $i = k + 1$ and repeat (ii).

EXAMPLE 5   Search for 43 on the list

$$10, 15, 20, 28, 37, 39, 43, 58, 67, 77, 81, 99$$

Solution   We shall give the steps of the binary search algorithm.

(i) There are twelve numbers on the list, so we set $i = 1$ and $j = 12$. Then $k = \lfloor (1 + 12)/2 \rfloor = 6$.

(ii) $a(6) = 39 < 43$, so set $i = 6 + 1 = 7$.

(iii) Now $k = \lfloor (7 + 12)/2 \rfloor = 9$.

(iv) $a(9) = 67 > 43$, so set $j = 8$.

(v) Now $k = \lfloor (7 + 8)/2 \rfloor = 7$.

(vi) $a(7) = 43$, so the algorithm terminates by finding 43. ●

A *combinatorial algorithm* is an algorithm which lists objects of a given kind. Algorithms which list integer vectors are particularly important; they give lists of permutations, for example. Many other combinatorial algorithms begin by associating integer vectors to the objects to be listed and then listing the integer vectors. We shall describe four algorithms for listing integer vectors. These actually belong to a whole family of such algorithms which we shall describe in the Advanced Exercises of the next section.

DEFINITION 8.1.2   Given a partially ordered set X, with the ordering denoted by $\leq$, we can define four partial orderings on X^n, the set of ordered n-tuples of elements of X, as follows (we will use the symbol $a \leq b$ to denote the partial ordering on X, and $a < b$ to mean that $a \leq b$ and $a \neq b$):

i. *lexicographical order*, or *lex order*:

$$(a_1, a_2, \ldots, a_n) < (b_1, b_2, \ldots, b_n)$$

in *lex order* if for some integer k with $0 \leq k < n$ we have $a_1 = b_1, a_2 = b_2,$ $\ldots$, and $a_k = b_k$, but $a_{k+1} < b_{k+1}$.

ii. *reverse lexicographical order*, or *revlex order*:

$$(a_1, a_2, \ldots, a_n) < (b_1, b_2, \ldots, b_n)$$

in *revlex order* if for some integer k with $0 \leq k < n$ we have $a_n = b_n,$ $a_{n-1} = b_{n-1}, \ldots$, and $a_{n-k+1} = b_{n-k+1}$, but $a_{n-k} < b_{n-k}$.

iii. If X is the set of non-negative integers define *graded lexicographical order*, or *grlex order*:

$$(a_1, a_2, \ldots, a_n) < (b_1, b_2, \ldots, b_n)$$

in *grlex order* if $a_1 + a_2 + \ldots + a_n < b_1 + b_2 + \ldots + b_n$ or if $a_1 + a_2 + \ldots + a_n = b_1 + b_2 + \ldots + b_n$ and $(a_1, a_2, \ldots, a_n) < (b_1, b_2, \ldots, b_n)$ in lex order.

iv. If X is the set of non-negative integers define *graded reverse lexicographical order*, or *grevlex order*:

$$(a_1, a_2, \ldots, a_n) < (b_1, b_2, \ldots, b_n)$$

in *grevlex order* if $a_1 + a_2 + \ldots + a_n < b_1 + b_2 + \ldots + b_n$ or if $a_1 + a_2 + \ldots + a_n = b_1 + b_2 + \ldots + b_n$ and $(a_1, a_2, \ldots, a_n) < (b_1, b_2, \ldots, b_n)$ in revlex order.

EXAMPLE 6   List the permutations of $\{1, 2, 3\}$ in lex order.

Solution

$$(1, 2, 3)$$
$$(1, 3, 2)$$

$$(2, 1, 3)$$
$$(2, 3, 1)$$
$$(3, 1, 2)$$
$$(3, 2, 1)$$

EXAMPLE 7   List the permutations of $\{1, 2, 3\}$ in revlex order.

Solution

$$(3, 2, 1)$$
$$(2, 3, 1)$$
$$(3, 1, 2)$$
$$(1, 3, 2)$$
$$(2, 1, 3)$$
$$(1, 2, 3)$$

Notice that this list is *not* the corresponding lex list in reverse order. There is a simple relation between lists in lex order and lists in revlex order. Can you describe it?

The orderings which we have defined give orderings for integer vectors, and they can also be used to list subsets of an ordered set. For example, to list the k-subsets of $\{1, 2, 3, \ldots, n\}$, order the elements of the subset in increasing order, view the resulting subset as an ordered k-tuple, and use any of the orderings which we have just defined.

EXAMPLE 8   List the 3-element subsets of $\{1, 2, 3, 4, 5\}$ in lex order, revlex order, grlex order, and grevlex order.

Solution   We give the lists below. Notice that the lists are all different. Notice, in particular, the difference between the list in lex order and the list in grlex order.

Lex Order	Revlex Order	Grlex Order	Grevlex Order
$\{1, 2, 3\}$	$\{1, 2, 3\}$	$\{1, 2, 3\}$	$\{1, 2, 3\}$
$\{1, 2, 4\}$	$\{1, 2, 4\}$	$\{1, 2, 4\}$	$\{1, 2, 4\}$
$\{1, 2, 5\}$	$\{1, 3, 4\}$	$\{1, 2, 5\}$	$\{1, 3, 4\}$
$\{1, 3, 4\}$	$\{2, 3, 4\}$	$\{1, 3, 4\}$	$\{1, 2, 5\}$
$\{1, 3, 5\}$	$\{1, 2, 5\}$	$\{1, 3, 5\}$	$\{2, 3, 4\}$
$\{1, 4, 5\}$	$\{1, 3, 5\}$	$\{2, 3, 4\}$	$\{1, 3, 5\}$
$\{2, 3, 4\}$	$\{1, 4, 5\}$	$\{1, 4, 5\}$	$\{2, 3, 5\}$
$\{2, 3, 5\}$	$\{2, 3, 5\}$	$\{2, 3, 5\}$	$\{1, 4, 5\}$
$\{2, 4, 5\}$	$\{2, 4, 5\}$	$\{2, 4, 5\}$	$\{2, 4, 5\}$
$\{3, 4, 5\}$	$\{3, 4, 5\}$	$\{3, 4, 5\}$	$\{3, 4, 5\}$

Error in this table

The combinatorial algorithms which we have described are very useful for listing permutations and combinations. Each of these algorithms can be used to give an ordering for labelled trees, using the Prüfer code. Several theorems in combinatorics about subsets of a set are proved by first choosing an ordering for the subsets: sometimes revlex order is used.

In computer algebra, the orderings of integer vectors which we have given are used to give orderings for monomials in several indeterminants

$$X_1^{n_1} X_2^{n_2} \ldots X_r^{n_r}$$

which can be used, for example, to give an analog of the Euclidean Algorithm for polynomials in several variables called *Buchberger's algorithm*. This algorithm is used to prove the *Gröbner Basis Algorithm*, which is fundamental in computer algebra. When computer algebra programs were first introduced, the grlex order was used to formulate these algorithms. The grevlex order is now used instead, because it was discovered that most algorithms in computer algebra run faster if the grevlex order is used.

APPLICATION   *The Man, Dog, Goat, and Cabbage Problem*

The Standard Gray Code is a basic example of a listing algorithm. Gray codes and their variants are very useful, and they occur in applications in ways that can be unexpected. To illustrate this we shall discuss the Man, Dog, Goat, and Cabbage Problem.

A man must carry a dog, a goat, and a box of cabbages across a river in a boat which will carry only the man and one other object. The man cannot leave the dog with the goat, or the goat with the box of cabbages.

There are two solutions to the problem with the minimum number of crossings, which is seven. In one, the man crosses with the goat, leaves the goat, returns and crosses with the dog, returns with the goat, crosses with the box of cabbages, and finally returns and crosses with the goat.

Can you find the other solution?

Alcuin of York (A.D. 735–804) is said to have stated this problem in a letter to Charlemagne, his student. Versions of the problem have been given by the Tigre of Ethiopia, the Bamileke of Cameroon, and the Kabijlie of Algeria, among many others.

Let us use the symbols M = man, D = dog, G = goat, and C = cabbage. Let us represent the solution to the problem which we have just described by symbols representing whatever objects are left on the near bank of the river at each crossing:

$$MDGC \quad DC \quad MDC \quad C \quad MGC \quad G \quad MG \quad \emptyset$$

The solution to the problem is given by a code on these four letters:

$$MDGC$$
$$DC$$
$$MDC$$
$$C$$
$$MGC$$
$$G$$

$$M\,G$$
$$\emptyset$$

Now let us replace these codewords by binary vectors of length four, with the entries denoting M, D, G, C in order, and where 1 means that the symbol is present and 0 means that the symbol is not present. This gives the code

$$(1, 1, 1, 1)$$
$$(0, 1, 0, 1)$$
$$(1, 1, 0, 1)$$
$$(0, 0, 0, 1)$$
$$(1, 0, 1, 1)$$
$$(0, 0, 1, 0)$$
$$(1, 0, 1, 0)$$
$$(0, 0, 0, 0)$$

Notice that the first two vectors in this list and the last two vectors have an even number of 1s, and that the middle four vectors have an odd number of 1s. Now deform the middle four vectors as follows:

Replace $(1, 1, 0, 1)$ *by* $(1, 1, 0, 0)$.
Replace $(0, 0, 0, 1)$ *by* $(1, 0, 0, 1)$.
Replace $(1, 0, 1, 1)$ *by* $(0, 0, 1, 1)$.
Replace $(0, 0, 1, 0)$ *by* $(0, 1, 1, 0)$.

Write the list again:

$$(1, 1, 1, 1)$$
$$(0, 1, 0, 1)$$
$$(1, 1, 0, 0)$$
$$(1, 0, 0, 1)$$
$$(0, 0, 1, 1)$$
$$(0, 1, 1, 0)$$
$$(1, 0, 1, 0)$$
$$(0, 0, 0, 0)$$

There are eight binary vectors of length four with an even number of 1s, and they form a linear subcode $E \subset Z_2^4$. All of these vectors occur in this list, and each pair of successive vectors differs in exactly two entries.

There is a one-to-one correspondence

$$f : Z_2^3 \to E$$

given by $f((a_1, a_2, a_3)) = (a_1, a_2, a_3, a_1 + a_2 + a_3)$, and the list of elements of E which we have given above becomes the Standard Gray Code for binary vectors of length three (except that the entries have been permuted in a circular fashion). We conclude that

The Standard Gray Code can be used to solve the Man, Dog, Goat, and Cabbage Problem.

EXERCISES

Sort the following lists using Mergesort, and then using Quicksort.

1. 4, 1, 7, 3, 2, 5, 9, 8

2. 9, 7, 2, 3, 9, 4, 8, 5

3. 1, 2, 5, 4, 9, 3, 7, 6

4. 15, 12, 1, 5, 7, 9, 6, 4, 19, 3, 16, 2, 8, 11, 14, 18

5. 7, 11, 13, 19, 18, 1, 5, 12, 4, 16, 17, 3, 15, 2, 14, 10

Use Binary Search to search for the number 37 on each of the following lists.

6. 2, 10, 16, 25, 31, 37, 40, 47

7. 5, 7, 8, 9, 37, 50, 90, 100

8. 10, 19, 23, 25, 35, 39, 40, 50

9. 7, 12, 17, 20, 30, 31, 37, 40, 50, 55, 62, 70, 71, 80, 89, 99

10. 1, 2, 3, 10, 14, 15, 20, 21, 25, 30, 36, 40, 41, 50, 60, 70

11. List the ordered pairs of elements of $\{1, 2, 3, 4\}$ in lex order.

12. List the ordered pairs of elements of $\{1, 2, 3, 4\}$ in revlex order.

13. List the ordered pairs of elements of $\{1, 2, 3, 4\}$ in grlex order.

14. List the ordered pairs of elements of $\{1, 2, 3, 4\}$ in grevlex order.

15. List the twenty three-element subsets of $\{1, 2, 3, 4, 5, 6\}$ in lex order.

16. List the three-element subsets of $\{1, 2, 3, 4, 5, 6\}$ in revlex order.

17. List the three-element subsets of $\{1, 2, 3, 4, 5, 6\}$ in grlex order.

18. List the three-element subsets of $\{1, 2, 3, 4, 5, 6\}$ in grevlex order.

ADVANCED EXERCISES

Remember the Tower of Hanoi problem for n disks and three pegs, which we described at the end of Section 1.2. The Standard Gray Code can be used to give a solution to this problem. Recall the Standard Gray Code for binary vectors of length three:

$$(0, 0, 0)$$
$$(0, 0, 1)$$
$$(0, 1, 1)$$
$$(0, 1, 0)$$
$$(1, 1, 0)$$
$$(1, 1, 1)$$
$$(1, 0, 1)$$
$$(1, 0, 0)$$

1. The Standard Gray Code for binary vectors of length three can be used to solve the Tower of Hanoi problem with three disks and three pegs. The idea is to indicate which disk is to be moved according to which coordinate changes as one moves through the Gray code. Give a precise description of the solution.

2. The Standard Gray Code for binary vectors of length n can be used to solve the Tower of Hanoi problem with n disks and three pegs. Give a precise description of the solution.

COMPUTER EXERCISES

Write programs to implement Mergesort and Quicksort, and use them to sort the following lists.

1. 21, 12, 14, 5, 47, 50, 33, 17, 1, 94, 51, 77, 85, 15, 19, 23, 59, 16, 93, 20, 55, 18, 12, 35, 28, 66, 24, 90, 80, 70, 44, 56, 34, 27, 30, 31, 18, 67, 73, 58.

2. 14, 92, 15, 95, 19, 20, 56, 40, 40, 40, 55, 81, 73, 82, 99, 37, 39, 50, 87, 90, 39, 33, 12, 2, 3, 57, 87, 37, 15, 16, 35, 39, 61, 73, 82, 83, 10, 11, 7, 8.

3. Use binary search to search for 35 on the list in Exercise 1.

4. Use binary search to search for 34 on the list in Exercise 1.

5. Use binary search to search for 61 on the list in Exercise 2.

Write a program which will list ordered or unordered subsets of a finite set in lex order, revlex order, grlex order, or grevlex order. Use your program to print the following lists.

6. List the four-element subsets of $\{1, 2, 3, 4, 5, 6, 7, 8, 9, 10\}$ in lex order.

7. List the four-element subsets of $\{1, 2, 3, 4, 5, 6, 7, 8, 9, 10\}$ in revlex order.

8. List the four-element subsets of $\{1, 2, 3, 4, 5, 6, 7, 8, 9, 10\}$ in grlex order.

9. List the four-element subsets of $\{1, 2, 3, 4, 5, 6, 7, 8, 9, 10\}$ in grevlex order.

8.2   GRAPH ALGORITHMS

The Beaver brought paper, portfolio, pens,

And ink in unfailing supplies:

While strange creepy creatures came out of their dens,

And watched them with wondering eyes.

Algorithms involving graphs are of basic importance in discrete mathematics. We have already stated several important graph algorithms in Chapter 4.

In this section, we shall discribe three algorithms involving graphs. First we shall recall the statement of Prim's Algorithm, which we considered in Chapter 4. Then we shall discuss a closely related algorithm, *Kruskal's Algorithm*, which was described by J. B. Kruskal in a paper published in 1956. Kruskal's Algorithm is an example of a *greedy algorithm*: a greedy algorithm repeatedly chooses the nearest, or farthest, or largest, or smallest object under consideration.

We shall contrast Prim's Algorithm and Kruskal's Algorithm.

Another class of graph algorithms finds shortest paths of various types in a graph. A basic example is *Dijkstra's Algorithm*, which was described by E.W. Dijkstra in a paper published in 1959.

As an application of the ideas in this section, we shall describe *network flows*, which have numerous applications. We shall state the basic *Max-Flow-Min-Cut Theorem*. This theorem is closely related to the theorems which we discussed in the previous chapter.

PRIM'S ALGORITHM   The input of this algorithm is a connected weighted graph G, with non-negative weights. The output is a minimum-weight spanning tree T of G.

   i. First choose any node v of G.

  ii. If the tree T has been constructed, add a new edge e to T such that one node of e belongs to T, the other does not, and the weight $w(e)$ is minimal subject to this condition.

 iii. Continue until (ii) is no longer possible.

Lemma   Prim's Algorithm terminates by finding a minimum-weight spanning tree T of G.

Proof   We will prove by induction on the number of nodes $n = |V(T)|$ of T that T is a subtree of G of minimum weight containing the node v.

The statement is trivially true if $V(T) = \{v\}$, and this starts the induction. Assume by induction that the statement is true for some $n = |V(T)| < |V(G)|$. The algorithm adds an edge e to T in such a way that $|V(T \cup \{e\})| = |V(T)| + 1$, and that the weight of e is minimal subject to this condition. This clearly implies that $T \cup \{e\}$ has minimum weight among the subtrees of G containing v. Therefore, when $n = |V(G)|$, T must be a spanning tree of G of minimum weight. ●

KRUSKAL'S ALGORITHM   The input of this algorithm is a finite weighted graph G. The output is a minimum-weight spanning forest of G.

   i. Choose an edge e_1 of G of minimum weight, and regard this edge and its two vertices as a forest F.

  ii. If the forest F has been constructed, with edges $e_1, e_2, \ldots, e_n$, add a new edge e_{n+1} to F in such a way that the addition of e_{n+1} does not create a cycle, and such that $w(e_{n+1})$ is minimal subject to this condition.

 iii. Continue with (ii) as long as possible.

Kruskal's Algorithm terminates by producing a spanning tree if and only if G is connected.

COROLLARY 8.2.1   A graph G is connected if and only if it has a spanning tree.

In Figure 8.1 we have a weighted graph, and the sequence of subtrees that Kruskal's algorithm generates, ending in a minimum-weight spanning tree. In Figure 8.2 we show how Prim's Algorithm gives a minimum-weight spanning tree for the same graph. Notice that the steps in the two algorithms are different, but that they give the same minimum-weight spanning tree as output.

In this example it happens that all the forests constructed by Kruskal's Algorithm are trees. This need not happen. In Figure 8.3 we have an example where Kruskal's Algorithm gives a minimum-weight spanning tree, but where one of the forests the

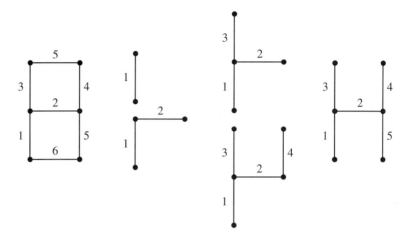

Figure 8.1   An Application of Kruskal's Algorithm

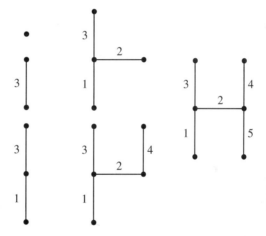

Figure 8.2   Prim's Algorithm Applied to the Same Graph

algorithm constructs is not a tree. Prim's Algorithm, applied to the same graph, in Figure 8.4, gives the same minimum-weight spanning tree, but the algorithm gives trees at every stage.

We have omitted a detail in the implementation of these two algorithms. When choosing an edge of minimum weight which does not create a cycle in Kruskal's Algorithm, or an edge of minimum weight in Prim's Algorithm, it may happen that there is more than one edge of minimum weight. As we have described the two algorithms, one simply chooses any appropriate edge of minimum weight.

It is possible to introduce a convention for breaking ties among edges of minimum weight in these two algorithms. It can be proved, for example, that given any minimum-weight spanning tree T of a weighted graph, there is a convention for breaking ties such that Kruskal's Algorithm will have T as output.

EXAMPLE 1   A subway system is to be constructed in a certain city. The positions of the subway stations and the distances between the stations are known. It is desired to build tracks

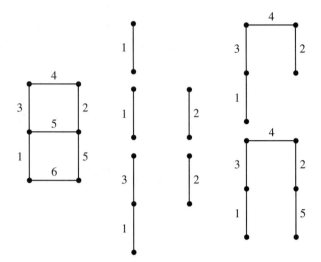

Figure 8.3   An Application of Kruskal's Algorithm

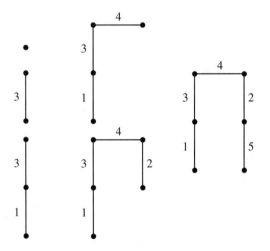

Figure 8.4   Prim's Algorithm Applied to the Same Graph

in such a way that any station can be reached from any other station, and such that the total amount of track is minimal. How can this be done?

Solution   We can model the configuration of stations, and possible tracks between them, by a connected weighted graph. By applying Kruskal's Algorithm or Prim's Algorithm to this graph we can find a minimal-weight spanning tree, and this gives the desired configuration of tracks.   ●

Another basic graph algorithm was found by the Dutch mathematician Edsger Wybe Dijkstra in 1959. It finds the shortest path between two nodes in a weighted graph. If *A* and *B* are nodes in a connected simple weighted graph *G*, where all the weights are non-negative real numbers, the *length* $L(A, B)$ of a path from A to B is the sum of the weights of the edges of the path. A *shortest path* joining A and B is a path whose weight is minimal among all paths joining A and B.

> **DIJKSTRA'S ALGORITHM**   This algorithm finds a shortest path between two nodes A and B in a connected simple weighted undirected graph G.
> **i.** Begin a path with node A.
> **ii.** Construct a set of nodes as follows. First, set $S_1 = \{A\}$. If S_k has been defined, label all nodes adjacent to elements of S_k with the shortest path to that node consisting of elements of S_k, except for that node. If w is the node with the smallest label, set $S_{k+1} = S_k \cup \{w\}$.
> **iii.** When $w = B$, stop.

We give an example of the iteration of Dijkstra's Algorithm in Figure 8.5. The relabelling is crucial to the implementation of the algorithm. For each iteration of the algorithm, we circle the node which has been added to S_k.

As with Kruskal's Algorithm, and with many other algorithms, there are details of the implementation of Dijkstra's Algorithm which can only be explained if one knows more about data structures. For example, there is a data structure called the *Fibonacci Heap* which can be used to improve the implementation of Dijkstra's algorithm. Although we cannot describe the Fibonacci Heap here, Lamé's Theorem is the basic result which is needed to do this.

APPLICATION   *Network Flows*

A basic application of our ideas about graphs is to the study of *networks*. Intuitively, a network is a system through which objects of any kind can be transmitted. The objects might be bits passing through a communications network, oil through a network of pipes, or trucks passing along a system of roads.

> **DEFINITION 8.2.1**   A *network* N is a weighted digraph such that the weight $c(e)$ of each edge e is a non-negative real number. $c(e)$ is called the *capacity* of the edge e.

We shall assume that a network has one initial node A (A is an initial node of each edge containing it), and one terminal node B (B is the terminal node of each edge containing it).

For each node v of a network, the *in-degree* of v is the sum of the weights of the edges coming into v, and the *out-degree* of v is the sum of the weights of the edges coming out of v.

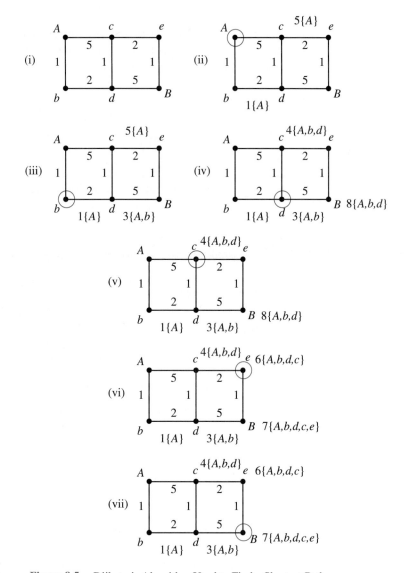

Figure 8.5   Dijkstra's Algorithm Used to Find a Shortest Path

DEFINITION 8.2.2   If N is a network, a *flow in N* is a non-negative function $f(e)$ on the edges e of N such that:

i. $f(e) \leq c(e)$ for all edges e of N.

ii. When N is regarded as a network using f instead of c, the in-degree and out-degree of each node of N are equal, except for the nodes A and B.

Intuitively, the capacity $c(e)$ of an edge is the maximum number of bits that can flow through part of a communications network, the maximum amount of oil that will flow through part of a pipeline, and so on.

It is easy to prove that given a flow f, the sum of the $f(e)$ for the edges leaving A is equal to the sum of the $f(e)$ for the edges entering B. This number is called the *value* of the flow. A flow is *maximal* if it is as large as possible for the given network.

A *cut set* C for a network N is a set of edges of N such that any path from A to B contains an edge in C, or equivalently such that removing the edges of C from N will disconnect A from B. The *capacity* of a cut C is the sum of the capacities $c(e)$ for edges e in C. A cut set C is *minimal* if the capacity of C is minimal among all cut sets of N.

The following theorem was proved by L. R. Ford, Jr., and D. R. Fulkerson, in a paper which was published in 1956.

The Max-Flow-Min-Cut Theorem   If N is a network, the maximal flow through N is equal to the minimal capacity of a cut C of N.

Given a network N, there is an algorithm for finding a minimal cut C for N, but we shall not state it. For simple networks, minimal cuts can be found by inspection.

For example, in Figure 8.6 we have a network N, where the capacities $c(e)$ of the edges are the numbers next to each edge.

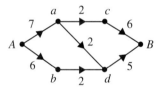

Figure 8.6   A Network

By inspection, we find that a minimal cut set C for N is

$$C = \{(a, c), (a, d), (b, d)\}$$

which has capacity $2+2+2 = 6$, so the maximal flow from A to B through this network is 6.

The Max-Flow-Min-Cut Theorem is closely related to the theorems that we discussed in the previous chapter: Hall's Theorem, Menger's Theorem, and Dilworth's Theorem, and they in turn are closely related to each other. Hall's Theorem is logically equivalent to Menger's Theorem. Menger's Theorem, with some additional arguments, can be used to prove the Max-Flow-Min-Cut Theorem. The Max-Flow-Min-Cut Theorem can be used to prove Hall's Theorem. Hall's Theorem follows from Dilworth's Theorem.

The Max-Flow-Min-Cut Theorem has been applied to the routing of logging trucks, to the routing of garbage trucks in New York City, and in hundreds of other ways.

The theory of network flows really began with the work of the German physicist Kirchhoff. It was while Kirchhoff was working on electrical networks that he proved the Matrix-Tree Theorem, which we stated in Chapter 4. *Kirchhoff's Laws* for electrical circuits involve an early notion of flow through a network. It is an interesting question, which seems never to have been studied, to what extent Kirchhoff anticipated the Max-Flow-Min-Cut Theorem.

EXERCISES

Use Kruskal's Algorithm and then Prim's Algoritm to find a minimum-weight spanning tree for each of the following graphs.

1.

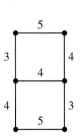

2.

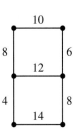

3.

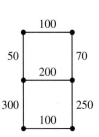

4.

5.

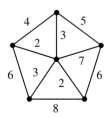

Use Dijkstra's Algorithm to find a shortest path between nodes A and B in the following graphs.

6.

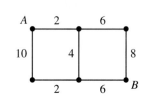

7.

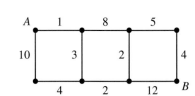

8.

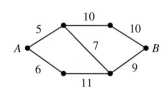

9.

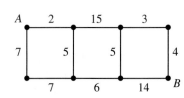

Show that Kruskal's Algorithm will give the indicated minimum-weight spanning tree of the given graph if it is implemented suitably (that is, if appropriate edges of minimum weight are chosen at each step).

10.

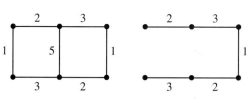

11.

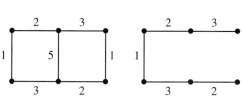

12.

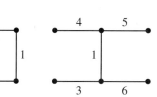

Find a minimal cut by inspection for each of the following networks, and determine the maximal flow from A to B.

13. *Has incorrect answer in back*

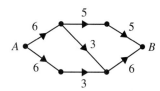

14.

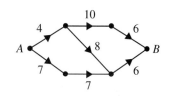

15.

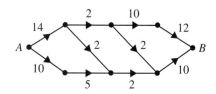

8.3   THE COMPLEXITY OF AN ALGORITHM

> *The Beaver had counted with scrupulous care,*
>
> *Attending to every word:*
>
> *But it fairly lost heart, and outgrabe in dispair,*
>
> *When the third repetition occurred.*

A basic problem in computer science is the problem of determining whether an algorithm will terminate, and if it does, how long it will take to terminate. The second of these two problems is the problem of determining the *complexity* of an algorithm.

> **DEFINITION 8.3.1**   The *complexity* of an algorithm is the number of steps that the algorithm takes to terminate, as a function of the length of the *input* of the algorithm.

As the definition implies, the complexity of an algorithm is measured by functions $f(n)$, where n is the length of the input.

The complexity of an algorithm is best studied by giving efficient estimates for the number of steps that the algorithm takes to terminate. This is done using *Big O notation*, *Omega notation*, and *Theta notation*.

We shall begin our discussion of complexity by discussing the Standard Gray Code Algorithm, which terminates in *exponential time*, and the Euclidean Algorithm, which terminates in *logarithmic time*.

The aim of this section is to give an outline of complexity theory. In order to study the complexity of algorithms effectively, we need to explain some results about recursions. We shall do this in the first section of the next chapter. The *Master Theorem*, which we shall state there, will allow us to give precise estimates for the complexity of many algorithms.

It frequently happens that an algorithm which may take many steps to terminate in the *worst case* will terminate much more rapidly in most cases that one encounters. This means that besides determining the complexity of an algorithm one should also consider its *average-case complexity*.

Using the ideas introduced in this section, we can describe *classes* of problems. We shall describe two particularly important classes, the class P of problems which can be solved by an algorithm which runs in *polynomial time*, and the class NP of problems whose solutions can be checked to be correct in polynomial time. Is $P = NP$? This is one of the basic problems in theoretical computer science. We shall also describe *NP-complete problems*.

DEFINITION 8.3.2   Given a function $g(n)$, we define three classes of functions $f(n)$:

i. The class "*big O*," $O(g(n))$ of functions $f(n)$ such that there are positive constants C and n_o such that

$$0 \le f(n) \le Cg(n)$$

for all $n \ge n_o$.

ii. The class "*Omega*," $\Omega(g(n))$ of functions $f(n)$ such that there are positive constants C and n_o such that

$$0 \le Cg(n) \le f(n)$$

for all $n \ge n_o$.

iii. The class "*Theta*," $\Theta(g(n))$ of functions $f(n)$ such that there are positive constants C, D, and n_o such that

$$0 \le Cg(n) \le f(n) \le Dg(n)$$

for all $n \ge n_o$.

Notice that the inequalities are only assumed to hold for sufficiently large n: they are *asymptotic* inequalities. Big O notation allows us to bound the number of steps that an algorithm takes from above; omega notation allows us to bound the number of steps from below; and theta notation allows us to bound the number of steps from above and below. The expression

$$f(n) = O(g(n))$$

means that $f(n)$ is in the class $O(g(n))$; $f(n) = \Omega(g(n))$ means that $f(n)$ is in the class $\Omega(g(n))$, and $f(n) = \Theta(g(n))$ means that $f(n)$ is in the class $\Theta(g(n))$.

Big O notation was introduced by the German mathematician Paul Bachmann in a book about number theory, published in 1892.

To determine the complexity of an algorithm which takes $f(n)$ steps to terminate, three things should be done:

   I.   Find a function $g(n)$ such that $f(n) = O(g(n))$, and the values of the function $g(n)$ are as small as possible. This will give the "worst-case complexity" of the algorithm.

  II.   If needed, describe $f(n)$ further by using omega notation and theta notation.

 III.   Find the average-case complexity of the algorithm.

The average-case complexity is determined after eliminating certain bad cases. We will be a little vague about exactly what this means, but we will give examples.

We described the Standard Gray Code in Section 1.3. Gray codes are named for Frank Gray, who discovered and patented them while working at the Bell Telephone Laboratories. The original purpose of this code was to avoid errors in signals transmitted by pulse code modulation.

EXAMPLE 1   Describe the complexity of the Standard Gray Code Algorithm.

Solution   Given a positive integer n, the Standard Gray Code Algorithm gives a list of the binary vectors of length n. In general a Gray Code is a code that gives a list of binary vectors of a given length, where each vector in the list differs from the preceding vector in one entry.

Gray Codes are far from unique. For example, it has been proved that there are 187,499,658,240 Gray Codes of length five.

In order to list the binary vectors of length n, the Standard Gray Code first lists the binary vectors of lengths $1, 2, \ldots, n$. If we start the algorithm with the empty vector, the total number of vectors to be listed is

$$1 + 2 + 2^2 + \ldots + 2^n = 2^{n+1} - 1$$

Since the Standard Gray Code Algorithm must list $2^{n+1} - 1$ binary vectors in order to list the binary vectors of length n its complexity is

$$O(2^{n+1})$$

Since $2^{n+1} = 2 \cdot 2^n$ we can also say that the complexity of the Standard Gray Code Algorithm is

$$O(2^n)$$

if we change the constant C in the definition from 1 to 2.

The Standard Gray Code Algorithm terminates in $f(n) = 2^{n+1} - 1$ steps. Since

$$2^n \leq 2^{n+1} - 1$$

if $n \geq 0$, we see that $f(n) = \Omega(2^n)$.

Putting our two estimates together, we see that $f(n) = \Theta(2^n)$.

The average-case complexity of the Standard Gray Code Algorithm is no different from the worst-case complexity, since the algorithm must always list 2^n binary vectors. ●

EXAMPLE 2   Describe the complexity of the Euclidean Algorithm.

Solution   The complexity of the Euclidean Algorithm is described by Lamé's Theorem which we proved in Section 2.2. Remember that Lamé's Theorem says that the Euclidean Algorithm will find the greatest common divisor of two integers $1 \leq n \leq m$ in no more than

$$\log_\phi(n) + 1$$

steps, where $\phi = (1 + \sqrt{5})/2$ is the Golden Ratio. ●

In complexity theory it is customary to work with logarithms to the base 2. The notation is

$$lgn = \log_2(n)$$

Remember the Base Change Formula for the logarithm:

$$\log_b(x) = \log_a(x)/\log_a(b)$$

We shall assume that both a and b are real numbers which are greater than 1.

Notice that it follows from this formula that if the complexity of an algorithm is $O(\log_a(n))$ then it is also $O(\log_b(n))$.

Since

$$\log_\phi(n) + 1 \leq 2\log_\phi(n)$$

for all $n \geq 2$, we can apply the Base Change Formula and conclude that the complexity of the Euclidean Algorithm is

$$O(\log_\phi(n)) = O(lgn)$$

The Standard Gray Code Algorithm is an example of an *exponential time algorithm*: because the exponential function 2^n grows rapidly such an algorithm will run very slowly. The Euclidean Algorithm is an example of a *logarithmic time algorithm*: because the logarithm function increases very slowly such an algorithm will run quite quickly.

Lamé's Theorem describes the worst-case complexity of the Euclidean Algorithm. When the Euclidean Algorithm is run on pairs of successive Fibonacci numbers $\log_\phi(n) + 1$ steps are always needed, so the statement of the theorem cannot be improved by any statement about complexity involving the logarithm.

However, Lamé's Theorem falls far short of describing the average-case complexity of the Euclidean Algorithm. The algorithm runs more rapidly for pairs of integers which are not successive Fibonacci numbers. So, if $f(n, m)$ is the number of steps which the Euclidean Algorithm takes to compute the greatest common divisor of $1 \leq n \leq m$, it is *not* true that $f(n, m) = \Omega(lgn)$, and it is *not* true that $f(n, m) = \Theta(lgn)$.

One of the simplest questions that one could ask about the Euclidean Algorithm is unsolved:

What is the average-case complexity of the Euclidean Algorithm?

The listing algorithms that we stated at the end of the previous section give lists of permutations or combinations of the elements of $\{1, 2, 3, \ldots, n\}$ of length k for any $1 \leq k \leq n$. Because

$$P(n, k) = n(n - 1)(n - 2) \ldots (n - k + 1) = n^k + \ldots$$
$$= P(n, k)/k! = (1/k!)n^k + \ldots$$

where the dots indicate lower order terms, the complexity of each of these algorithms is given by a polynomial of degree k.

Here is a list of the most common types of complexity:

Constant Time: $O(1)$
Doubly Logarithmic Time: $O(lglgn)$
Logarithmic Time: $O(lgn)$
Linear Time: $O(n)$
nlgn Time: $O(nlgn)$
Polynomial Time: $O(n^a)$, *a* a positive real number
Exponential Time: $O(2^n)$
Doubly Exponential Time: $O(2^{2^n})$

For example, the Euclidean Algorithm runs in logarithmic time, the listing algorithms for permutations and combinations run in polynomial time, and the Standard Gray Code Algorithm runs in exponential time.

EXAMPLE 3   If a computer can carry out a billion operations a second, and if the length of the input of an algorithm is $n = 100$, estimate the length of time that will be necessary to execute the algorithm if it has any of the above kinds of complexity.

Solution

For constant time, 10^{-9} seconds.
For *lglgn* time, $lglg100 \cdot 10^{-9} = 2.73 \cdot 10^{-9}$ seconds.
For *lgn* time, $lg100 \cdot 10^{-9} = 6.64 \cdot 10^{-9}$ seconds.
For linear time, $100 \cdot 10^{-9} = 10^{-7}$ seconds.
For *nlgn* time, $6.64 \cdot 10^{-7}$ seconds.
For polynomial time, for example $O(n^2)$ time, $10,000 \cdot 10^{-9} = 10^{-5}$ seconds.
For exponential time, $2^{100} \cdot 10^{-9} = 1.27 \cdot 10^{21}$ seconds, or about $4 \cdot 10^{13}$ years.
For doubly exponential time, the number of seconds is a number of about $3.82 \cdot 10^{29}$ digits.

Notice the enormous difference between the two exponential time algorithms and all the others. For this reason, computer scientists regard polynomial time and quicker algorithms as "fast" and exponential time algorithms as "slow."

Algorithms which terminate in constant time are quite rare, so the two examples that we have given, the Euclidean Algorithm and the Standard Gray Code Algorithm, in some sense represent opposite extremes. We have listed the different types of complexity in increasing order, and provided that the functions in the complexity statements cannot be replaced by smaller ones, the order is strict. That is, an exponential time algorithm cannot have polynomial time, an algorithm which is $\Theta(n^3)$ cannot be $O(n^2)$, and so on.

The function $lglgn$ increases extremely slowly, so an algorithm whose complexity is $O(lglgn)$ will terminate very rapidly. Such algorithms are rare. On the other hand, algorithms whose complexity is doubly exponential, and which terminate very slowly, are relatively common. Many of the algorithms in computer algebra have doubly exponential complexity.

What is the complexity of the sorting and searching algorithms which we discussed in Section 8.1?

The complexity of Mergesort is $O(nlgn)$.

The complexity of Quicksort is $O(n^2)$, but its average-case complexity is $O(nlgn)$, and the constant is small, so Quicksort often runs faster than Mergesort.

The complexity of Binary Search is $O(nlgn)$.

Now we shall prove two results about complexity.

THEOREM 8.3.1   If $f(X) = a_d X^d + a_{d-1} X^{d-1} + \ldots + a_1 X + a_0$ is a polynomial of degree d, then $f(X) = O(X^d)$.

Proof   Let C be the maximum of the absolute values of the coefficients a_i of $f(X)$. Then

$$|f(X)| \leq \sum_{i=0}^{d} |a_i X^i| \leq \sum_{i=0}^{d} |a_i X^d| \leq C(d+1)|X^d|$$

and this proves the theorem.   ●

We should also mention that

$$O(f+g) = \max\{O(f), O(g)\}$$
$$O(f \cdot g) = O(f) \cdot O(g)$$

THEOREM 8.3.2   The following two rules govern the growth of sums:

   (i) $1 + X + X^2 + \ldots + X^n = O(X^{n+1})$

If k is a positive integer, then

   (ii) $1^k + 2^k + \ldots + n^k = O(n^{k+1})$

Proof   The assertion (i) follows from the identity

$$1 + X + X^2 + \ldots + X^n = (X^{n+1} - 1)/(X - 1)$$

and the assertion (ii) follows from Bernoulli's Formula for the sum of the first n kth powers which we gave at the end of Chapter 3, and the previous theorem:

$$1^k + 2^k + \ldots + n^k = \left(\frac{1}{k+1}\right)(B_{k+1}(n+1) - B_{k+1}),$$

and this proves the theorem.   ●

The precise estimate for the sum of powers given by Bernoulli's Formula actually proves more than the assertion that the sum of the first n kth powers is $O(n^{k+1})$. The expression given by the Bernoulli polynomial is accurate for small values of n as well as large ones.

EXAMPLE 4   Suppose that the kth step of an algorithm can be carried out in k^2 seconds. How long will it take to execute the first n steps of the algorithm?

Solution   By Bernoulli's Formula, the number of seconds required is

$$1^2 + 2^2 + \ldots + n^2 = \frac{n(n+1)(2n+1)}{6}$$

which is $O(n^3)$.   ●

The combinatorial algorithms which we described in Section 8.1 are very useful for listing permutations and integer vectors. Since there are $n!$ permutations of n objects and n^n n-tuples of n objects, we clearly have

$$n! \leq n^n$$

Taking logarithms we get the useful formula

$$\log(n!) \leq \log(n^n) = n\log(n)$$

This formula gives an approximation to $n!$. A more accurate approximation is given by *Stirling's Formula* (the wavy equal sign means "approximately equal to"):

$$n! \approx \sqrt{2\pi n}\left(\frac{n}{e}\right)^n$$

This formula is quite accurate; and it can be made as accurate as desired by taking more terms in the formula

$$n! = \sqrt{2\pi n}\left(\frac{n}{e}\right)^n\left(1 + \frac{1}{12(n+1)} + \frac{1}{288(n+1)^2} + \ldots\right)$$

Stirling's Formula is due to James Stirling (1692–1770). One of Stirling's contributions was to find the term $\sqrt{2\pi}$ in the formula above. This term is also crucially important in probability theory.

EXAMPLE 5   Find 9!, and then approximate 9! using Stirling's Formula, and again using three terms of the series above for $n!$.

Solution $9! = 362,880$, and Stirling's Formula gives

$$9! \approx \sqrt{18\pi} \left(\frac{9}{e}\right)^9 = (7.519884824\ldots)(47,811.48671\ldots) = 359,536.8733\ldots$$

and the relative error is less than 1 percent. Using the more exact formula, we have

$$9! \approx (359,536.8733)\left(1 + \frac{1}{12 \cdot 10} + \frac{1}{288 \cdot 100}\right)$$

$$= (359,536.8733\ldots)(1.008368055\ldots) = 362,545.4976\ldots$$

and the relative error is less than 0.1 percent.   ●

DEFINITION 8.3.3   We define three complexity classes of problems:

i. The complexity class P is the class of problems which can be solved by an algorithm which runs in polynomial time.

ii. The complexity class NP is the class of problems whose solutions can be verified to be correct by a polynomial time algorithm.

iii. The class of NP-*complete* problems consists of problems such that if the problem can be solved, then any problem in the class NP can be solved in polynomial time, which implies that $P = NP$.

It is clear that the class P is contained in the class NP. Computer scientists believe that $P \neq NP$ for many reasons, the first being that it is usually far simpler to verify that a solution to a problem is correct quickly than it is to find a solution to the problem quickly.

Another serious reason for believing that $P \neq NP$ is the existence of NP-complete problems. We will give a very partial list of NP-complete problems. Many other NP-complete problems are known.

Another possibility which should be mentioned is that the question as to whether $P = NP$ may not be decidable. That is, it may be true that $P = NP$ in some models of set theory but not in others.

Recall that we stated the Travelling Salesman Problem in Chapter 4. We discussed graph coloring in Chapter 7; the Graph Coloring Problem is the problem of coloring the nodes of a graph with the fewest colors. The Hamiltonian Circuit Problem is the problem of determining whether a graph has a Hamiltonian circuit. The Satisfiability Problem for Boolean functions is the problem of determining whether the Boolean function $f(x_1, x_2, \ldots, x_n)$ can be given the value 1 by a suitable assignment of values 0 and 1 to the variables.

Four *NP*-Complete Problems The following problems are NP-complete:

(i) The Travelling Salesman Problem

(ii) The Graph Coloring Problem

(iii) The Hamiltonian Circuit Problem

(iv) The Satisfiability Problem for Boolean Functions

APPLICATION *Kruskal's Algorithm and Prim's Algorithm*

If G is a finite weighted graph, we will use the notation $V = |V(G)|$ and $E = |E(G)|$.

To implement Kruskal's Algorithm, first we must inspect each edge to find an edge of minimal weight: this takes $O(E)$ time.

If the algorithm has constructed a tree with edges $e_1, e_2, \ldots, e_n$, we must resort the remaining edges so that the edge labelled e_{n+1} does not create a cycle, and so that e_{n+1} has minimum weight subject to this condition: this takes $O(E \lg E)$ time.

Thus, altogether, Kruskal's Algorithm takes $O(E \lg E)$ time.

Remember that Prim's Algorithm begins by choosing an arbitrary node of G. When the algorithm has constructed a tree T, the next time it is executed the algorithm must list the nodes which do not belong to T, which takes $O(V \lg V)$ time, and then find an edge of minimal weight joining one of these nodes to a node of T, which takes $O(E \lg V)$ time. Thus, altogether, Prim's Algorithm takes $O(V \lg V + E \lg V)$ time.

EXERCISES

Prove the following statements involving *big O* notation.

1. $x^2 + 2 = O(x^2)$.   **2.** $x^2 + 5x + 2 = O(x^2)$.

3. $x^3 + 2 = O(x^3)$.

4. $x^3 + x^2 + x + 1 = O(x^3)$.

5. $x \lg(x) = O(x^2)$.

Prove the following statements involving Ω notation.

6. $x^2 - x = \Omega(x^2)$.

7. $x^2 + 2x - 1000 = \Omega(x^2)$.

8. $x^3 - x^2 + x - 200 = \Omega(x^3)$.

9. $x \lg x - 1000x + 500 = \Omega(x)$.

10. $3^x - x - 1000 = \Omega(2^x)$.

Prove the following statements involving Θ notation.

11. $x + 100 = \Theta(x)$.   **12.** $x^2 + x + 1 = \Theta(x^2)$.

13. $x^2 - x + 1 = \Theta(x^2)$.

14. $x \lg x + 20 = \Theta(x \lg x)$.

15. $x \lg x + x + 100 = \Theta(x \lg x)$.

If a computer can carry out 10^9 operations per second, and if the kth step of the algorithm takes 2^k seconds to execute, how long will the computer take to execute the algorithm if the following number of steps are used? (Express your answer in seconds, hours, days, or years, as appropriate.)

16. $n = 20$.   **17.** $n = 30$.

18. $n = 40$.   **19.** $n = 50$.

20. $n = 60$.

The following sums express the number of seconds that a certain algorithm takes to execute n steps of a program, assuming that the computer can carry out 10^9 operation per second. Use Bernoulli's Formula to estimate the amount of time that the algorithm will to take to terminate, and again express your answer in seconds, hours, days, or years, as appropriate.

21. $1 + 2 + \ldots + 1,000,000$.

22. $1^2 + 2^2 + \ldots + 100^2$

23. $1^2 + 2^2 + \ldots + 10,000^2$

24. $1^3 + 2^3 + \ldots + 10,000^3$

25. $1^{10} + 2^{10} + \ldots + 1,000^{10}$

Approximate each of the following factorials using Stirling's Formula, and then by using three terms of the series for $n!$. Compute the percentage error for each of your approximations.

26. 5!   **27.** 8!   **28.** 12!

29. 20!   **30.** 100!

ADVANCED EXERCISES

1. Is it true that $3^x = \Theta(2^x)$?

2. If a and b are real numbers greater than 1, when is $a^x = \Theta(b^x)$?

3. Is $\lg\lg x = \Theta(\lg x)$?   **4.** Is $2^x = \Theta(2^{2^x})$?

5. Prove that if $f(X)$ and $g(X)$ are polynomials of the same degree, then $f(X) = \Theta(g(X))$.

COMPUTER EXERCISES

Check the following statements about complexity of the form $f(n) = O(g(n))$ by finding a value of n for which $f(n) \le g(n)$.

1. $1,000n^2 + 7,000n + 10,000 = O(n^3)$.

2. $500n^2 + 25,700n + 100,000 = O(n^3)$.

3. $10,000n^2 + 102,000n + 1,000,000 = O(n^3)$.

4. $500n\lg(n) = O(n^2)$.

5. $1,000n\lg(n) + 20,000n = O(n^2)$.

6. $500n^{10} + n^5 + 20,000 = O(2^n)$.

7. $n^{50} + n^{40} + n^{30} + n^{20} + n^{10} + 1 = O(2^n)$.

8. Describe an algorithm which finds the longest palindrome in a given finite sequence (remember that a palindrome is a sequence which is the same in reverse order). What is the complexity of your algorithm?

9. Given a finite sequence of numbers or letters, describe an algorithm which finds a longest sequence common to the given sequence and the same sequence in reverse order. What is the complexity of your algorithm?

10. Prove that among the longest sequences common to a sequence and the same sequence in reverse order, there is always a palindrome. Is *every* common sequence of maximal length a palindrome?

8.4   CHAPTER SUMMARY AND SUPPLEMENTARY EXERCISES

> *It felt that, in spite of all possible pains,*
>
> *It had somehow contrived to lose count,*
>
> *And the only thing now was to rack its poor brains*
>
> *By reckoning up the amount.*
>
> *"Two added to one—if that could but be done,"*
>
> *It said "with one's fingers and thumbs!"*
>
> *Recollecting with tears how, in earlier years,*
>
> *It had taken no pains with its sums.*

Section 8.1

In this section, we described the problems of sorting, searching, and listing a set of objects. We described two sorting algorithms, Mergesort and Quicksort. We described Binary Search, and a number of listing algorithms for integer vectors.

Section 8.2

In this section, we discussed graph algorithms. We described Prim's Algorithm, Kruskal's Algorithm, Dijkstra's Algorithm, and the Max-Flow-Min-Cut Theorem.

Section 8.3

In this section, we described the complexity of an algorithm. We introduced big O notation, omega notation, and theta notation. We described conventions about functions which are used in complexity theory. We described the complexity of several algorithms. We described the classes P and NP. We described the complexity of Kruskal's Algorithm and Prim's Algorithm.

EXERCISES

Use Kruskal's Algorithm, and then Prim's Algorithm, to find a minimum weight spanning tree for each of the following graphs:

1.

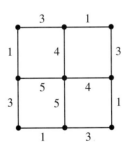

2.

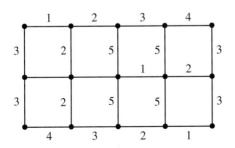

Use Dijkstra's Algorithm to find a shortest path from node A to node B in each of the following graphs:

3.

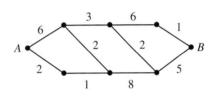

4.

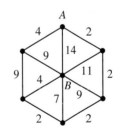

Use the Max-Flow-Min-Cut Theorem to find the maximal flow through each of the following networks:

5.

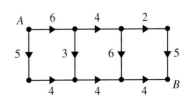

6.

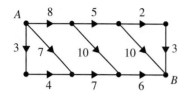

COMPUTER EXERCISES

At the end of Chapter 3, we stated Jakob Bernoulli's formula for the sum of the first n kth powers. Euler discovered a remarkable analog of this formula for the sum of all inverse even powers, which he stated in the *Introductio in Analysin Infinitorum*:

$$\sum_{n=1}^{\infty} n^{-2k} = \frac{(2\pi)^{2k}(-1)^{k+1} B_{2k}}{2 \cdot (2k)!}$$

where B_{2k} is the $2k$th Bernoulli number.

For example,

$$\sum_{n=1}^{\infty} n^{-2} = \frac{\pi^2}{6}$$

$$\sum_{n=1}^{\infty} n^{-4} = \frac{\pi^4}{90}$$

and so on.

In the following exercises, you are asked to evaluate the given sums and their bounds, and show how close the sums come to the bounds.

1. $1^{-2} + 2^{-2} + \ldots + n^{-2} < \frac{\pi^2}{6}$, for $n = 10$, $n = 100$, $n = 1000$.

2. $1^{-4} + 2^{-4} + \ldots + n^{-4} < \frac{\pi^4}{90}$, for $n = 10$, $n = 100$, $n = 1000$.

3. $1^{-6} + 2^{-6} + \ldots + n^{-6} < \frac{\pi^6}{945}$, for $n = 10$, $n = 100$, $n = 1000$.

Notice that Euler's Formula involves only even inverse powers. It is very remarkable that no one has been able to find a formula for the sum of all the inverse odd powers for $k \geq 3$, although this sum is finite.

4. Find the smallest bound you can for the sums

$$1^{-3} + 2^{-3} + \ldots + n^{-3}$$

by evaluating this sum for $n = 10$, $n = 100$, $n = 1000$. If you can find a simple expression for

$$\sum_{n=1}^{\infty} n^{-3}$$

you will have solved an old problem in number theory.

9 Combinatorics

9.1 RECURSIONS AND THEIR SOLUTION

> *Taking Three as the subject to reason about—*
> *A convenient number to state—*
> *We add Seven, and Ten, and then multiply out*
> *By One Thousand diminished by Eight.*
> *The result we proceed to divide, as you see,*
> *By Nine Hundred and Ninety and Two:*
> *Then subtract Seventeen, and the answer must be*
> *Exactly and perfectly true.*

We have mentioned François-Édouard-Anatole Lucas several times in this book. Lucas wrote that

> *The theory of recurrent series is an inexhaustible mine which contains*
> *all the properties of numbers.*

Lucas and earlier writers such as Euler used the phrase "recurrent series" to denote what is now called a "linear recursive sequence." We defined and discussed such sequences in Section 5 of Chapter 1. The concept of a linear recursive sequence is of basic importance, and we now give it a more general definition.

> **DEFINITION 9.1.1**   A sequence $\{a_n\}_{n=0}^{\infty}$ is *linear recursive of degree r* if there are constants $c_1, c_2, \ldots, c_r$ and a positive integer $N \geq r$ such that
>
> $$a_n = c_1 a_{n-1} + c_2 a_{n-2} + \ldots + c_r a_{n-r}$$
>
> if $n \geq N$.

In the earlier definition, we took $N = r$, but here we allow the possibility that the sequence may have a longer initial sequence which is not described by the recursion. A sequence which is linear recursive of a given degree may also be linear recursive with a smaller degree, but the *minimal degree* of the sequence is an invariant of the sequence.

Lucas outlined an extensive theory based on the properties of linear recursive sequences in which the Fibonacci numbers and the Lucas numbers play central roles. He gave an exposition of this theory and its applications in two long papers which were published in the first volume of the *American Journal of Mathematics* in 1879.

Modern algorithms for primality testing are still strongly influenced by Lucas's ideas and methods. Lucas had the very original idea that doubly periodic functions called *elliptic functions* can be studied by using linear recursive sequences of degree three or four.

In this section, we shall give a brief description of the theory of linear recursive sequences. Introducing the *generating function* of a sequence will allow us to describe recursions more effectively. The *Master Theorem* is part of this theory, and we shall show how it can be used to describe the complexity of some of the algorithms that we have discussed. As an application, we shall use linear recursions to describe *AVL trees*.

In the second section, we shall give a brief introduction to probability theory, and explain why it is useful in discrete mathematics. We shall show how probability theory can be used to describe the *reliability* of a network.

In the third section of this chapter, we shall discuss the concept of a *group*, and describe three important counting rules which are expressed in terms of groups. We shall use *Polya's Theorem* to count rooted trees, and we shall give a formula for the number of unlabelled trees.

Why are linear recursive sequences important?

 I. Linear recursive sequences count objects of many kinds: graphs, trees, spanning trees, organic chemicals, and more.

 II. Linear recursive sequences can be used to study problems which are important both in mathematics and computer science, such as the primality problem.

 III. The theory of linear recursive sequences can be used to determine the complexity of an algorithm.

DEFINITION 9.1.2   A *polynomial* is an expression

$$f(x) = \sum_{k=1}^{n} a_k x^k = a_0 + a_1 x + \ldots + a_n x^n$$

A *power series* is an expression

$$f(x) = \sum_{k=0}^{\infty} a_k x^k = a_0 + a_1 x + a_2 x^2 + \ldots + a_n x^n + \ldots$$

A *rational function* $f(x)$ is a ratio of polynomials $P(x)$ and $Q(x)$:

$$f(x) = \frac{P(x)}{Q(x)}$$

The coefficients of a polynomial or a power series can be real numbers, complex numbers, or elements of other number systems such as Z_n. Polynomials and power series can be added and multiplied, like numbers in other number systems.

We will be particularly concerned with power series which can be expressed as rational functions, as in the next example and theorem.

EXAMPLE 1   Prove that the *geometric series* $\sum_{k=0}^{\infty} x^n$ can be expressed as

$$\frac{1}{1-x} = 1 + x + x^2 + \ldots + x^n + \ldots$$

Solution   If we multiply both sides of the equation by $1 - x$ we have

$$1 = (1 - x)(1 + x + x^2 + \ldots + x^n + \ldots)$$
$$= 1 + (x - x) + (x^2 - x^2) + \ldots + (x^n - x^n) + \ldots = 1$$

It is very important that when we consider identities like the previous one, we do not mean that the identity holds for every value of the variable x. For example, if we substitute $x = 2$ in the identity for the geometric series, we get

$$-1 = \frac{1}{1-2} = 1 + 2 + 2^2 + \ldots = \infty$$

which of course is nonsense. The series converges to $\frac{1}{1-x}$ if $-1 < x < 1$, but we will be concerned with the identity only as a formal expression, without regard for the values of x for which the series converges.

> *Power series in elementary combinatorics are regarded as formal expressions. When we work with them, we are not concerned with questions of convergence.*   ●

THEOREM 9.1.1 If $n = 1, 2, \ldots$ is a positive integer, we have the identity

$$\frac{1}{(1-x)^n} = \sum_{k=0}^{\infty} \binom{n+k-1}{k} x^k$$

which defines the *inverse binomial series*.

Proof We shall prove this by induction on n. If $n = 1$, the inverse binomial series is the geometric series, and the identity is the identity which we have just proved.

Now assume that the identity holds for a given n. To complete the proof by induction we must show that the coefficient of x^k in

$$\frac{1}{(1-x)^{n+1}}$$

is

$$\binom{(n+1)+k-1}{k} = \binom{n+k}{k}$$

We have

$$\frac{1}{(1-x)^{n+1}} = \left(\frac{1}{(1-x)^n}\right)\left(\frac{1}{1-x}\right)$$

$$= \left(\sum_{k=0}^{\infty} \binom{n+k-1}{k} x^k\right)\left(\sum_{k=0}^{\infty} x^k\right)$$

$$= \ldots + \left(\binom{n+0-1}{0} + \binom{n+1-1}{1} + \ldots + \binom{n+k-1}{k}\right) x^k + \ldots$$

$$= \ldots + \binom{n+k}{k} x^k + \ldots$$

by the Parallel Summation identity which we stated in Section 3.3, and this completes the proof by induction. ●

If $n = 2$, we have

$$\frac{1}{(1-x)^2} = \sum_{k=0}^{\infty} \binom{k+1}{k} x^k = 1 + 2x + 3x^2 + 4x^3 + \ldots$$

Notice that the geometric series and all the inverse binomial series are inverses of polynomials. There is a simple algorithm which describes the inverse of a polynomial or more generally of a power series.

THEOREM 9.1.2 Let

$$f(x) = \sum_{k=0}^{\infty} a_k x^k$$

be a power series such that $a_0 \neq 0$. Then $f(x)$ has a unique multiplicative inverse $g(x)$, $f(x)g(x) = 1$, where

$$g(x) = \sum_{k=0}^{\infty} b_k x^k$$

and where the coefficients of $g(x)$ are determined by the following equations:

$$a_0 b_0 = 1$$
$$a_0 b_1 + a_1 b_0 = 0$$
$$a_0 b_2 + a_1 b_1 + a_2 b_0 = 0$$
$$\dots$$
$$a_0 b_k + a_1 b_{k-1} + \dots + a_k b_0 = 0$$
$$\dots$$

Proof   If $g(x)$ is the multiplicative inverse of $f(x)$, then $f(x)g(x) = 1$, so the product of the constant terms of $f(x)$ and $g(x)$ must be equal to 1, and the coefficients of all higher powers of x in $f(x)g(x)$ must be zero. That is the meaning of these equations. Since

$$b_k = a_0^{-1}(-a_1 b_{k-1} - a_2 b_{k-2} - \dots - a_k b_0)$$

the equations give a recursion which determines the coefficients of $g(x)$.   ●

EXAMPLE 2   Use this algorithm to find the inverse of the polynomial $(1 - x)^2$.

Solution   The polynomial which we wish to invert is $(1 - x)^2 = 1 - 2x + x^2$. The recursion for the coefficients of the inverse $g(x)$ gives

$$b_0 = 1$$
$$b_1 - 2 = 0, \text{ and if } k \geq 2,$$
$$b_k - 2b_{k-1} + b_{k-2} = 0, \text{ or } b_k = 2b_{k-1} - b_{k-2}.$$

We shall prove by induction that $b_k = k + 1$. This is true for $k = 0$ and $k = 1$ by the formulas above. If $k \geq 2$, $b_k = 2b_{k-1} - b_{k-2} = 2(k) - (k - 1) = k + 1$, and this completes the proof by induction.   ●

Is the multiplicative inverse of a non-zero power series $f(x)$ always a power series? This is not quite true. The problem is that the constant term of a power series may be zero, so that the algorithm above cannot be applied. However, this is a minor problem. By factoring out the highest possible power of x, we can write $f(x)$ in the form

$$f(x) = x^n h(x)$$

where $h(x)$ has non-zero constant term. $h(x)$ has an inverse $k(x)$, which is a power series, so the inverse of $f(x)$ can be written

$$x^{-n} k(x)$$

A series of this kind is called a *Laurent series*. What we have shown is that the inverse of any non-zero power series is a Laurent series.

Power series are important in discrete mathematics because there is a power series associated with every sequence of numbers. The properties of the sequence are mirrored by the properties of the power series. This power series is called the *generating function* of the sequence. In fact, we shall define two kinds of generating functions of a sequence, the generating function and the exponential generating function.

DEFINITION 9.1.3   Let

$$a_0, a_1, a_2, \ldots$$

be a sequence of numbers. The *generating function* of the sequence is the power series

$$f(x) = \sum_{k=0}^{\infty} a_k x^k = a_0 + a_1 x + a_2 x^2 + \ldots + a_k x^k + \ldots$$

and the *exponential generating function* of the sequence is the power series

$$f(x) = \sum_{k=0}^{\infty} (a_k/k!) x^k = a_0 + a_1 x + \ldots + (a_k/k!) x^k + \ldots$$

The numbers in the sequence can be real numbers, complex numbers, or elements of another number system such as Z_n.

The generating functions of many well-known sequences are rational functions.

EXAMPLE 3   Describe the generating function of the Fibonacci numbers.

Solution   Recall that $F_0 = 0$, $F_1 = 1$, and if $n \geq 2$, then $F_n = F_{n-1} + F_{n-2}$. If $f(x)$ is the generating function, we write $f(x)$, $xf(x)$, and $x^2 f(x)$.

$$f(x) = 0 + x + F_2 x^2 + \ldots + F_n x^n + \ldots$$
$$xf(x) = 0 + 0 + F_1 x^2 + \ldots + F_{n-1} x^n + \ldots$$
$$x^2 f(x) = 0 + 0 + F_0 x^2 + \ldots + F_{n-2} x^n + \ldots$$

so

$$f(x) - xf(x) - x^2 f(x) = f(x)(1 - x - x^2) = x$$

and

$$f(x) = \frac{x}{1 - x - x^2}$$

EXAMPLE 4   Describe the generating function of the Lucas numbers.

Solution   Euler described this sequence in the *Introductio ad Analysin Infinitorum*. Recall that $L_0 = 2$, $L_1 = 1$, and if $n \geq 2$, $L_n = L_{n-1} + L_{n-2}$. Then

$$f(x) = 2 + x + L_2 x^2 + \ldots + L_n x^n + \ldots$$
$$x f(x) = 0 + 2x + L_1 x^2 + \ldots + L_{n-1} x^n + \ldots$$
$$x^2 f(x) = 0 + 0 + L_0 x^2 + \ldots + L_{n-2} x^n + \ldots$$

so

$$f(x) - x f(x) - x^2 f(x) = f(x)(1 - x - x^2) = 2 - x$$

and

$$f(x) = \frac{2 - x}{1 - x - x^2}$$

●

EXAMPLE 5   Describe the generating function of Euler's sequence.

Solution   Euler discussed this sequence in the *Introductio ad Analysin Infinitorum*. The sequence is

$$a_0 = 1, 1, 2, 2, 3, 3, 4, 4, \ldots$$

and satisfies the recursion $a_n = a_{n-1} + a_{n-2} - a_{n-3}$ if $n \geq 3$. By a calculation just like the ones in the previous two examples, we see that

$$f(x) = \frac{1}{1 - x - x^2 + x^3}$$

●

Here are a few examples of exponential generating functions. Recall that the *Taylor expansion* of a function $f(x)$ at the origin is given by

$$f(x) = \sum_{k=0}^{\infty} (f^{(k)}(0)/k!) x^k$$

In other words, the Taylor expansion can be viewed as the exponential generating function of the sequence

$$f^{(k)}(0)$$

where $f^k(0)$ is the kth derivative of $f(x)$ evaluated at $x = 0$.

EXAMPLE 6   Describe the Taylor series expansions of $\cos(x)$, $\sin(x)$, e^x, $\tan(x)$, and $\sec(x)$.

Solution

$$\cos(x) = 1 - x^2/2! + x^4/4! - x^6/6! + \ldots$$
$$\sin(x) = x - x^3/3! + x^5/5! - x^7/7! + \ldots$$

$$e^x = 1 + x + x^2/2! + x^3/3! + \ldots + x^n/n! + \ldots$$

$$\tan(x) = x + 2x^3/3! + 16x^5/5! + 272x^7/7! + \ldots$$

$$= \sum_{k=0}^{\infty} A_{2k+1} x^{2k+1}/(2k+1)!$$

$$\sec(x) = 1 + x^2/2! + 5x^4/4! + 61x^6/6! + \ldots$$

$$= \sum_{k=0}^{\infty} A_{2k} x^{2k}/(2k)!$$

The Taylor expansions of $\cos(x)$, $\sin(x)$, and e^x are the exponential generating functions of the following sequences

$$1, 0, -1, 0, 1, 0, -1, 0, \ldots$$

$$0, 1, 0, -1, 0, 1, 0, -1, \ldots$$

$$1, 1, 1, 1, 1, 1, 1, 1, \ldots$$

Since these sequences are periodic, we can easily find all their terms.

By constrast, the numbers A_{2k+1} and A_{2k} which occur in the Taylor expansions of $\tan(x)$ and $\sec(x)$ are called *tangent numbers* and *secant numbers* (secant numbers are also called *Euler numbers*), and are by no means easy to compute. The obvious algorithm for computing them, by taking higher derivatives of $\tan(x)$ and $\sec(x)$, is quite slow. In the Advanced Exercises, we shall give a better algorithm for computing tangent numbers and secant numbers.

Notice that the generating functions of the sequences of Fibonacci numbers, Lucas numbers, and Euler numbers are all rational functions.

THEOREM 9.1.3   The following conditions are equivalent:

(i)   The sequence $\{a_k\}_{k=0}^{\infty}$ is linear recursive.

(ii)   The generating function $f(x) = \sum_{k=0}^{\infty} a_k x^k$ is a rational function $P(x)/Q(x)$, where $Q(x)$ has the form

$$Q(x) = 1 - c_1 x - c_2 x^2 - \ldots - c_r x^r$$

Proof   First, assume that $\{a_k\}_{k=0}^{\infty}$ is a linear recursive sequence. Recall that this means that there are positive integers r and N and constants $c_1, c_2, \ldots, c_r$ such that

$$a_n = c_1 a_{n-1} + c_2 a_{n-2} + \ldots + c_r a_{n-r}$$

if $n \geq N$.

This is the same as saying that the coefficient of x^n in

$$f(x)(1 - c_1 x - c_2 x^2 - \ldots - c^r x^r)$$

is equal to 0 if $n \geq N$. But then $f(x)Q(x) = P(x)$, where $P(x)$ is a polynomial, so $f(x) = P(x)/Q(x)$ is a rational function.

Now assume that the generating function $f(x)$ is a rational function, with the given denominator $Q(x)$. Since $f(x) = P(x)/Q(x)$, where $P(x)$ is a polynomial, $P(x) =$

$f(x)Q(x)$. But this means that the terms of the sequence satisfy the given recursion for $n \geq N$.   ●

Notice that the roots of the denominator $Q(X)$ in the previous theorem are never equal to zero. By the Fundamental Theorem of Algebra, we can find a root of $Q(x)$, and then factor $Q(x)$ into linear factors by induction:

$$Q(x) = C(x - a_1)(x - a_2)\ldots(x - a_r)$$

Since the roots a_k of $Q(x)$ are all non-zero, we can rewrite the factorization as

$$Q(x) = D(1 - a_1^{-1}x)(1 - a_2^{-1}x)\ldots(1 - a_r^{-1}x)$$

This factorization of $Q(x)$ in terms of *inverse roots* is very convenient. Of course some of the roots, or inverse roots, may occur more than once. The number of times that a root occurs is called its *multiplicity*.

Remember that when we are given a rational function $P(x)/Q(x)$ we can always divide $Q(x)$ into $P(x)$ using synthetic division, and write $P(x) = S(x)Q(x) + R(x)$, where $S(x)$ is a polynomial and $R(x)$ is a polynomial of degree less than the degree of $Q(x)$.

We will state the next theorem, about partial fractions, without proof. It may be familiar to you from a second calculus course, where it is shown that the partial fraction decomposition of a rational function allows us to integrate the function.

THEOREM 9.1.4   The partial fraction decomposition of a rational function.

Let $P(x)/Q(x)$ be a rational function, where the degree of $P(x)$ is strictly smaller than the degree of $Q(x)$. Then $P(x)/Q(x)$ is a finite sum of terms of the form

$$\frac{A_1}{(1 - ax)}, \frac{A_2}{(1 - ax)^2}, \ldots, \frac{A_n}{(1 - ax)^n}$$

for each inverse root a of $Q(x)$, where n is the multiplicity of a.   ●

The inverse roots which occur in the above expression may be complex numbers. If we wish to avoid this and use only real numbers, we must consider more complicated expressions with powers of irreducible quadratic polynomials in the denominator. You probably have heard a discussion of these problems when partial fractions were discussed in a second calculus course.

As Euler pointed out, we can expand the terms above using the Inverse Binomial Series, and reach the following important conclusion.

Binet's Formula, General Form   There is an explicit formula for the terms of any linear recursive sequence.

We shall illustrate this for the Fibonacci sequence, the Lucas sequence, and for Euler's sequence.

EXAMPLE 7   Find Binet's Formula for the Fibonacci sequence.

Solution   The generating function is $f(x) = \frac{x}{1-x-x^2}$. Recall the Golden Ratio and its conjugate

$$\phi = \frac{1 + \sqrt{5}}{2}, \phi' = \frac{1 - \sqrt{5}}{2}$$

and notice that

$$1 - x - x^2 = (1 - \phi x)(1 - \phi' x)$$

The partial fraction decomposition of the generating function is

$$\sum_{n=0}^{\infty} F_n x^n = \frac{x}{1 - x - x^2} = \frac{1/\sqrt{5}}{(1 - \phi x)} - \frac{1/\sqrt{5}}{(1 - \phi' x)}$$

Expanding terms on the right-hand side gives

$$F_n = \frac{1}{\sqrt{5}}((\phi)^n - (\phi')^n)$$

which is Binet's Formula for the Fibonacci numbers. We proved in Chapter 2 that this formula holds by induction. Our method here not only proves that the formula holds, but produces the formula explicitly. ●

EXAMPLE 8   Find Binet's Formula for the Lucas sequence.

Solution   The generating function is

$$f(x) = \sum_{n=0}^{\infty} L_n x^n = \frac{2 - x}{1 - x - x^2}$$

The partial fraction decomposition of the generating function is

$$\sum_{n=0}^{\infty} L_n x^n = \frac{2 - x}{1 - x - x^2} = \frac{1}{(1 - \phi x)} + \frac{1}{(1 - \phi' x)}$$

Expanding terms on each side gives

$$L_n = (\phi)^n + (\phi')^n$$

and we have seen that this is Binet's Formula for the Lucas numbers. ●

EXAMPLE 9   Find Binet's Formula for Euler's sequence.

Solution   The partial fraction decomposition of the generating function is

$$\frac{1}{1 - x - x^2 + x^3} = \frac{1}{(1 - x)^2(1 + x)} = \frac{1/2}{(1 - x)^2} + \frac{1/4}{(1 - x)} + \frac{1/4}{(1 + x)}$$

Notice that the denominator of the generating function contains a squared term. Expanding the series above, the nth term of the sequence is equal to

$$\frac{2n + 3 + (-1)^n}{4}$$ ●

Here is an example where the Fibonacci sequence is used to solve a problem about graphs. The *ladder graph* L_n consists of two parallel rows of n nodes connected as follows:

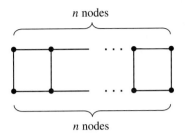

A *one-factor* of a graph G is a subgraph F of G with $V(F) = V(G)$ such that the edges of F are disjoint and every node of G belongs to an edge of F. There is one one-factor of L_1, namely L_1 itself:

There are two one-factors of L_2:

There are three one-factors of L_3:

There are five one-factors of L_4:

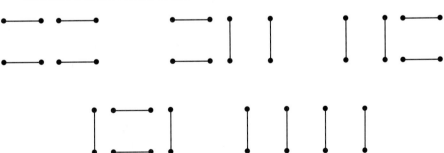

EXAMPLE 10   Prove that the number of one-factors of the ladder graph L_n is the Fibonacci number F_{n+1}.

Solution   We checked above that L_1 has $F_2 = 1$ one-factor, and that L_2 has $F_3 = 2$ one-factors. If $n > 2$, each one-factor of L_n must end in

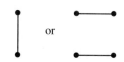

or

In the first case, the one-factor can be completed by any one-factor of L_{n-1}, and by induction there are F_n of these. In the second case, the one-factor can be completed by any one-factor of L_{n-2}, and by induction there are F_{n-1} of these. Thus the total number of one-factors of L_n is

$$F_n + F_{n-1} = F_{n+1}$$

and this completes the proof by induction.   ●

APPLICATION   *Applications to Complexity*

Suppose that we have an algorithm which works by dividing a problem of size n into a subproblems, each of size n/b, and that the time it takes to do this is measured by a function $f(n)$.

This corresponds to a recursion of the form

$$T(n) = aT(n/b) + f(n)$$

A recursion of this kind is called a *divide-and-conquer relation*.
Suppose that we define $T(1) = 1$ with the recursion

$$T(n) = 3T(n/4) + n$$

if $n > 1$.

If we iterate this recursion, we get

$$\begin{aligned}
T(n) &= n + 3T(n/4) \\
&= n + 3(n/4 + 3T(n/16)) \\
&= n + 3(n/4 + 3(n/16 + 3T(n/64))) \\
&= n + 3n/4 + 9n/16 + 27n/64 + \ldots + 3^{\log_4(n)} T(1)
\end{aligned}$$

if $n = 4^k$ for some integer k.

Thus

$$T(n) \leq n \sum_{i=0}^{\infty} (3/4)^i + n^{\log_4(3)}$$

$$T(n) \leq n \left(\frac{1}{1 - 3/4} \right) + o(n)$$

$$T(n) \leq 4n + o(n)$$

$$T(n) = O(4n) = O(n)$$

(Here we have used the fact that $\log_4(3) < 1$.)

If $T(1) = 1$ and $T(n) = aT(n/b)$, an exactly similar argument shows that

$$T(n) = n^{\log_b(a)} + \sum_{j=0}^{\log_b(a)-1} a^j f(n/b^j)$$

if $n = b^k$ for some integer k.

It turns out that, with very mild assumptions on T, the assumption that n is a power of b can be omitted.

Analysis of the previous equation gives a proof of the Master Theorem.

The Master Theorem   Let $a \geq 1$ and $b > 1$ be constants, $f(n)$ a function, and let $T(n)$ be defined by the recurrence

$$T(n) = aT(n/b) + f(n)$$

where n/b is understood to be the integer nearest to n/b. Then

(i)   If $f(n) = O(n^{\log_b(a)-\epsilon})$ for some positive constant $\epsilon > 0$, then
$T(n) = \Theta(n^{\log_b(a)})$.

(ii) If $f(n) = \Theta(n^{\log_b(a)})$, then $T(n) = \Theta(n^{\log_b(a)}\lg n)$.

(iii) If $f(n) = \Omega(n^{\log_b(a)+\epsilon})$ for some constant $\epsilon > 0$, and if $af(n/b) \leq cf(n)$ for some constant $c < 1$ and all sufficiently large n, then $T(n)$
$= \Theta(f(n))$.   ●

The theorem compares $f(n)$ with $n^{\log_b(a)}$. Roughly speaking, the Master Theorem says that the larger of the two functions determines the complexity of $T(n)$. We should stress that there are a few recursions which are not described by any of the three cases of the theorem.

EXAMPLE 11   Use the Master Theorem to describe the complexity of $T(n)$ if $T(1) = 1$ and $T(n) = 2T(n/2) + 1$.

Solution   Here $a = b = 2$, so $\log_b(a) = 1$, and $f(n) = 1$. Clearly $f(n) = 1 = O(n^{\log_b(a)-1/2}) = O(n^{1/2})$, so

$$T(n) = \Theta(n)$$ ●

EXAMPLE 12   Use the Master Theorem to describe the complexity of $T(n)$ if $T(n) = 2T(n/2) + n$.

Solution   Again $a = b = 2$, so $\log_b(a) = 1$, and $f(n) = n$. Since $f(n) = n = n^{\log_b(a)}$, the condition in the second case of the Master Theorem is clearly satisfied, so

$$T(n) = \Theta(n\lg n)$$ ●

EXAMPLE 13   Use the Master Theorem to find the complexity of $T(n)$ if $T(1) = 1$ and $T(n) = 2T(n/2) + n^2$.

Solution   Here $a = b = 2$, so $\log_b(a) = 1$, and $f(n) = n^2$. Clearly $f(n) = n^2 = \Omega(n^{\log_b(a)+1/2})$ and $2f(n/2) = n^2/2 < (3/4)n^2$ for all n and in particular for all sufficiently large n.

By the third case of the Master Theorem,

$$T(n) = \Theta(n^2)$$

    ●

APPLICATION   *AVL Trees*

The structure of trees is a basic theme in computer science and in this book, and the structure of binary trees is particularly important. The most important example of a binary tree is the full binary tree. However, it is inconvenient to use only this tree as a model for a process, or as a data structure. Therefore one wants "good" binary trees, trees whose properties do not differ drastically from the properties of the full binary tree. Such a tree should be "balanced." That is, the left and right subtrees at each node should differ only slightly.

The Russian mathematicians G. M. Adel'son-Vel'ski and E. M. Landis proposed a definition of a "good" binary tree which has been widely accepted. The *height* of a rooted tree is the maximal distance from the root to a terminal node.

> **DEFINITION 9.1.4**   An *AVL tree* is a rooted binary tree with the property that for each node the height of the left and right subtrees at that node differ at most by one.

In Figure 9.1 we have two AVL trees.

Figure 9.1   Two AVL Trees

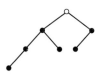

Figure 9.2
A Leftmost AVL Tree

What "goodness" property should we expect from such a tree? In applications, one *searches* through a tree. A depth-first search will run quickly if the height of the tree is small in comparison with the number of nodes. The following theorem explains why it is advantageous to search through an AVL tree.

The theorem states that a search through an AVL tree will be quick even if the tree is as "bad" as possible. A *leftmost* AVL tree is a tree such that at each node the height of the left subtree is one more than the height of the right subtree. We picture such a tree in Figure 9.2.

THEOREM 9.1.5   The height of a leftmost AVL tree is proportional to the log of the number of nodes in the tree.

Proof   Let N_H be the number of nodes in a leftmost binary tree T of height H. Because T is a leftmost AVL tree

$$N_H = 1 + N_{H-1} + N_{H-2}$$

and $N_1 = 1$ and $N_2 = 2$.

This is almost the recurrence for the Fibonacci numbers. In fact

$$N_H = F_{H+2} - 1$$

if $H > 1$.

Using our approximation for the Fibonacci numbers

$$N_H = (\phi)^{H+2}/\sqrt{5} - 1$$

Now taking logs $(\lg(x) = \log_2(x))$

$$\lg(N_H + 1) = (H + 2)\lg(\phi) - (1/2)\lg(5)$$

or

$$H \approx 1.44\lg(N_H)$$ ●

This result means that any depth-first search through an AVL tree will terminate quickly.

EXERCISES

Each of the following sequences of real numbers is linear recursive. Find the recursion which generates each sequence.

1. $a_0 = 1, 2, 3, 4, 5, \ldots$.

2. $a_0 = 1, 2, 1, 2, 1, 2, \ldots$.

3. $a_0 = 1, 3, 1, 3, 1, 3, \ldots$.

4. $a_0 = 1, 2, 3, 1, 2, 3, 1, 2, 3, \ldots$.

5. $a_0 = 1, 4, 9, 16, 25, \ldots$.

6. $a_0 = 1, 2, 3, 5, 8, 13, \ldots$.

7. $a_0 = 2, 2, 4, 6, 10, 16, \ldots$.

8. $a_0 = 1, 2, 3, 6, 11, 20, 37, \ldots$.

9. $a_0 = 1, -1, 0, 0, -1, -1, -2, -4, \ldots$.

10. $a_0 = 1, 1, 1, 4, 7, 13, 28, \ldots$.

Find the power series which are the inverses of the following polynomials. You may use the formula for the inverse binomial series or the algorithm for generating the inverse of a power series.

11. $1 - x^2$ **12.** $1 + x^2$

13. $1 - x^3$ **14.** $(1 + x)^2$

15. $(1 - x)^3$ **16.** $(1 + x)^3$

17. $1 + x + x^2$ **18.** $1 - x^4$

19. $2 - x^2$ **20.** $1 + x + x^3$

Give the generating function for each of the following sequences. Give the partial fraction decomposition for the generating function, and use it to give an explicit formula for the general term a_n of the sequence.

21. $a_0 = 1, 2, 4, 8, 16, 32, \ldots$.

22. $a_0 = 1, 0, 1, 0, 1, 0, 1, 0, \ldots$.

23. $a_0 = 3, 5, 9, 17, \ldots$.

24. $a_0 = 1/2, -3/4, 7/8, -15/16, \ldots$.

25. $a_0 = 1, 1, 1, 2, 2, 2, 3, 3, 3, \ldots$.

Use the Master Theorem to find the complexity of the following recursions, where in each case $T(1) = 1$.

26. $T(n) = 2T(n/3) + 2$.

27. $T(n) = 2T(n/3) + \sqrt{n}$.

28. $T(n) = 3T(n/3) + n$.

29. $T(n) = 4T(n/2) + n^2$.

30. $T(n) = 2T(n/3) + n\lg n$.

ADVANCED EXERCISES

Here are two sequences which can be viewed as generalizations of the Fibonacci sequence. The first is due to J. H. Conway, and the second to D. Hofstadter.

Conway's sequence is defined by $a(1) = a(2) = 1$ and the recursion

$$a(n) = a(a(n-1)) + a(n - a(n-1))$$

for $n > 2$.

The first few terms of Conway's sequence are 1, 1, 2, 2, 3, 4, 4, 4, ...

Hofstadter's sequence is defined by $Q(1) = Q(2) = 1$ and the recursion

$$Q(n) = Q(n - Q(n-1)) + Q(n - Q(n-2))$$

for $n > 2$.

The first few terms of Hofstadter's sequence are 1, 1, 2, 3, 3, 4, 5, 5, ...

1. Compute the first thirty terms of Conway's sequence.
2. Is Conway's sequence always monotone increasing?
3. Compute the first thirty terms of Hofstadter's sequence.
4. Is Hofstadter's sequence always monotone increasing?

The coefficients of the Taylor expansions of even the simplest functions often have interesting combinatorial properties. For example, the Taylor expansions of the tangent and the secant are

$$\tan(x) = x + 2x^3/3! + 16x^5/5! + 272x^7/7! + \dots$$

$$= \sum_{k=0}^{\infty} A_{2k+1} x^{2k+1}/(2k+1)!$$

$$\sec(x) = 1 + x^2/2! + 5x^4/4! + 61x^6/6! + \dots$$

$$= \sum_{k=0}^{\infty} A_{2k} x^{2k}/(2k)!$$

The numbers A_{2k+1} are called *tangent numbers* and the numbers A_{2k} are called *secant numbers*.

The Russian mathematician V. I. Arnold discovered a remarkable recursion which generates the tangent numbers and the secant numbers. Consider the following Pascal Triangle:

					1					
				0	→	1				
			1	←	1	←	0			
		0	→	1	→	2	→	2		
	5	←	5	←	4	←	2	←	0	
0	→	5	→	10	→	14	→	16	→	16
61	←	61	←	56	←	46	←	32	←	16	← 0

The recursion computes the entries of each row by adding up the entries in the previous row, up to the given entry, in the sense of the arrows in the given row.

The numbers on the left-hand diagonal of this Pascal Triangle are the secant numbers. The numbers on the right-hand diagonal are the tangent numbers.

1. Compute the first six derivatives of $\tan(x)$ and $\sec(x)$. Use these to compute the first six tangent numbers and secant numbers.
2. Use the recursion which we have described to add four rows to the Pascal Triangle above. Use this to express the Taylor expansion of the tangent and secant to ten places.

COMPUTER EXERCISES

1. Use your computer algebra program to compute the first ten derivatives of the tangent and the secant. Express the Taylor expansion of the tangent and the secant to ten places.

2. Use your computer algebra program to compute as many derivatives of the tangent and secant as possible. Give the corresponding Taylor expansion of the tangent and the secant.

3. Write a program to generate the Pascal Triangle for the tangent and secant numbers which we described in the Advanced Exercises. Use it to express the Taylor expansion of the tangent and secant to as many places as possible.

4. Which algorithm for computing the Taylor expansion of the tangent and the secant is best: computing and evaluating the derivatives or using the Pascal Triangle which we have described?

9.2   PROBABILITIES

> *"'Tis the voice of the Jubjub!" he suddenly cried.*
>
> *(This man, that they used to call "Dunce.")*
>
> *"As the Bellman would tell you," he added with pride,*
>
> *"I have uttered that sentiment once.*
>
> *" 'Tis the note of the Jubjub! Keep count, I entreat;*
>
> *You will find I have told it you twice.*
>
> *'Tis the song of the Jubjub! The proof is complete,*
>
> *If only I've stated it thrice."*

On February 16, 1709, Rémond de Montmort sent a copy of his book *On Games of Chance* to Isaac Newton (1642–1727):

> *I do myself the honor Sir of sending you a book with the title* On Games
> of Chance. *I would consider myself infinitely happy if this work could
> meet with your approbation.*

Montmort's book was found later among Newton's papers. The close relationship between probability and combinatorics is suggested by the title of one of the chapters, *On Combinations*.

Long before receiving Montmort's book, Newton had helped and encouraged another French mathematician, Abraham DeMoivre (1667–1754), who moved to England in 1685. DeMoivre began to study probabilities, and Newton encouraged him in this. When he was asked questions about probabilities, Newton would often say

> *Go to Mr. DeMoivre. He knows these things better than I do.*

DeMoivre's book *The Doctrine of Chances* exerted a lasting influence on the theory of probabilities. In *The Doctrine of Chances*, in Montmort's *On Games of Chance*, and in Jakob Bernoulli's *Ars Conjectandi*, the theory of probabilities began to take shape.

Let us remember that Boole hoped to unify combinatorics, logic, and the theory of probabilities, as did Leibniz.

Probability theory can be used to show that interesting objects exist without describing them explicitly. In 1948, Claude E. Shannon published the paper *A Mathematical theory of communication*, which contains a result now called *Shannon's Theorem*. This theorem asserts that given an information channel, there is a code which will allow information to pass through the channel in the best possible way, arbitrarily close to the capacity of the channel and with an arbitrarily small probability of error.

Shannon's Theorem is at the origin of coding theory. The theorem is proved using probabilities, and while it proves that "good codes exist," it gives no hint about how to construct them. In the theory of codes, methods were developed to construct the codes whose existence is proved by Shannon's Theorem.

The theory of probabilities involves three basic concepts: *sample spaces*, *random variables*, and *limit theorems*. Each random variable X has a *distribution*. The *binomial distribution* and the *normal distribution* are the most important examples of probability distributions. The *DeMoivre-Laplace Theorem* and the *Central Limit Theorem* are examples of limit theorems.

In this section, we shall give a brief exposition of these concepts. We shall illustrate them with examples from the history of the theory, such as *De Méré's Paradox* and the *Dice Game of Newton and Pepys*. We shall describe the *tails* of the binomial distribution, which have applications to describing the complexity of an algorithm.

As an application, we shall show how probabilities can be used to study the *reliability* of a network.

In the Advanced Exercises, we shall briefly describe *inversion formulas*, which play a fundamental role in combinatorics. Each finite poset has an inversion formula associated with it. We shall discuss the inversion formula which is associated with the n-cube, *binomial inversion*.

DEFINITION 9.2.1   The *sample space S* is the set of all outcomes of an experiment or process.

DEFINITION 9.2.2   A *random variable X* is a real-valued function on the sample space.

EXAMPLE 1   What is the sample space for the experiment of tossing a coin three times?

Solution   Let us denote "heads" by H and "tails" by T. The sample space S is

$$
\begin{array}{cccc}
 & HHT & HTT & \\
HHH & HTH & THT & TTT \\
 & THH & TTH &
\end{array}
$$

If we set $X =$ the number of heads, X is a random variable.

EXAMPLE 2   What is the sample space for the experiment of tossing a coin four times?

Solution   The sample space S is

$$
\begin{array}{ccccc}
 & & HHTT & & \\
 & HHHT & HTHT & HTTT & \\
HHHH & HHTH & HTTH & THTT & TTTT \\
 & HTHH & THHT & TTHT & \\
 & THHH & THTH & TTTH & \\
 & & TTHH & &
\end{array}
$$

If we set $X =$ the number of heads, then X is a random variable.

DEFINITION 9.2.3   An *event* is a subset of the sample space.

For example, if we toss a coin three times, then

$$E = \{HTT, THT, TTH\}$$

is the event that heads comes up exactly once.

If we toss a coin four times, then

$$E = \{HTTT, THTT, TTHT, TTTH\}$$

is the event that heads comes up exactly once.

If the sample space S is finite, we can define the *probability* of an event E as follows

$$P(E) = |E|/|S|$$

Here it is assumed that all elements in the sample space have an equal likelihood of occurring. Probabilities can also be assigned if this condition is not satisfied. We shall give an example later.

EXAMPLE 3   If we toss a coin three times, what is the probability that exactly one head will come up? If we toss a coin four times, what is the probability that exactly one head will come up?

Solution   The sample space S for tossing a coin three times has eight elements, and the event E that exactly one head comes up has three, so the probability is

$$P(E) = |E|/|S| = 3/8 = .3750 = 37.50\%.$$

For the second question,

$$P(E) = |E|/|S| = 4/16 = .2500 = 25.00\% \qquad \bullet$$

EXAMPLE 4   What is the probability that if we toss two fair dice the sum of the numbers on the dice will be seven or eleven?

Solution   The sample space for tossing two fair dice is a six-by-six grid, which we pictured in Section 3.2. We noticed there that there are six ways for the sum of the numbers on the dice to be seven, and two ways for the sum to be eleven. Thus there are eight possibilities in all for this event E out of thirty-six, so

$$P(E) = |E|/|S| = 8/36 = .2222 = 22.22\% \qquad \bullet$$

The probability function P has two simple properties.

   I. $0 \leq P(E) \leq 1$.

   II. $P(E \cup F) = P(E) + P(F) - P(E \cap F)$.

Property I is clear since probabilities are non-negative and cannot be larger than $P(S) = |S|/|S| = 1$. Property II is simply the Principle of Inclusion-Exclusion for two sets, also known as *finite additivity*. The probability function also obeys the Principle of Inclusion-Exclusion for any finite number of sets.

> **DEFINITION 9.2.4**   Two events E and F are
> **i.** *disjoint*, if $E \cap F = \emptyset$.
> **ii.** *independent*, if $P(E \cap F) = P(E)P(F)$.

THEOREM 9.2.1   If E and F are disjoint events, then

> **(i)** $P(E \cup F) = P(E) + P(F)$.

It is always true that

> **(ii)** $P(E^c) = 1 - P(E)$.

Proof   (i) follows by finite additivity, since $P(E \cap F) = P(\emptyset) = 0$. (ii) follows because $P(E \cup E^c) = P(S) = 1$ and $P(E \cap E^c) = P(\emptyset) = 0$.   ●

The Binomial Distribution   Suppose that we toss a coin n times and that the outcomes of each toss are independent (the outcome of each toss does not affect the outcome of any other toss). The probability of getting heads is $P(H) = p$ and the probability of getting tails is $P(T) = q$, where $0 \le p, q \le 1$, and $p + q = 1$. Let X be the number of heads that comes up. Then

$$P(X = k) = \binom{n}{k} p^k q^{n-k}$$

for $k = 0, 1, \ldots, n$.

The coin is fair (it is equally likely that heads and tails come up) if $p = q = 1/2$. For a fair coin, the expression of the binomial distribution can be simplified somewhat:

$$P(X = k) = \binom{n}{k}(1/2)^n = \binom{n}{k}/2^n$$

EXAMPLE 5   A fair coin is tossed seven times, and X is the number of heads. Find the following probabilities:

> **(i)** $P(X = 3)$.
> **(ii)** $P(X \le 2)$.
> **(iii)** $P(3 \le X \le 5)$.

Solution

> **(i)** $P(X = 3) = \binom{7}{3}/2^7 = 35/128 = .2734 = 27.34\%$.
> **(ii)** $P(X \le 2) = P(X = 0) + P(X = 1) + P(X = 2) =$
> $\left(\binom{7}{0} + \binom{7}{1} + \binom{7}{2} \right)/2^7 = (1 + 7 + 21)/128 = .2266 = 22.66\%$.

(iii) $P(3 \leq X \leq 5) = P(X = 3) + P(X = 4) + P(X = 5) =$
$\left(\binom{7}{3} + \binom{7}{4} + \binom{7}{5}\right)/2^7 = (35 + 35 + 21)/128 = .7109 = 71.09\%.$ ●

Now we will give two examples to illustrate the behavior of the binomial distribution in the case where $p \neq q$.

In the summer of 1653, the Duke of Roannez introduced Pascal to Antoine Gombaud, Chevalier de Méré. De Méré was a member of the nobility and a gambler, and he asked Pascal the following question.

De Méré's Paradox   Player A and Player B are playing the following game with dice. Player A tosses four dice and tries to get at least one six, while Player B tosses twenty-four pairs of dice and tries to get at least one pair of sixes. Since $4/6 = 24/36$ the game should be fair. However, experience shows that Player A has an advantage. Why does this happen?

First notice that if the dice are fair, the probability of getting a six is $p = 1/6$, while the probability of not getting a six is $q = 5/6$. We can use the rule for complements, and the binomial distribution, to compute and compare the probabilities that the two players succeed.

Pascal answered de Méré's question as follows. This was one of the first applications of the binomial distribution.

$$P(A \text{ succeeds}) = 1 - P(A \text{ fails}) = 1 - (5/6)^4 = .5177$$

while

$$P(B \text{ succeeds}) = 1 - P(B \text{ fails}) = 1 - (35/36)^{24} = .4914$$

In other words, Player A has an advantage of about 2.6%. This agrees very well with de Méré's observation.

Shown in Table 9.1 is a dice game of the kind which occurs in de Méré's Paradox. Player A tosses n dice and tries to get at least one six, while Player B tosses $6n$ pairs of dice and tries to get at least one pair of sixes. Notice that Player A always has an advantage.

Samuel Pepys, the author of the famous *Diary*, asked Isaac Newton a question about a gambling game.

Mr. Pepys to Mr. Isaac Newton, Wednesday, November 22, 1693:

> *Sir: ... Now so it is, that the late project (of which you cannot but heard) of Mr. Neale the Groom-Porter his lottery, has almost extinguished for some time at all places of public conversation in this town, especially among men of numbers, every other talk but what relates to the doctrine of determining between the true proportions of the hazards incident to this or that given chance or lot.*

Pepys went on to ask Newton the following question. Suppose that Player A tosses six dice and tries to get at least one six, while Player B tosses twelve dice and tries to get at least two sixes. Is this a fair game? Newton replied to Pepys:

Table 9.1 A Table for de Méré's Paradox

Player A tosses n dice	Player A's Chance	Player B's Chance
$n = 1$	.1667	.1555
$n = 2$	.3056	.2868
$n = 3$	.4213	.3977
$n = 4$	.5177	.4914
$n = 5$	.5981	.5705
$n = 6$	.6651	.6372
$n = 7$	.7209	.6937
$n = 8$	.7674	.7413
$n = 9$	.8062	.7816

Mr. Isaac Newton to Mr. Pepys, Cambridge, November 26, 1693:

Sir: I was very glad to hear of your good health by Mr. Smith, and to have any opportunity given me of showing how ready I should be to serve you or your friends upon any occasion, and wish that something of greater moment would give me a new opportunity of doing it so as to become more useful to you than in solving only a mathematical question.

Newton explained that the game is not fair. We shall give his argument restated in modern language.

We shall analyze this game just as we analyzed de Méré's Paradox, by computing the probabilities that each player succeeds.

$$P(A \text{ succeeds}) = 1 - P(A \text{ fails}) = 1 - (5/6)^6 = .6651.$$

Remember that Player B can fail in two ways: by getting no sixes or only one six.

$$P(B \text{ succeeds}) = 1 - P(B \text{ fails})$$
$$= 1 - (5/6)^{12} - 12 \cdot (1/6)(5/6)^{11} = 1 - .1122 - .2692 = .6186$$

In other words, Player A has an advantage of more than 4.5%, a very large advantage in a gambling game. As Newton put it,

I say, that A has an easier task than B.

Table 9.2 shows the Dice Game of Newton and Pepys. Player A tosses n dice and tries to get at least one six, while Player B tosses $2n$ dice and tries to get at least two sixes.

Notice how the probabilities in this game behave differently from those in the game based on de Méré's Paradox. At first, Player A has an advantage, but as more dice are tossed, the advantage shifts to Player B.

Remember that a random variable X is a function on the sample space, which we assume to be finite. Every random variable has a *distribution*

$$f(x) = P(X = x)$$

Table 9.2   A Table for the Dice Game of Newton and Pepys

Player A tosses n dice	Player A's Chance	Player B's Chance
$n = 1$	.1667	.0278
$n = 2$	.3056	.1319
$n = 3$	.4213	.2632
$n = 4$	.5177	.3953
$n = 5$	.5981	.5155
$n = 6$	.6651	.6187
$n = 7$	.7209	.7040
$n = 8$	.7674	.7728
$n = 9$	.8062	.8272

The distribution records the probabilities that the random variable will take on a given value. The binomial distribution is an example.

Three quantities are associated with a random variable X:

The *mean*, or *expected value*

$$\mu = \mu_X = E(X) = \sum x \cdot f(x)$$

The *variance*

$$\text{var}(X) = E((X - \mu_X)^2) = E(X^2) - \mu^2$$

The *standard deviation*

$$\sigma_X = \sqrt{\text{var}(X)}$$

The standard deviation measures the "expected distance" from the random variable to its mean.

EXAMPLE 6   A fair coin is tossed four times, and X is the number of heads. Find the mean, variance, and standard deviation of X.

Solution   The mean is

$$\mu = E(X) = 0 \cdot (1/16) + 1 \cdot (4/16) + 2 \cdot (6/16) + 3 \cdot (4/16) + 4 \cdot (1/16) = 32/16 = 2$$

The variance is

$$\begin{aligned}\text{var}(X) &= E(X^2) - \mu^2 \\ &= 0^2 \cdot (1/16) + 1^2 \cdot (4/16) + 2^2 \cdot (6/16) + 3^2 \cdot (4/16) + 4^2 \cdot (1/16) - 2^2 \\ &= 80/16 - 4 = 1\end{aligned}$$

The standard deviation is

$$\sigma_X = \sqrt{1} = 1 \qquad \bullet$$

In general, the mean and standard deviation for the binomial distribution can be evaluated as follows.

THEOREM 9.2.2   If a coin is tossed n times and X is the number of heads, then

$$\mu = \mu_X = np$$
$$\sigma = \sigma_X = \sqrt{npq}$$

●

Here is an example from Montmort's *On Games of Chance* which involves expectations.

EXAMPLE 7   Baron de la Hontan, the French explorer, visited a Canadian Indian tribe where the following game was played. Player A and Player B bet an equal amount. Player A tosses eight counters in the air. The counters are black on one side and white on the other. Player A wins if an odd number of counters turns up black; Player A wins double the amount if the counters turn up all black or all white; and otherwise Player B wins. Is this a fair game?

Solution   We will answer this question by computing Player A's expectation. The game is fair if the expectation of each player is zero. The probabilities involved are the entries of the eighth row of the Pascal Triangle, divided by $2^8 = 256$. Let X be the amount which Player A will receive in the various cases ($X = 1, 2,$ or -1 times the amount that each player has bet).

$$E(X) = 2/256 + 8/256 - 28/256 + 56/256 - 70/256 + 56/256 - 28/256$$
$$+ 8/256 + 2/256 = 6/256$$

The game is not fair, and Player A has a slight advantage.   ●

Two random variables X, Y are *independent* if

$$P(X = x \text{ and } Y = y) = P(X = x)P(Y = y)$$

The following theorem is the basic result about the distribution of sums of independent random variables. Notice that if two random variables have the same distribution, then they have the same mean μ and standard deviation σ.

Central Limit Theorem   Let $X_1, X_2, \ldots, X_n$ be independent random variables with the same distribution and mean $\mu = 0$. Then

$$P(z_1 \leq (X_1 + \ldots + X_n)/\sigma\sqrt{n} \leq z_2) \to \left(1/\sqrt{2\pi}\right) \int_{z_1}^{z_2} e^{-z^2/2} dz$$

as $n \to \infty$.   ●

The function

$$f(x) = \left(1/\sqrt{2\pi}\right) e^{-z^2/2}$$

describes the *standard normal distribution*, and the Central Limit Theorem says that the mean of a sum of independent random variables with the same distribution has a distribution which is closer and closer to a normal distribution as n increases. The integral computes the area under the graph of the standard normal distribution between z_1 and z_2.

In the case of the coin tossing game, where $X_k = 1$ if the coin comes up heads on the kth toss and $X_k = -1$ if the coin comes up tails on the kth toss, the Central Limit Theorem is known as the *DeMoivre-Laplace Theorem.*

The DeMoivre-Laplace Theorem expresses the very important fact, which was first recognized by DeMoivre, that the binomial distribution approaches the normal distribution as the number of trials n increases.

DeMoivre published the first description of the normal distribution in 1733. He published in an obscure journal, and the result was not widely noticed. It was then rediscovered by Laplace, Gauss, and others. A proof of the Central Limit Theorem was first given by the Russian mathematician V. I. Kolmogorov in 1930.

In general, if X is a random variable which is normally distributed, we can form the *z-score*

$$z = (X - \mu)/\sigma$$

which has the effect of making the mean of X equal to zero and the standard deviation of X equal to one. The distribution of z is then the standard normal distribution. If we wish to compute the probability that $x_1 \leq X \leq x_2$, we can do this by computing the z-scores z_1 and z_2 of x_1 and x_2 and then computing

$$P(x_1 \leq X \leq x_2) = P(z_1 \leq z \leq z_2) = (1/\sqrt{2\pi}) \int_{z_1}^{z_2} e^{-z^2/2} dz$$

by using a normal table.

Because the binomial distribution closely approximates the normal distribution when n is large, we can use the normal distribution to study the binomial distribution. In fact what happens is rather elegant: just as it becomes inconvenient to compute the terms of the binomial distribution, because the binomial coefficients become large, it is no longer necessary to do so because the normal approximation can be used.

In the coin tossing problem, if $p = q = 1/2$, the binomial distribution is well approximated by the normal distribution if $n \geq 10$. We will give three examples to show how the normal approximation to the binomial distribution can be used.

The last two examples show how the normal distribution can be used to approximate the *tails* of the binomial distribution, that is, to find the probabilities of a large or small number of heads occurring. This kind of computation comes up frequently in algorithmic problems.

In computing the tails of the binomial distribution, we make two simplifying assumptions: that the coin is fair ($p = q = 1/2$) and that the number of tosses is large (we shall choose $n = 100$). Other more complicated formulas can be used to describe the tails of the binomial distribution if these assumptions are not satisfied.

EXAMPLE 8   A fair coin is tossed 100 times. What is the probability that the number of heads is between 50 and 60?

Solution   By the theorem about the mean and standard distribution of the binomial distribution, $\mu = np = 100(1/2) = 50$, and $\sigma = \sqrt{npq} = \sqrt{100(1/2)(1/2)} = 5$. The z-scores are

$z_{50} = (50 - 50)/5 = 0$ and $z_{60} = (60 - 50)/5 = 2$. From a normal table,

$$P(0 \leq z \leq 2) = .4772 = 47.72\%$$ ●

EXAMPLE 9   A fair coin is tossed 100 times. What is the probability that the number of heads is 60 or more?

Solution   As in the previous problem, $z_{60} = 2$. Half the area under the standard normal curve is .5000, so the probability is

$$P(z \geq 2) = .5000 - .4772 = .0228 = 2.28\%$$ ●

EXAMPLE 10   A fair coin is tossed 100 times. What is the probability that the number of heads is 35 or less?

Solution   The z-score of $X = 35$ is $z_{35} = (35 - 50)/5 = -3$. Then from a normal table,

$$P(z \leq -3) = P(z \geq 3) = .5000 - .4987 = .0013 = .13\%$$ ●

APPLICATION   *The Reliability of a Network*

We discussed networks and their properties in the previous chapter.

There are several ways to formulate the reliability problem for a network G. We may be concerned with the possibility that all nodes can communicate with each other, or with the possibility that there is an operational path between two fixed nodes, and so on. Here we will concern ourselves only with the *all-terminal reliability problem*, the problem of determining when all nodes can communicate. In this problem, the network G is viewed as an undirected graph.

In our model, nodes never fail but edges may fail. The nodes may be computers or stations in a communications network. The edges may be communication channels, which are subject to failure due to overload, storms, and so on.

We shall make a further assumption, that all edges of G have the same probability p of failure. For example, if 10% of the channels in a communications system fail in any given year, we can use $p = .9$ as the probability that the channels will be operational, so we would assign $p = .9$ to the edges in the corresponding graph.

In order for the system modelled by the graph G to be operational in spite of the failure of some channels, when we remove the edges corresponding to the failed channels, we must have a subgraph G' left which is operational, that is, G' must be a connected graph with $E(G') \subset E(G)$. If N_i is the number of operational subgraphs of G with i edges, and if $e = |E(G)|$, then the *reliability polynomial*

$$\text{Rel}(p) = \sum_{i=0}^{e} N_i p^i (1 - p)^{e-i}$$

measures the probability that G will be operational.

For example, consider the network in Figure 9.3. An operational subgraph must contain at least three edges, so $N_0 = N_1 = N_2 = 0$. We see that $N_3 = 8$, $N_4 = 5$, and

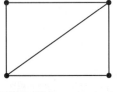

Figure 9.3
A Network

$N_5 = 1$, so

$$\text{Rel}(p) = 8p^3(1-p)^2 + 5p^4(1-p) + p^5$$

Suppose that we have two networks G_1 and G_2 with reliability polynomials $\text{Rel}_1(p)$ and $\text{Rel}_2(p)$. Then we say that G_2 is more reliable than G_1 if

$$\text{Rel}_1(p) < \text{Rel}_2(p)$$

The reliability of a network may vary with the probability of edge failure, and we will give an example to illustrate this. Consider the two networks in Figure 9.4.

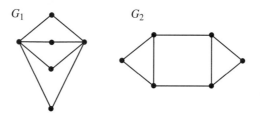

Figure 9.4   Two Networks

If the network on the left is G_1 and the network on the right is G_2, the reliability polynomials are

$$\text{Rel}_1(p) = 32p^5(1-p)^3 + 24p^6(1-p)^2 + 8p^7(1-p) + p^8$$
$$\text{Rel}_2(p) = 30p^5(1-p)^3 + 25p^6(1-p)^2 + 8p^7(1-p) + p^8$$

The difference $\text{Rel}_1(p) - \text{Rel}_2(p) = 2p^5(1-p)^3 - p^6(1-p)^2$ is zero when $p = 2/3$, negative when $p > 2/3$, and positive when $p < 2/3$. Thus G_1 is more reliable when $p < 2/3$, and G_2 is more reliable when $p > 2/3$.

EXERCISES

Two fair dice are tossed, and X is the sum of the numbers which come up. Find the following probabilities.

1. $P(X = 6)$ **2.** $P(X = 7)$
3. $P(X \leq 4)$ **4.** $P(X \geq 8)$

A fair coin is tossed n times, and X is the number of heads that come up. Find the following probabilities.

5. $n = 5$, $P(X = 3)$ **6.** $n = 6$, $P(X = 3)$
7. $n = 7$, $P(X = 5)$ **8.** $n = 8$, $P(X = 4)$
9. $n = 5$, $P(X \leq 2)$ **10.** $n = 6$, $P(X \leq 2)$
11. $n = 7$, $P(2 \leq X \leq 4)$
12. $n = 8$, $P(X \leq 3)$ **13.** $n = 8$, $P(3 \leq X \leq 5)$
14. $n = 8$, $P(X \geq 6)$

A coin is tossed n times, for which $p = 0.3$ and $q = 0.7$. Find the following probabilities.

15. $n = 5$, $P(X = 3)$ **16.** $n = 5$, $P(X = 4)$
17. $n = 7$, $P(X \leq 2)$ **18.** $n = 7$, $P(X \geq 5)$
19. $n = 8$, $P(X \leq 2)$ **20.** $n = 8$, $P(3 \leq X \leq 5)$
21. $n = 8$, $P(X \geq 6)$

22. Give your own computation of Player A's chance and Player B's chance in the game associated with de Méré's Paradox when Player A tosses eight dice and Player B tosses 48 pairs of dice.

23. Give your own computation of Player A's chance and Player B's chance in the Dice Game of Newton and Pepys when Player A tosses eight dice and Player B tosses sixteen dice.

A fair coin is tossed 100 times, and X is the number of heads that comes up. Use the normal approximation to the binomial distribution to compute the following probabilities.

24. $P(50 \le X \le 55)$ **25.** $P(50 \le X \le 65)$

26. $P(40 \le X \le 60)$ **27.** $P(42 \le X \le 52)$

28. $P(38 \le X \le 44)$ **29.** $P(X \le 40)$

30. $P(X \le 38)$ **31.** $P(X \le 43)$

32. $P(X \ge 45)$ **33.** $P(X \ge 55)$

34. $P(X \ge 58)$ **35.** $P(X \ge 63)$

ADVANCED EXERCISES

Here we shall give a very brief description of *inversion formulas*, which play a fundamental role in combinatorics.

Let S be a finite set, and suppose that $A_1, A_2, \ldots, A_n$ are subsets of S. How many elements of S belong to exactly m of the subsets A_i?

If we define

$$B_m = \cup(A_{i_1} \cap A_{i_2} \cap \ldots \cap A_{i_m})$$

then B_m is the set of elements which belong to m or more of the subsets A_i, and $B_m - B_{m+1}$ is the set of elements which belong to exactly m of the A_i. The number of elements of this set is counted by a variant of the Principle of Inclusion-Exclusion, called the *Binomial Inversion Formula*.

Binomial Inversion   The number of elements of S belonging to exactly m of the subsets A_i is

$$|B_m - B_{m+1}| = \sum |A_{i_1} \cap \ldots \cap A_{i_m}|$$

$$- \sum \binom{m+1}{m}|A_{i_1} \cap \ldots \cap A_{i_{m+1}}|$$

$$+ \ldots + (-1)^{n-m}\binom{n}{m}|A_1 \cap \ldots \cap A_n|$$

where the sums are over intersections of the given type.

Binomial inversion is inversion on the n-cube. There is another inversion formula, *Möbius Inversion*, which is inversion on the poset $\mathrm{Div}(n)$. Since $\mathrm{Div}(m)$ is the n-cube if m is square-free with n prime factors, the Möbius Inversion Formula generalizes the Binomial Inversion Formula. There is an inversion formula for any finite poset, a generalization of Binomial Inversion and Möbius Inversion.

The statement of Binomial Inversion for a probability function is just the same as the statement for the counting function.

Inversion formulas are the discrete analogs of the *Fourier transform*, which is important in all parts of mathematics and physics.

1. Use the Binomial Inversion formula to prove the following theorem.

Theorem 9.2.3

Let $A_1, \ldots, A_n \subset S$ be finite sets such that the probability of any k-fold intersection is $1/(k + 1)$. Then

$$P(B_m - B_{m+1}) = 1/(n + 1)$$

for $m = 0, 1, \ldots, n$ (we assume that $A_{n+1} = \emptyset$).   ●

Hint: Use the Inverse Expansion Identity which we stated in Section 3.3.

Notice how this theorem generalizes the statement of the Towns with Clubs problem from Chapter 1. In that problem, there are three clubs (three subsets of S), such that half the elements of S belong to each subset, one-third of the elements of S to each pair of subsets, and one-fourth of the elements of S to all three subsets.

COMPUTER EXERCISES

1. Verify the conclusion of the theorem in the Advanced Exercises if the number of subsets is 10.

9.3   GROUPS AND COUNTING

They sought it with thimbles, they sought it with care;

They pursued it with forks and hope;

They threatened its life with a railway share;

They charmed it with smiles and soap.

In many parts of mathematics, and in particular in discrete mathematics, it is important to take account of the *symmetries* of the objects which are being studied. The theory of *groups* attempts to give a systematic account of symmetries and their properties.

An object may have obvious symmetries, like the symmetries of an equilateral triangle which we shall describe below. Other symmetries may be hidden. For example, the equations which describe the motion of a wave in a shallow canal have hidden symmetries which can be used to describe the properties of the wave.

As an example of hidden symmetries, we shall show how permutations can be used to describe the marriage laws of the Murngin, a tribe of Northern Australia. In the Advanced Exercises, we shall describe the eighteenth-century game of *Treize* in terms of permutations. In the Supplementary Exercises, we shall describe the Victorian game of *Mousetrap* in a similar way.

Lagrange's Theorem, *Burnside's Lemma*, and *Polya's Theorem* are three powerful counting rules which exploit symmetries. We shall show how each of these rules can be used to solve counting problems.

The theory of groups began in work of Joseph-Louis Lagrange (1736–1813) on permutations and C. F. Gauss (1777–1855) on quadratic forms. The theory was developed by Niels Henrik Abel (1802–1829) and Evariste Galois (1811–1832), and applied to the problem of finding the roots of a polynomial equation. *Galois theory* has important applications to the theory of codes.

One way to understand the importance of groups in counting problems is to realize that it is always easier to count labelled objects than it is to count unlabelled objects. If we count labelled objects of some kind, we can use the symmetries described by groups to "remove the labels" and count the resulting unlabelled objects. As an application of these ideas, we shall show how to count rooted trees and unlabelled trees.

DEFINITION 9.3.1   A *group* G is a set with a binary law of composition $GoG \to G$ which satisfies the following axioms

$$G1. \ ao(boc) = (aob)oc$$

for all $a, b, c \in G$.

G2. There is an element $e \in G$ such that

$$eoa = aoe = a$$

for all $a \in G$.
  G3. For each $a \in g$ there is an element $a^{-1} \in G$ such that

$$a^{-1}oa = aoa^{-1} = e$$

The first axiom says that the law of composition is *associative*. The second axiom says that the group has an *identity element*. The third axiom says that each element of the group has an *inverse*.

It can be proved that identity elements and inverses in a group are unique. Once it has been made clear what the law of composition is, we shall omit the symbol "o." That is, we shall write $aob = ab$, and so on.

The term "group" was first used by Galois. The *Traité des substitutions et des équations algébriques* (Treatise on permutations and algebraic equations) of Camille Jordan (1838–1922), which was published in 1870, was an important early text in the theory of groups.

As the name of Jordan's book implies, groups were first thought of as sets of permutations which are closed under composition. A permutation π of a set X can be thought of as a one-to-one correspondence $\pi : X \rightarrow X$. Permutations can also be described using *cycle notation*.

DEFINITION 9.3.2   Let X be a set. A *k-cycle* is an expression

$$(a_1, a_2, \ldots, a_k)$$

where $a_i \in X$, and the cycle denotes the permutation $\pi(a_i) = a_{i+1}$ if $i < k$, and $\pi(a_k) = a_1$.

In particular, a *fixed point* i of a permutation π is given by a cycle of length one: (i). A cycle of length two (i, j) is called a *transposition*. Two cycles are *disjoint* if they have no elements in common.

We can picture a k-cycle like this:

$$a_1 \rightarrow a_2 \rightarrow \ldots \rightarrow a_{k-1} \rightarrow a_k \rightarrow a_1$$

The following theorem describes how permutations can be decomposed into cycles. We shall omit the proof. When we write a product of cycles, it is understood that the cycles are composed from right to left.

THEOREM 9.3.1   Let X be a finite set.

  (i)   Every permutation of X can be expressed as a finite product of disjoint cycles.

(ii)   The identity

$$(a_1, a_2, \ldots, a_k) = (a_1, a_k)(a_1, a_{k-1}) \ldots (a_1, a_2)$$

shows that every cycle of length k can be expressed as a product of $k - 1$ transpositions.

(iii)   A permutation of X can be expressed as a product of either an even number of transpositions or an odd number of transpositions, but not both.   ●

Disjoint cycles commute, so it does not matter what order the cycles in **(i)** are written in.

A permutation is called *even* if it factors into an even number of transpositions, and *odd* if it factors into an odd number of transpositions. The last part of the theorem says that a permutation cannot be both even and odd.

It is easy to compose permutations in cycle notation. If we are given a product of cycles, start at the right and keep track of the effect of each cycle on each element. For example, given

$$(1, 2, 3)(1, 2, 4)$$

start on the right with the element 1. The cycle on the right maps 1 to 2, and the next cycle maps 2 to 3, so we write $(1, 3)$. The cycle on the right leaves 3 fixed and the next cycle maps 3 to 1, so we close up the cycle and write $(1, 3)$. Similarly, the cycle on the right maps 2 to 4, and 4 to 1 which is then mapped to 2, so we get another cycle $(2, 4)$. Altogether,

$$(1, 2, 3)(1, 2, 4) = (1, 3)(2, 4) = (2, 4)(1, 3).$$

Similarly,

$$(1, 2, 4)(1, 2, 3) = (1, 4)(2, 3)$$

Notice that $(1, 2, 3)(1, 2, 4) \neq (1, 2, 4)(1, 2, 3)$. In other words, composition of permutations need not be commutative.

A subset H of a group G is called a *subgroup* if it satisfies the group axioms for the same law of composition.

DEFINITION 9.3.3   The set of all permutations of the set $\{1, 2, \ldots, n\}$ is called the *symmetric group* S_n. The set of even permutations of $\{1, 2, \ldots, n\}$ is called the *alternating group* A_n.

It is easy to verify that S_n is a group. More generally, the set of all one-to-one correspondences of any set with itself is a group. It is also easy to check that A_n is a group, and that A_n is a subgroup of S_n.

A group G is *finite* if the set G is finite, and the number of elements of G is called the *order* of G and denoted $|G|$.

In Chapter 3 we saw that

$$|S_n| = n!$$

and it follows from Lagrange's Theorem that

$$|A_n| = n!/2$$

EXAMPLE 1   List the elements of the group S_3 and its subgroup A_3.

Solution   The elements of S_3 are the identity e and the cycles $(1, 2), (1, 3), (2, 3), (1, 2, 3), (1, 3, 2)$. 3-cycles factor into two transpositions and hence are even, so the elements of A_3 are e, $(1, 2, 3), (1, 3, 2)$.   ●

Examples of Groups

   I.   The set of integers Z with addition is a group.

   II.   The set of integers (mod n) Z_n with addition of congruence classes is a group.

   III.   The set Z_n^* of elements of Z_n with multiplicative inverses and multiplication of congruence classes is a group.

   IV.   The symmetric group S_n.

   V.   The alternating group A_n.

The symmetries of an object form a group. In Figure 9.5 we have an equilateral triangle and a square. There are six symmetries of the triangle: the identity, rotation counterclockwise through 120 or 240 degrees, and reflections about any of the three axes of symmetry. These symmetries all correspond to permutations of the three vertices, so the group of symmetries of the triangle is the symmetric group S_3.

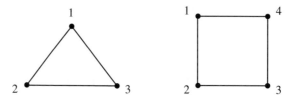

Figure 9.5   A Triangle and a Square

The symmetries of a regular polygon with n sides form a group called the *dihedral group* D_n. This group always has $2n$ elements: n rotations, and reflections about the n axes of symmetry. It happens that $D_3 = S_3$, but if $n > 3$, D_n is always a proper subgroup of S_n.

The symmetries of a square are the elements of the group D_4. The rotations are the identity e, $(1, 2, 3, 4)$, $(1, 3)(2, 4)$, and $(4, 3, 2, 1)$, and the reflections are $(2, 4)$, $(1, 2)(3, 4)$, $(1, 3)$, and $(1, 4)(2, 3)$.

We noticed earlier that permutations need not commute. Here is a very curious application of this fact.

While writing his thesis on kinship structures, the French ethnologist Claude Levi-Strauss studied the marriage laws of a number of tribes. The marriage laws of primitive tribes can be quite complicated.

The marriage laws of the Murngin, a tribe of Northern Australia, are particularly complicated, and it seemed to Levi-Strauss that there was something mathematical about them. Levi-Strauss asked his friend André Weil whether it was possible to give them a mathematical description. Weil answered Levi-Strauss' question in a remarkable paper *Sur l'étude algébrique de certains types de lois de mariage* (On the algebraic study of certain types of marriage laws), which was published in 1949.

EXAMPLE 2   Describe André Weil's model for the marriage laws of the Murngin.

Solution   Weil found that the marriage laws of the Murngin could be analyzed in the following way. First, the men and women in the tribe who are allowed to marry are partitioned into disjoint sets

$$M_1, M_2, \ldots, M_n$$

the *types* of marriages which are possible. Now the marriage laws of the Murngin can be expressed by the following three conditions:

(A)   Each man or woman of the tribe can contract only one type of marriage. That is, if the man or woman belongs to M_i, there is a unique type M_j consisting of the persons whom that man or woman is allowed to marry.

(B)   For each man or woman, the type of the woman or man whom he or she is allowed to marry is uniquely determined by the person's sex and the type of marriage of the person's parents.

(C)   Each man is allowed to marry the daughter of the brother of his mother.

The first two conditions mean that the possible types of marriages correspond to functions $\pi(M_i) = M_j$, that is, to permutations of the n classes $M_1, M_2, \ldots, M_n$. Thus each woman A and man B of the tribe have permutations π_A and π_B associated with them.

Condition (C), when it is analyzed, turns out to mean that

$$\pi_A \pi_B = \pi_B \pi_A$$

In other words, woman A and man B are allowed to marry if and only if the corresponding permutations commute!   ●

It is possible to describe when two permutations commute, in terms of their cycle decompositions: the criterion is by no means simple. It is quite remarkable that the Murngin found a condition of comparable subtlety, expressed in their marriage laws.

If $H \subset G$ is a subgroup and $a \in G$, the *left coset* of H by a is the set

$$aH = \{ah : h \in H\}$$

The number of left cosets of a subgroup H of a group G is called the *index* of H in G, and denoted by $(G : H)$.

Lagrange's Theorem   Let G be a finite group and $H \subset G$ a subgroup. Then

$$|G| = |H|(G : H)$$

Proof   First, we claim that for any subgroup H of any group G, the left cosets give a partition of G. To prove this we must show, just as in the proof of the Partition Theorem, that two left cosets aH and bH are disjoint if they are not equal. We can prove this by showing that if two left cosets have an element in common, then they must be equal.

Suppose that $x \in aH \cap bH$. Then $x = ah_1 = bh_2$, where $h_1, h_2 \in H$, so $a = bh_2h_1^{-1}$. Then if $ah \in aH$, $ah = bh_2h_1^{-1}h \in bH$, so $aH \subset bH$. By symmetry, $bH \subset aH$, so $aH = bH$.

Now assume that G and hence H and $(G : H)$ are finite. The mapping

$$f : H \to aH$$

defined by $f(h) = ah$ is a one-to-one correspondence. It is clearly surjective, and if $f(h_1) = ah_1 = f(h_2) = ah_2$, then $a^{-1}ah_1 = h_1 = h_2 = a^{-1}ah_2$, so f is injective. Thus $|aH| = |H|$ for all left cosets. Since the left cosets partition G into disjoint subsets each with the same number of elements, the theorem follows.   ●

Here is an application of Lagrange's Theorem.

The Key Ring Problem   Four keys are placed on a key ring. How many configurations of keys are possible, if two are considered to be the same if they can be interchanged by moving the key ring in any way?

Solution   The set of all possible configurations of the four keys is the set of all permutations of four objects, S_4, which is also a group. The motions of the key ring form a subgroup, D_4. The number of different configurations is the index of D_4 in S_4, which is

$$(S_4 : D_4) = 24/8 = 3$$

We show three configurations in Figure 9.6.   ●

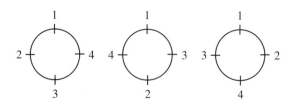

Figure 9.6   The Key Ring Problem

The intuitive idea that the symmetries of an object form a group can be made precise by defining an *action of a group on a set.*

DEFINITION 9.3.4   An *action* of a group G on a set X is a function $G \times X \to X$ which satisfies the following two conditions. We will denote the image of (g, x) by gx.

   I. $ex = x$ for all $x \in X$.

   II. $(g_2 g_1)x = g_2(g_1 x)$ for all $x \in X$ and all pairs of elements $g_1, g_2 \in G$.

There are three types of sets associated with any group action:

   I. $X_g = \{x \in X : gx = x\}$

   II. $Gx = \{gx : g \in G\}$

   III. $G_x = \{g \in G : gx = x\}$

The set Gx is called the *orbit* of x. The set G_x is called the *stabilizer* of x, and is a subgroup of G. If G is finite it is easy to prove that

$$|Gx| = (G : G_x);$$

that is, that the order of the orbit of any element $x \in X$ is equal to the index of its stabilizer. The stabilizers G_y of elements of a given orbit Gx all have the same number of elements as G_x.

The following result was stated by William Burnside (1852–1927) and is named for him, but it was known earlier to several mathematicians and was stated earlier by Ferdinand Georg Frobenius (1849–1917).

Burnside's Lemma   Let G be a finite group acting on a finite set X. If R is the number of orbits of this action, then

$$R \cdot |G| = \sum_{g \in G} |X_g|$$

Proof   Let $E = \{(g, x) : gx = x\}$, and let $B = \{Gx : x \in X\}$ be the set of orbits of the action. Define a function $f : E \to B$ by setting $f((g, x)) = Gx$. Clearly

$$|E| = \sum_{g \in G} |X_g|$$

and for any $x \in X$,

$$f^{-1}(Gx) = \{(g, y) : g \in G_y \text{ and } y \in Gx\}$$

Because the stabilizers G_y all have the same number of elements as G_x, this latter set has $|G_x|(G : G_x) = |G|$ elements. Because the function f is surjective, Burnside's Lemma now follows by the Surjective Function Rule.   ●

EXAMPLE 3   The vertices of a square are colored with any of three colors, blue (B), red (R), or white (W). The square is then allowed to rotate or flip in any way, and two colorings are considered to be equivalent if they can be exchanged by such a motion. How many different colorings are there?

Solution   To solve this problem using Burnside's Lemma we must count the number of elements in the sets X_g for each element $g \in D_4$. Remember that there are eight elements in D_4.

There are $3^4 = 81$ possible colorings and the identity element fixes them all, so $|X_e| = 81$. For a coloring to be fixed under $(1, 2, 3, 4)$ or $(4, 3, 2, 1)$ all the colors must be the same, so for these two elements $|X_g| = 3$. For a coloring to be fixed under $(1, 3)(2, 4)$ opposite vertices must have the same color, so for this element $|X_g| = 9$. For the two reflections in an axis through two opposite vertices, $|X_g| = 27$, and for the two reflections in an axis through midpoints of opposite sides, $|X_g| = 9$. Altogether,

$$R|D_4| = 8R = 81 + 3 + 3 + 9 + 27 + 27 + 9 + 9 = 168$$

so the number of inequivalent colorings is $R = 168/8 = 21$.   ●

Now we shall describe *Polya's Theorem*, a powerful counting rule which, among other things, can be viewed as a generalization of both Lagrange's Theorem and Burnside's Lemma. This theorem was originally stated in Polya's monograph *Combinatorial Enumeration of Groups, Graphs, and Chemical Compounds*, which was published in 1937 (the original monograph was in German). Polya's Theorem gives what is in a sense the ultimate expression of the role of symmetries in counting problems. It is particularly useful in the enumeration of trees.

Recall the idea which we stated at the beginning of this section, that the role of groups in counting problems is to allow us to count unlabelled objects by first counting labelled objects and then "removing the labels." The statement of Polya's Theorem which we shall give here is adapted to counting colored objects which are not equivalent under some group of transformations.

While Polya's Theorem can be stated in many ways, all versions of the theorem depend on the concept of the *cycle index*.

DEFINITION 9.3.5   Let G be a subgroup of S_n. The *cycle index* of G is

$$P_G = \frac{1}{|G|} \sum f_{i_1 i_2 \dots i_n} x_1^{i_1} x_2^{i_2} \dots x_n^{i_n}$$

where

$$f_{i_1 i_2 \dots i_n}$$

is the number of elements of G which decompose into i_1 cycles of length one, i_2 cycles of length two, ..., and i_n cycles of length n.

By convention, each term is chosen in such a way that $\sum_j i_j = n$. This means for example that the identity is viewed as the cycle $(1)(2)\ldots(n)$ and is represented by x_1^n.

EXAMPLE 4   Compute the cycle index of S_3.

Solution   The elements of S_3, decomposed into cycles according to the rule which we have given, are $(1)(2)(3)$, $(1,2)(3)$, $(2)(1,3)$, $(1)(2,3)$, $(1,2,3)$, and $(3,2,1)$. Thus

$$P_{S_3} = (1/6)(x_1^3 + 3x_1x_2 + 2x_3)$$ ●

EXAMPLE 5   Compute the cycle index of D_4.

Solution   The elements of D_4, decomposed according to the rule which we have given, are $(1)(2)(3)(4)$, $(1,2,3,4)$, $(1,3)(2,4)$, $(4,3,2,1)$, $(1,3)(2)(4)$, $(1,2)(3,4)$, $(1)(3)(2,4)$, and $(1,4)(2,3)$. Thus

$$P_{D_4} = (1/8)(x_1^4 + 3x_2^2 + 2x_1^2x_2 + 2x_4)$$ ●

Now we shall state Polya's Theorem, which in a sense gives the ultimate expression of counting rules involving symmetry. There are several statements of Polya's Theorem. We shall give one which is adapted to counting the G-inequivalent colorings of some geometric object X. We should mention that Burnside's Lemma is used in the proof of Polya's Theorem.

The proof of Polya's Theorem is too difficult to be given here. However, the idea of the theorem is quite simple: the number of inequivalent colorings is to be obtained by "substitution in the cycle index."

Polya's Theorem   Let $G \subset S_n$ be the group of symmetries of a geometric object X. The number of G-inequivalent r-colorings of X is

$$P_G(r, r, \ldots, r)$$ ●

EXAMPLE 6   Find the number of inequivalent 2-colorings of the vertices of an equilateral triangle.

Solution   Here $G = S_3$, and we know that the cycle index is $P_G = (1/6)(x_1^3 + 3x_1x_2 + 2x_3)$. Hence the number of 2-colorings is

$$P_{S_3}(2, 2, 2) = (1/6)(2^3 + 3 \cdot 2 \cdot 2 + 2 \cdot 2) = (1/6)(24) = 4.$$

We list the different 2-colorings in Figure 9.7.   ●

EXAMPLE 7   Find the number of inequivalent 3-colorings of the vertices of a square.

Solution   Using Burnside's Lemma, we saw that the number of inequivalent colorings is 21. Now we take $P_{D_4} = (1/8)(x_1^4 + 3x_2^2 + 2x_1^2x_2 + 2x_4)$ and substitute:

$$P_{D_4} = (1/8)(3^4 + 3 \cdot 3^2 + 2 \cdot 3^2 \cdot 3 + 2 \cdot 3) = (1/8)(168) = 21$$ ●

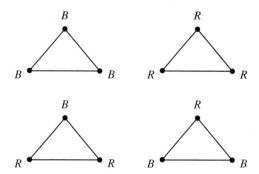

Figure 9.7   Two-Colorings of a Triangle

APPLICATION   *Rooted Trees and Unlabelled Trees*

We have already given the formula n^{n-2} for the number of labelled trees with n nodes, which is due to Arthur Cayley. It is an important question in combinatorics to count or list trees of various types. Two questions of this kind are particularly important:

I.   What is the number T_n of rooted trees with n nodes?

II.   What is the number t_n of unlabelled trees with n nodes?

Cayley gave a formula which can be used to compute T_n in his paper *On the theory of the analytical forms called trees*, which was published in 1857. If we remove the root from a rooted tree, we can label the adjacent nodes of the resulting trees as roots, giving a finite number of rooted trees. Cayley realized that this observation, if it is interpreted carefully enough, gives a recursion for the numbers T_n. He was led to the formula

$$(1 - x)^{-1}(1 - x^2)^{-T_1}(1 - x^3)^{-T_2}(1 - x^4)^{-T_3} = 1 + T_1 x + T_2 x^2 + T_3 x^3 + \ldots$$

Cayley stated this formula but gave no proof for it. The first proof was given by Polya, using Polya's Theorem, in his monograph of 1937.

Cayley's formula expresses a deep insight into the structure of rooted trees, but it is rather hard to understand. It simplifies matters a great deal if we realize that Cayley's equation is equivalent to the following explicit recursion for the numbers T_n. Because it is equivalent to Cayley's equation, we shall call it *Cayley's Formula*.

Cayley's Formula   The number T_n of rooted trees with n nodes satisfies the recursion

$$T_{n+1} = \frac{1}{n} \sum_{k=1}^{n} \left(\sum_{d \mid k} d T_d \right) T_{n-k+1}$$

where $d \mid k$ means that d divides k.

EXAMPLE 8   Compute T_5 using Cayley's Formula.

Solution   It is conventional to set $T_0 = 1$. Clearly $T_1 = T_2 = 1$. $T_3 = 2$ and $T_4 = 4$ because there are four rooted trees with four vertices:

Now using Cayley's Formula

$$T_5 = (1/4)[T_1T_4 + (T_1 + 2T_2)T_3 + (T_1 + 3T_3)T_2 + (T_1 + 2T_2 + 4T_4)T_1]$$
$$= (1/4)[1 \cdot 4 + (1 + 2 \cdot 1) \cdot 2 + (1 + 3 \cdot 2) \cdot 1 + (1 + 2 \cdot 1 + 4 \cdot 4) \cdot 1]$$
$$= (1/4)[4 + 6 + 7 + 19] = (1/4)[36] = 9$$

and in fact there are nine rooted trees with five nodes.   ●

Let $T(x) = \sum_{k=0}^{\infty} T_k x^k$ be the generating function for the sequence of T_k, and $t(x) = \sum_{k=0}^{\infty} t_k x^k$ be the generating function for the sequence of t_k. *Otter's Formula* was published by R. Otter in 1948.

Otter's Formula   The generating functions $t(x)$ and $T(x)$ are related by the equation

$$t(x) = T(x) - \frac{1}{2}[T(x)^2 - T(x^2)]$$ ●

EXERCISES

Factor the following products of cycles into disjoint cycles (remember to start on the right, and note the effect of each cycle as you move from right to left). You may omit 1-cycles.

1. $(1, 2, 3)(2, 3)$
2. $(1, 2, 3)(1, 2)$
3. $(1, 2, 3)(1, 3)$
4. $(1, 3)(1, 2)$
5. $(1, 2)(1, 3)$
6. $(1, 2, 3, 4)(1, 3)(2, 4)$
7. $(1, 3)(2, 4)(1, 2, 3, 4)$
8. $(1, 2, 3, 4)(1, 3, 2, 4)$
9. $(1, 3, 2, 4)(1, 2, 3, 4)$
10. $(1, 2, 3, 4, 5)(1, 3, 4)(1, 2, 5)$
11. State whether each of the permutations in 1 through 10 above is even or odd.
12. List the 24 elements of S_4 in cycle notation.
13. List the 12 elements of A_4 in cycle notation.
14. List the 10 elements of D_5 in cycle notation.
15. List the 12 elements of D_6 in cycle notation.
16. Use Lagrange's Theorem to solve the Key Ring Problem for five keys, and give examples of all the inequivalent configurations of keys.

17. Use Lagrange's Theorem to solve the Key Ring Problem for six keys, and give examples of all the inequivalent configurations of keys.
18. Use Burnside's Lemma to count the inequivalent 2-colorings of the vertices of a triangle (remember that we found four of these).
19. Use Burnside's Lemma to count the inequivalent 3-colorings of the vertices of a triangle.
20. Use Burnside's Lemma to count the inequivalent 4-colorings of the vertices of a triangle.
21. Use Burnside's Lemma to count the inequivalent 5-colorings of the vertices of a square.
22. Use Burnside's Lemma to count the inequivalent 2-colorings of the vertices of a regular pentagon.

Recall that we discussed the regular tetrahedron, or 3-simplex, in Chapter 4. This is a figure whose faces are four equilateral triangles. The group of motions of the tetrahedron is the alternating group A_4: the identity e; two rotations about each of the four axes of symmetry through a vertex and the center of the opposite face; and three rotations halfway about an axis from the center of one edge through the center of the opposite edge.

23. Use Burnside's Lemma to find the number of inequivalent 3-colorings of the vertices of a tetrahedron.

We have often discussed the 3-cube in this book. The group of motions of the 3-cube is the symmetric group S_4. The 3-cube has four long diagonals from a given vertex to the furthest opposite vertex, and it can be proved that the group of motions of the 3-cube can be identified with the group of permutations of these four long diagonals.

24. Use Burnside's Lemma to find the number of inequivalent 3-colorings of the vertices of the 3-cube.

25. Use Polya's Theorem to find the number of inequivalent 3-colorings of the vertices of an equilateral triangle.

26. Use Polya's Theorem to find the number of inequivalent 5-colorings of the vertices of a square.

27. Use Polya's Theorem to find the number of inequivalent 2-colorings of the vertices of a regular pentagon.

28. Use Polya's Theorem to find the number of inequivalent 2-colorings of the vertices of a regular hexagon.

29. Use Polya's Theorem to find the number of inequivalent 3-colorings of the vertices of a tetrahedron.

30. Use Polya's Theorem to find the number of inequivalent 3-colorings of the vertices of the 3-cube.

ADVANCED EXERCISES

In Montmort's book *On Games of Chance* he analyzes the chances of winning some of the gambling games which were played in France at the time (the early 1700s). Games like Pharaoh, Lansquenet, Bassete, Piquet, Treize, and others are studied in his book.

The game of Treize is played as follows ("treize" is the French word for "thirteen"). There are thirteen cards in a deck, numbered from 1 to 13. The deck is shuffled, and the cards are dealt one at a time. As each card is dealt, the player counts the number of cards which have been dealt. If this number ever agrees with the number on the card, the player wins the game.

The game can be played with any number of cards. Shuffling the deck is the same as permuting the cards, and it is clear that the player will lose the game just in case this permutation has no fixed points. A permutation without fixed points is called a *derangement*, and the number of derangements of a set with n elements is denoted $d(n)$.

1. Prove that if $n > 2$, then $d(n) = (n-1)(d(n-2) + d(n-1))$.

2. Prove that if $n > 1$, then $d(n) = nd(n-1) + (-1)^n$.

3. Prove that

$$n! = \sum_{k=0}^{n} \binom{n}{k} d(n-k)$$

4. Use the Principle of Inclusion-Exclusion to prove Euler's formula for $d(n)$:

$$d(n) = n! \left(1 - \frac{1}{1!} + \frac{1}{2!} - \ldots + (-1)^n \frac{1}{n!} \right)$$

5. Conclude from the formula in (4) that $d(n)$ is the nearest integer to $n!/e$.

6. Conclude from (5) that the probability of winning the game of Treize is almost exactly $(e-1)/e = .6321 = 63.21\%$.

7. Let $S = S_n$, the set of all permutations of n objects, be regarded as a sample space. Let X be the number of fixed points of a permutation. Use the identity in (3) to prove that

$$E(X) = 1;$$

that is, that the expected number of fixed points of a permutation is exactly one. Notice that this means that one *expects* to win the game of Treize.

8. Use the identity in (3) to prove that

$$\sigma_X = 1;$$

that is, that the standard deviation of the number of fixed points of a permutation is also exactly one.

COMPUTER EXERCISES

1. Use the recursion $d(n) = (n-1)(d(n-2)+d(n-1))$ to compute as many values of $d(n)$ as you can.

2. Use the recursion $d(n) = nd(n-1)+(-1)^n$ to compute as many values of $d(n)$ as you can.

3. Use Cayley's Formula to compute the number of rooted trees with n nodes T_n for as many values of n as you can.

4. Use your result in (3) and Otter's Formula to compute the number of unlabelled trees t_n for as many values of n as you can.

9.4   CHAPTER SUMMARY AND SUPPLEMENTARY EXERCISES

> *"There is Thingumbob shouting!" the Bellman said.*
>
> *"He is shouting like mad, only hark!*
>
> *He is waving his hands, he is wagging his head,*
>
> *He has certainly found a Snark!"*

Chapter Summary

Section 9.1

In this section, we discussed the basic notion of recursiveness, particularly linear recursiveness. We defined the generating function and the exponential generating function of a sequence and gave several examples. We discussed divide-and-conquer relations. We showed how the Master Theorem can be used to determine the complexity of an algorithm which is defined using a divide-and-conquer technique. As an application of these ideas, we described AVL trees.

Section 9.2

In this section, we discussed probabilities and their applications in discrete mathematics. We introduced the binomial distribution and gave several examples of its use. We stated the Central Limit Theorem and pointed out that it implies in particular that in the limit the binomial distribution is normal. Using this observation, we showed how to compute the tails of the binomial distribution. We showed how probabilities can be used to study the reliability of a network.

Section 9.3

In this section, we discussed the notion of symmetry in discrete mathematics, expressed in terms of groups. We showed how Lagrange's Theorem, Burnside's Lemma, and

Polya's Theorem can be used to solve problems involving symmetries. We discussed Cayley's Formula for the number of rooted trees with a given number of nodes, and Otter's Formula, which allows one to compute the number of unlabelled trees with a given number of nodes.

EXERCISES

In this book, we have discussed several examples of counting numbers, for example the Bernoulli numbers. The *Stirling numbers of the first kind* $s(n, k)$ and the *Stirling numbers of the second kind* $S(n, k)$ are important examples of counting numbers. They are named for James Stirling (1692–1770). These numbers are defined by the recursions

$$x(x - 1)(x - 2) \ldots (x - n + 1)$$
$$= s(n, n)x^n - s(n, n - 1)x^{n-1}$$
$$\quad + s(n, n - 2)x^{n-2} + \ldots + (-1)^n s(n, 1)x$$
$$x^n = S(n, 1)x + S(n, 2)x(x - 1)$$
$$\quad + S(n, 3)x(x - 1)(x - 2) + \ldots$$
$$\quad + S(n, n)x(x - 1)(x - 2) \ldots (x - n + 1)$$

Newton noticed that power series in x^n sometimes converge more quickly when expressed as series in the *falling factorials* $x(x - 1)(x - 2) \ldots (x - n + 1)$. It was for this reason that Stirling made tables of the coefficients in the equations above, and this is why the coefficients are named for him.

1. Write the two equations above for $n = 1, 2, 3, 4, 5$. The Stirling numbers satisfy the recursions

$$s(n + 1, k) = s(n, k - 1) + ns(n, k)$$

and

$$S(n + 1, k) = S(n, k - 1) + kS(n, k)$$

Notice how these recursions differ only in the coefficient in the second term. Notice too how each is the recursion for the Pascal Triangle, perturbed by the addition of that coefficient.

2. Use the recursion for the Stirling numbers of the first kind to generate a table of those numbers for $n \le 8$.

3. Use the recursion for the Stirling numbers of the second kind to generate a table of those numbers for $n \le 8$.

The Stirling numbers of the first kind $s(n, k)$ count the number of permutations of n things which decompose into exactly k cycles (one-cycles must be included).

4. Verify this for permutations of $n = 4$ things.

5. Verify this for permutations of $n = 5$ things.

The Stirling numbers of the second kind $S(n, k)$ count the number of partitions of a set with n elements into k non-empty subsets.

6. Verify this for partitions of a set with $n = 4$ elements.

7. Verify this for partitions of a set with $n = 5$ elements.

8. Prove that the coefficient of x^n in the expansion of

$$\frac{1 + x}{(1 + x^2)(1 - x)^2}$$

is either $n + 1$ or $n + 2$, and give a rule to determine which one occurs.

In the previous section, we showed how to use Polya's Theorem to count colorings of the vertices of a tetrahedron or a cube. This involved finding their groups of symmetries, which turn out to be A_4 in the first case and S_4 in the second case. Colorings of the vertices of the octahedron can be counted in just the same way, since the group of symmetries of the octahedron is also S_4.

How can we count colorings of the vertices of the other two three-dimensional regular solids, the icosahedron and the dodecahedron? We must find their groups of symmetries, which turn out to be the same.

It is possible to inscribe five tetrahedra in either the icosahedron or the dodecahedron in such a way that the group of symmetries of each of these regular solids is the group of even permutations of the five tetrahedra, that is, the group A_5. Once we know this, colorings of the vertices of the icosahedron or the dodecahedron can be counted using either Burnside's Lemma or Polya's Theorem.

Before doing the computations, it is convenient to divide the permutations in A_5 into types.

9. Prove that the even permutations of $\{1, 2, 3, 4, 5\}$ can be described as follows. We have the identity e, 20 3-cycles (i, j, k), 24 5-cycles (i, j, k, l, m), and 15 products of disjoint transpositions $(i, j)(k, l)$.

10. Using the information in the previous problem, find the number of inequivalent 2-colorings of the vertices of the icosahedron using Burnside's Lemma.

11. Compute the cycle index of A_5.

12. Use Polya's Theorem to find the number of inequivalent 2-colorings of the vertices of the icosahedron.

ADVANCED EXERCISES

In the Advanced Exercises for the previous section, we described derangements (permutations with no fixed points), and showed how they arise in the eighteenth-century game of Treize. Now we shall describe a Victorian card game called *Mousetrap*. The analysis of this game involves making lists of permutations and naturally extends the analysis of the game of Treize.

Arthur Cayley discussed the game of Mousetrap in two papers, one published in 1857, and the other in 1878.

The game of Mousetrap is played with n cards, where n can be any positive integer. The cards are shuffled, and arranged in a circle (or in a row, where we view the last card as being adjacent to the first card). Beginning with the first card, the player counts "one, two, three, ..." When the number on any card agrees with the count, that card is removed. The player then begins counting "one, two, three, . . . , " beginning with the next card. The game continues until no more cards can be eliminated. The player wins if all cards are eliminated.

For example, suppose that the cards in order are 4213. We count "one, two," and remove the card 2, leaving 413. We count "one," and remove card 1. Then beginning with card 3 we count "one, two, three, four," and remove card 4. Finally, we count "one, two, three" on card 3, and eliminate card 3.

By contrast, you will see that if the cards in order are 3214 that only cards 2 and 1 are removed.

Clearly the outcome of the game depends on the permutation represented by the cards. Notice that it is exactly the derangements for which no cards are removed.

Here is a table of the outcomes of the game of Mousetrap for four cards. This is a corrected version of Cayley's original table, which contained mistakes!

Permutation	Cards Removed
1234	1
1243	1342
1324	12
1342	1
1423	1234
1432	1423
2134	3214
2143	
2314	4
2341	
2413	
2431	3142
3124	4
3142	
3214	21
3241	23
3412	
3421	
4123	
4132	4
4213	2134
4231	2
4312	
4321	

Notice that there are six cases in which all the cards are removed, that is, six cases in which the player wins the game.

1. Reconstruct this table yourself. Give a brief explanation for each entry.

COMPUTER EXERCISES

In the game of Mousetrap with n cards, we can ask for the number of cases in which any number of cards i are removed. This gives the following Pascal Triangle, which was computed by R. K. Guy and R. J. Nowakowski. In the table, the number of cards thrown out is given in the top row, and the number of cards in the left-hand column.

n	0	1	2	3	4	5	6	7
1	0	1						
2	1	0	1					
3	2	2	0	2				
4	9	6	3	0	6			
5	44	31	19	11	0	15		
6	265	180	105	54	32	0	84	
7	1854	1255	771	411	281	138	0	330

Notice that the row sums are equal to $n!$, and that the numbers in the first column are the derangement numbers.

1. Explain why the entries on the second diagonal are all 0.
2. Use your computer algebra program to generate three more rows of this table.

No explicit recursion is known for the numbers on the right-hand diagonal. No explicit two-term recursion is known for the entries in this Pascal Triangle. Finding such recursions would be of great interest.

Models of Computation

10.1 LANGUAGES AND GRAMMARS

Just the place for a Snark! I have said it twice:

That alone should encourge the crew.

Just the place for a Snark! I have said it thrice:

What I tell you three times is true.

In this chapter, we shall describe and relate three topics: languages, logic, and machines.

The methods used to record data and to enter data in a computer of some kind have undergone a long evolution. For most of recorded history, data was simply recorded on tablets of some kind, or on paper. An example of a different idea was given in the early nineteenth century by the *Jacquard loom*, which used punched cards to describe the weaving patterns carried out by the machine. The idea of using punched cards to record data was applied in other situations, notably in the United States Census of 1890, in which more than fifty million punched cards were used. The method was then adopted by banks and companies, and punched cards remained an important method of recording data until the late 1960s.

As computer science developed, and as electronic computers gradually replaced the analog computers which had been used, there emerged a new and uniform concept of entering data, or instructions, in a machine. The data would be entered as a sequence of symbols, often binary bits.

From its beginnings, the development of computer hardware was accompanied by insights about the theoretical structure of machines. The idea developed that a machine is driven by strings on elements of a set called an *alphabet,* the strings being called *words,*

327

and the set of all words involved being called a *language*. A machine of a given type *accepts* words from a language of a given type. The study of machines of a given type is *equivalent* to the study of the languages which the machines will accept.

It is natural to seek to describe languages whose words are generated by some simple rule. We shall describe a class of languages of this kind, the *regular languages*.

Just as there are grammatical rules in English or French which determine what words are in the language, so we can formulate the notion of a *grammar*. Each grammar has a language associated with it. We will describe which grammars give rise to regular languages. We will describe *context-free* grammars and their associated languages. At the end of the section, we shall describe the *Chomsky hierarchy of languages*.

We shall begin the second section of this chapter with a brief history of computer science. Then we shall begin to describe the relationship between languages and machines. *Kleene's Theorem*, which asserts that *finite-state machines* are exactly the machines which accept a regular language as input, is a model for theorems of this kind. The *Pumping Lemma* will help us to decide whether a language is regular. We shall describe *Turing machines*, which give what seems to be the most general model for computation. *Church's Thesis* makes this intuitive idea precise. The concepts of *recursiveness* and *recursive enumerability* are essential here. At the end of the section, we shall describe *DNA computing*.

In the third section, we shall discuss the relationship between logic and computability. We shall discuss the Gödel Incompleteness Theorem, Turing's theorem about the Decision Problem, and the Tenth Hilbert Problem. Gödel's Theorem can be expressed in terms of Turing machines. When Gödel's Theorem is expressed in this way, it sets limits to computation in ways which are still not well understood.

We shall end the third section, and the book, with an idea about *The Hunting of the Snark*.

DEFINITION 10.1.1

i. An *alphabet* A is any set. The elements of A are called *letters*.
ii. A *word* is a finite string of letters from A, called *words on A*. There is a unique *empty string*, denoted by ϵ. The set of all words on A is denoted A^*.
iii. Given an alphabet A, a *language L* is any set of words on A.

The alphabet for the English language is

$$A = \{a, b, c, d, e, f, \ldots, z\}$$

and the English language is a language, in our sense, on this alphabet.

The set of binary bits forms an alphabet

$$A = \{0, 1\}$$

We could choose alphabets consisting of a certain number of letters:

$$\{a\}$$
$$\{a, b\}$$
$$\{a, b, c\}$$
$$\{a, b, c, d\}$$

and so on.

Recall that the DNA sequences which we have discussed are sequences of chemicals T = thymine, A = adenine, G = guanine, and C = cytosine. Thus we can view DNA sequences as words on the alphabet

$$A = \{T, A, G, C\}$$

In RNA, thymine is replaced by U = uracil, so RNA sequences can be viewed as words on the alphabet

$$A = \{U, A, G, C\}$$

First we must define some commonly used notation regarding languages.

I. $\emptyset$, ϵ, and a represent, respectively, a set with no strings, a set consisting only of the string of length zero, and a set consisting only of the string of length 1 a.

II.   (a) (L) represents the same set of strings as L.

   (b) LR represents the set of strings ab obtained by *concatenating* a string a of L with a string b of R.

   (c) $L + R$ represents the union of L and R.

   (d) L^*, the *Kleene star* of L represents the language obtained by concatenating zero or more strings of L, that is,

   $$L^* = \epsilon + L + LL + LLL + \dots$$

   We shall write L^n for the concatenation of L n times, so that

   $$L^* = \epsilon + L + L^2 + L^3 + \dots$$

III. $()$ takes precedence over $*$, $*$ takes precedence over concatenation, and concatenation takes precedence over union. For example,

$$LR + S^*$$

means "take the $*$ of S and the concatenation of L and R, and take the union of the results."

The operation of concatenation is associative. For three words a, b, c we have

$$(ab)c = a(bc)$$

The empty string acts as an identity element:

$$\epsilon a = a\epsilon$$

A set of strings which is closed under concatenation and contains ϵ, such as L^*, forms an algebraic structure called a *monoid*, or a *semi-group*. These structures are closely related to the groups which we studied in the previous chapter.

DEFINITION 10.1.2   A *regular expression* over an alphabet A is defined recursively as follows:

i. The empty set $\emptyset$ and ϵ are regular expressions.
ii. Each a in A is a regular expression.
iii. If L and R are regular expressions, then (L), LR, $L + R$, and L^* are also regular expressions.
iv. Nothing else is a regular expression.

We shall write $aa = a^2$, $aaa = a^3$, $aaaa = a^4$, and so on.

EXAMPLE 1   If we have two languages $L_1 = \{ab, ac\}$ and $L_2 = \{b, bc\}$ on the alphabet $A = \{a, b, c\}$, what is the concatenation $L_1 L_2$?

Solution

$$L_1 L_2 = \{ab^2, ab^2c, acb, acbc\}$$ ●

Every regular expression defines a set of words.

EXAMPLE 2   If $A = \{a, b\}$, what set of words is described by $(a + b)^*$?

Solution $(a + b)^*$ is the set of all words in a and b. ●

EXAMPLE 3   If $A = \{a, b, c\}$, what set of words is described by $c(a + b + c)^*c^2$?

Solution $c(a + b + c)^*c^2$ is the set of words in a, b, c which begin with c and end with cc. ●

DEFINITION 10.1.3   If R is a regular expression on some alphabet A, the *regular language* $L(R)$ is the set of words determined by R. A language L is *regular* if $L = L(R)$ for some regular expression R.

For example, the sets of words in the two previous examples form regular languages. Regular languages on the alphabet $A = \{0, 1\}$ are particularly important.

EXAMPLE 4   If $R = (0 + 1)^*$, what is $L(R)$?

Solution $L(R)$ is the set of all binary strings, which is a regular language.   ●

EXAMPLE 5   If $R = 0^3(1)^*0^4$, what is $L(R)$?

Solution $L(R)$ is the set of all binary strings which consist of three zeros, then any number of ones, and then four zeros.   ●

EXAMPLE 6   If $R = 0^*1001^*$, what is $L(R)$?

Solution $L(R)$ is the set of binary strings which consist of any number of zeros, then the string 100, then any number of ones.   ●

We can also describe a language directly as a set. For example,

$$L = \{ab^n, n \geq 0\}$$

consists of all words on the alphabet $A = \{a, b\}$ which consist of an a followed by any number of bs.

The language

$$L = \{a^n b^n, n > 0\}$$

consists of all words on the alphabet $A = \{a, b\}$ which consist of a seqeuence of as of length at least one followed by the same number of bs.

When a language is described as a set of words, how can we decide whether it is regular? For example, which of the last two examples is a regular language?

The language in the first of these two examples is regular, since it is really described by the regular expression ab^*. The language in the second example is not regular. We will see how to prove this in the next section, when we describe the *Pumping Lemma*.

Languages such as English and French can be described by their *grammars*: sets of rules which describe how the words in the language are formed. In a similar way, languages accepted as input by various types of machines can be described by grammars.

DEFINITION 10.1.4   A *grammar* G consists of the following:

  i. A finite set V of *variables* or *nonterminals*.
  ii. A finite set T of *terminals*, with $V \cap T = \emptyset$.
  iii. An element $s \in V$ called the *start symbol*.
  iv. A finite set P of *productions*. Productions are denoted $A ::= B$.

It is a convention that the "trivial production" $s ::= \epsilon$ is always included in the set P of productions.

Clearly a grammar is determined by its productions.

> **DEFINITION 10.1.5**   The *language* $L(G)$ of a grammar G consists of all words in the terminals which can be obtained from the start symbol using the productions.

The inclusion of the trivial production $s ::= \epsilon$ in G means that we always have $\epsilon \in L(G)$.

The set of productions of G will usually contain some of the form $s ::= A$, where the word A will usually contain variables. Other productions will allow terminals to be substituted for these variables, giving the words of the language.

Here are several examples of grammars and their associated languages.

EXAMPLE 7   If the grammar G has $V = \{s, A, B\}$ and $T = \{a, b\}$, and the productions are $s ::= AB$, $A ::= a$, and $B ::= b$, what is $L(G)$?

Solution $L(G)$ will always contain the empty string ϵ, by convention. The only ways that strings in the terminals can be formed is $s ::= AB ::= aB ::= ab$, or $s ::= AB ::= Ab ::= ab$, so $L(G) = \{\epsilon, ab\}$.   ●

EXAMPLE 8   If the grammar G has $V = \{s, A, B\}$ and $T = \{a, b\}$, and the productions are $s :: a$, $s ::= AsB$, $A ::= a$, $B ::= b$, what is $L(G)$?

Solution   The production $s ::= AsB$ may be applied n times, giving $s ::= A^n s B^n$. We must then substitute a for A, a for s, and b for B (we may substitute for A or B earlier, but this makes no difference). The result is a word consisting of $n + 1$ as followed by n bs, and the language consists of these words and the empty string ϵ.   ●

EXAMPLE 9   If the grammar G has the productions $s ::= AB$, $A ::= a^2$, $A ::= \epsilon$, $B ::= a^2 A$, $B ::= b$, what is $L(G)$?

Solution   The words in $L(G)$ consist of ϵ, words consisting of an even number of as, and words consisting of an even number of as followed by a b.   ●

> **DEFINITION 10.1.6**   A grammar G is
> i. *Left-regular* if all the productions not involving the start symbol are of the form $A ::= a$ or $A ::= aB$, and *right-regular* if all productions not involving the start symbol are of the form $A ::= a$ or $A ::= Ba$.
> ii. *Context-free* if all the productions not involving the start symbol, and which replace a word involving variables by a word in the terminals, are of the form $A ::= w$, where w is a word in the terminals of G. A language is context-free if it is of the form $L(G)$ for a context-free grammar G.
> iii. *Context-sensitive* if all the productions not involving the start symbol, and which replace a word involving variables by a word in the terminals, are of

the form $xAy ::= xwy$, where x, y, w are words in the terminals of G. A language is context-sensitive if it is of the form $L(G)$ for a context-sensitive grammar G.

Notice that every context-free language can be regarded as a context-sensitive language by taking $x = y = \epsilon$ in the definition of context-sensitivity.

Regular languages can be characterized by regular grammars.

THEOREM 10.1.1   The follwing conditions are equivalent:

  (i) The language L is regular.

  (ii) $L = L(G)$ for some left-regular grammar G.

  (iii) $L = L(G)$ for some right-regular grammar G.   ●

In a context-sensitive grammar, the word w can be substituted for the variable A only in the "context" of the words x and y.

EXAMPLE 10   Give an example of a context-free grammar, and its associated language.

Solution   The grammar with productions $s ::= ab$ and $s ::= asb$ is context-free, and will generate the language

$$L = \{a^n b^n, n > 0\}$$

that we considered above.   ●

EXAMPLE 11   Give an example of a context-sensitive grammar, and its associated language.

Solution   Let the productions be $s ::= BAB$, $B ::= a$, $aAa ::= aba$. This grammar is context-sensitive because of the last production. The words of the associated language are ϵ and *aba*.   ●

Now we shall define two closely related types of sets of positive integers. Both are important in the theory of computation, and both are related to the types of languages which we are studying.

DEFINITION 10.1.7 A set of positive integers S is

 i. *Recursive* if the elements of S are generated by a recursion, that is, if $S = \{a_1, a_2, \ldots, a_n, \ldots\}$, then there is a recursion which computes a_n in terms of $a_1, a_2, \ldots, a_{n-1}$ provided that n is sufficiently large.

 ii. *Recursively enumerable* if there is an algorithm which will determine whether any positive integer a is a member of S in finitely many steps.

For example, the set of Fibonacci numbers is recursive, and also recursively enumerable.

A recursive set is always recursively enumerable. There are recursively enumerable sets which are not recursive, and there are also explicit sets of positive integers which are not recursively enumerable. We shall discuss these matters in the next two sections.

Finally we shall describe a very general class of languages. The definition involves a kind of machine which we shall describe in the next section, where we shall also describe the concept of a machine accepting a language as input.

> **DEFINITION 10.1.8**   A language L is *recursively enumerable* if there is a Turing machine M such that M accepts L as input.

APPLICATION   *The Chomsky Hierarchy*

In this section, we have discussed four classes of languages:

  I.   Regular languages

  II.   Context-free languages

  III.   Context-sensitive languages

  IV.   Recursively enumerable languages

This scheme for classifying languages is called the *Chomsky Hierarchy*. It is named for the linguist N. Chomsky, who described it in a series of papers, the first of which was published in 1956.

It is a basic fact that each of these classes of languages contains the class before it:

  I.   Every regular language is a context-free language.

  II.   Every context-free language is a context-sensitive language.

  III.   Every context-sensitive language is recursively enumerable.

In the next two sections, we shall show how the Chomsky Hierarchy is important in the study of machines and computation.

It is very curious that the Chomsky Hierarchy has not been applied to the structure of the genetic code. Let us remember that short sequences in the genetic code, such as the sequences which code for proteins, are actually sequences of triplets on the alphabet T = thymine, A = adenine, G = guanine, C = cytosine.

Let the alphabet B be the set of 64 triplets of letters T, A, G, C. Then we ask:

> *Where do protein sequences fit into the Chomsky Heirarchy on the alphabet B? For example, are protein sequences words in a regular language on this alphabet?*

EXERCISES

Fix an alphabet $A = \{0, 1, a, b, c, d\}$. For each of the following regular expressions, describe the corresponding set of words on A, that is, the corresponding regular language.

1. $(a + b)^*ab$

2. $(0 + 1)^*11(0 + 1)^*$

3. $(00 + 11)^*$

4. $(01)^*(10)^*$

5. $(a + b)^*(c)^*$

6. $(a)^*(b)^*$

7. $(01 + 10)^*$

8. $(a + b)^*(c + d)^*$

9. $(a + bc)^*$

10. $(a + b + cd)^*$

For a grammar G with the following productions, describe the language $L(G)$. State whether $L(G)$ is regular, context-free, or context-sensitive.

11. $s ::= AB, A ::= a, A ::= aB, B ::= b.$

12. $s ::= AB, A ::= ab, A ::= aB, B ::= b, B ::= aA.$

13. $s ::= AsB, A ::= a, A ::= aB, B ::= b, B ::= Ba.$

14. $s ::= AB, s ::= As, A ::= a, A ::= Ba, B ::= b.$

15. $s ::= A, s ::= B, A ::= a, A ::= aB, B ::= b.$

16. $s ::= A, s ::= B, A ::= a, A ::= aB, B ::= b, B ::= bA.$

17. $s ::= ABC, A ::= a, A ::= aB, A ::= bC, B ::= cC, B ::= b, C ::= c.$

18. $s ::= AB, s ::= ABs, A ::= a, aAa ::= aba, B ::= b.$

19. $s ::= b, s ::= aA, A ::= b, A ::= aA, A ::= bB, B ::= a, B ::= aB.$

20. $s ::= ABC, A ::= a, A ::= aA, B ::= b, B ::= bB, C ::= c, C ::= cC.$

10.2   FINITE-STATE MACHINES AND TURING MACHINES

> *"It's a Snark!" was the sound that first came to their ears,*
> *And seemed almost too good to be true.*
> *Then followed a torrent of laughter and cheers:*
> *Then the ominous words "It's a Boo—"*
> *Then, silence. Some fancied they heard in the air*
> *A weary and wandering sigh*
> *That sounded like "-jum!" but the others declare*
> *It was only a breeze that went by.*

It is difficult, perhaps impossible, to give a precise history of computer science. Ideas for calculating machines occurred to many people, often nearly at the same time. Other ideas turn out to have been anticipated, at least to some extent. For example, it is traditional to give Pascal credit for the first calculating machine, which was designed and built in the 1640s. But a German named Schickard designed a similar machine, which was never built, in the 1620s.

We mentioned in Chapter 1 that Pascal was the first to express the Principle of Induction in its modern form, and that it was ignored until De Morgan published it in 1838. In a similar way, little effort was made to improve on Pascal's calculating machine until the nineteenth century. Then the English mathematician Charles Babbage (1792-1871) described and partly constructed a Difference Engine, whose purpose was to use difference tables to compute values of functions. He then began to build a more complicated machine, the Analytical Engine. In this work, he was partly assisted by Augusta Ada Byron, Countess of Lovelace (1815–1852), Lord Byron's daughter. Both of Babbage's machines were eventually built, and both exerted a lasting influence on designs for computers.

The punched card equipment which was used in the United States Census of 1890, which we mentioned in the previous section, was invented by an employee of the Census Bureau named Hermann Hollerith. Hollerith patented his equipment and formed a company, the Tabulating Machine Company. This became part of the Computing-Tabulating-Recording Company, which in turn developed into the International Business Machines Corporation (*IBM*).

From 1900 to 1920, calculating machines were developed and used on a large scale. Machines like the *Brunsviga*, *Felt's Comptometer*, and *Burrough's Adding and Listing Machine* were used for accounting purposes, but also to a small extent for scientific computing.

In the 1930s, there was steady progress in what we would now regard as computer science. Columbia University's Astronomical Computing Bureau, which had a close relationship with IBM, was one center where this change took place. In 1937, this facility was named for Thomas J. Watson of IBM. The machines involved were electromechanical, still not purely electronic computers. A joint project involving Harvard and IBM developed the Mark I computer. At MIT, a Differential Analyzer was built under the guidance of Vannevar Bush. At Bell Laboratories, the Complex Number Computer was designed and built.

Fully electronic computers originated in the late 1930s and early 1940s. John V. Atanasoff, of Iowa State University, was one of the first to emphasize the importance of a purely electronic computer. A fully electronic machine called the *Colossus* was built at Bletchley Park in England in 1943 and used for coding work. It contained 1500 vacuum tubes.

In 1943, the Moore School of Electrical Engineering of the University of Pennsylvania was given a contract by the Aberdeen Ballistics Research Laboratory to build a computer called the Electronic Numerical Integrator And Computer (the *ENIAC*), which was finished in 1946. This computer contained 18,000 vacuum tubes. Applied to ballistics work, it sharply reduced the time needed for computations. The success of the ENIAC made a deep impression, and inspired work on several new computers. The Universal Automatic Computer (the *UNIVAC*), the first general-purpose electronic computer, was built by a company which is now a part of the Sperry Rand Corporation, and delivered to the United States Census Bureau in 1951.

By the early 1950s, it was clear that future development of computer science lay with the fully electronic computer. In the succeeding decades, computer design was gradually improved. Input using punched cards or paper tape was replaced with input using a high-level language such as *FORTRAN*, *PASCAL*, or *C*. Silicon chips allowed vacuum tubes to be eliminated, allowing vast reductions in size. Hard disks with large capacity became available for the storage of data.

In this section, we shall give an introduction to the theory of machines. We shall describe *finite-state machines* and *Turing machines.* At the end of the section, we shall describe *DNA computing.*

The simplest model for computation is the *finite-state machine.* The following sort of example of a finite-state machine is often given.

Suppose that we have a vending machine which will dispense a candy bar when 15 cents is inserted. For simplicity, we will assume that the machine contains only one candy bar. The machine will accept dimes or nickels. The *initial state I* for the machine is when no money has been inserted. The machine is in state C when 15 cents or more has been inserted, and the machine has dispensed the candy bar. State C is the machine's only *accepting state.* There is also a state in which the machine has accepted 5 cents, and a state in which the machine has accepted 10 cents.

We can represent this machine by the digraph in Figure 10.1. We represent the initial state by an I, the accepting state by a C in a double circle, and the other states by letters in circles. The edges record the effect of inserting dimes or nickels into the machine. The loops express the fact that once the machine enters state C it does not leave it.

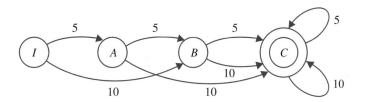

Figure 10.1   A Finite-State Machine

Now suppose that the candy bar costs 20 cents, and again that the machine will accept nickels or dimes. Figure 10.2 gives the digraph for this machine.

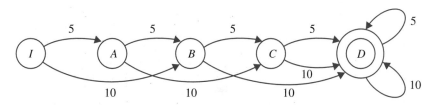

Figure 10.2   A Finite-State Machine

Here is a formal definition of a finite-state machine.

DEFINITION 10.2.1   A *finite-state machine* is a quintuple (A, S, T, I, f), where:

i. A is a finite alphabet.

> **ii.** S is a finite set of *internal states*.
> **iii.** $T \subset S$ is a set of *accepting states*.
> **iv.** $I \in S$ is the *initial state*.
> **v.** $f : A \times S \to S$ is the *next-state function*.

Finite-state machines are usually described by their digraphs, as in Figures 10.1 and 10.2. The accepting states are indicated by double circles. The edges are labelled with letters from A, which indicate how the next-state function works. In our two examples, there is only one accepting state, but in general there can be more than one.

In our first example, $A = \{5, 10\}$, $S = \{I, A, B, C\}$, $f(I, 5) = A$, $f(B, 5) = C$, and so on. The sequences $5, 10, 10, 5,$ or $5, 5, 5$ will put the machine in its state C whereas the sequence $5, 5$ will not. The state C is the only accepting state.

> **DEFINITION 10.2.2**   Let M be a finite-state machine. A word $w = a_1 a_2 \ldots a_n$ in A is *accepted* by M if the sequence $a_1, a_2, \ldots, a_n$ describes a path in the digraph for M which begins with the initial state I and ends with an accepting state of M. The *language L* which M accepts is the set of all words which M accepts.

EXAMPLE 1   Describe the language L accepted by the finite-state machine in Figure 10.1.

Solution   Let us set $5 = a$ and $10 = b$. A word which M accepts corresponds to a path from the initial state (in which the machine has accepted no money) to state C (in which the machine has accepted fifteen cents or more). The language L, therefore, consists of words in a and b which contain at least three as; of words in a and b which contain at least two bs; and of words in a and b which contain at least one a and one b.   ●

EXAMPLE 2   Describe the language L accepted by the finite-state machine in Figure 10.2.

Solution   Once again let us set $5 = a$ and $10 = b$. A word which M accepts corresponds to a path from the initial state (in which the machine has accepted no money) to state D (in which the machine has accepted twenty cents or more). The language L, therefore, consists of words in a and b which contain at least four as; of words in a and b which contain at least three as and at least one b; and of words in a and b which contain at least two as and at least one b.   ●

What is the practical importance of finite-state machines? Useful circuit elements and useful items of software can be modelled by finite-state machines. The same items of hardware and software can be studied by considering the input and output of these machines.

Let us take a basic example, the shift register. There are several ways to describe shift registers. Here we shall give a description which differs somewhat from the description of the example in Section 1.5.

In Figure 10.3, we give an example of a three-bit shift register, viewed as a finite-state machine. Each of the eight states is represented by one of the eight binary vectors of length three. The machine accepts the language consisting of all strings of 0s and 1s. A bit a acts by moving each bit in every vector one place to the left, and replacing the bit on the right by a. For example, $0 : (1, 0, 1) \rightarrow (0, 1, 0)$ and $1 : (0, 1, 0) \rightarrow (1, 0, 1)$. We can choose any state to be the initial state (the shift register can first be filled with any binary vector of length three), and all states are accepting states.

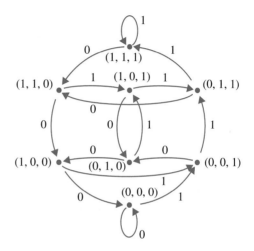

Figure 10.3   A Three-Bit Shift Register as a Finite-State Machine

Of course the diagram in Figure 10.3 is really the 3-cube, and it is clear that the n-cube will give a similar model of an n-bit shift register.

Notice the symmetry of this digraph: interchanging 0 and 1 gives an isomorphism of this digraph with itself.

In Figure 10.4, we give another example, which shows that the integers (mod n) can always be regarded as a finite-state machine. The states are the congruence classes (mod n): $[0], [1], [2], \ldots, [n-1]$. This machine accepts the language $\{a^k\}$, and a acts by $a : [k] \rightarrow [k+1]$, which describes the next-state function. We can regard any state as the initial state, and all states are accepting states. The existence of this example is one reason that we were careful to study the integers (mod n) in Chapter 2.

Our last example makes an important point about finite-state machines. Because a finite-state machine has only finitely many states, it cannot be used to count arbitrarily large integers.

> *We can use a finite-state machine to count the integers (mod n) for any n, but no finite-state machine will count integers of arbitrary size.*

This clearly shows the need to consider more complicated types of machines. Here is another point about finite-state machines, which also applies to the other machines that we will consider.

> *Finite-state machines can be modelled in either hardware or software.*

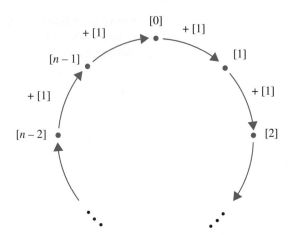

Figure 10.4   The Integers (mod n) as a Finite-State Machine

For example, it is possible to actually build a circuit element representing a shift register; or a program can be written which will give the shift register's output. Coding machines often contain several shift registers, and the usual practice is to actually construct them as circuit elements. Many computer codes are obtained from shift register output, but the shift register is usually modelled in software.

The following theorem was proved by S. C. Kleene in a paper which was published in 1956. The paper's original purpose was to study neural networks.

Kleene's Theorem   The following conditions are equivalent:

(i)   The language L is regular.

(ii)   There is a finite-state machine M which accepts L.   ●

Because the languages in the two previous examples are accepted by finite-state machines, they are each regular languages.

The Pumping Lemma   Suppose that a finite-state machine M accepts a language L over an alphabet A, and that M has k states. If w is a word of L which contains more than k letters from A, then $w = xyz$, where for every positive integer m, $w(m) = xy^m z$ is accepted by M.

The Pumping Lemma is a powerful tool for recognizing regular languages. The following example is a standard application of it.

EXAMPLE 3   Prove that the language

$$L = \{a^n b^n, n > 0\}$$

is not regular.

Solution   If L is regular, then there is a finite-state machine M which accepts L. Suppose that M has k states, and let $w = a^k b^k$. By the Pumping Lemma, $w = xyz$ where $y \neq \epsilon$, and

$w_2 = xy^2z$ is also accepted by M. Since $w \in L$, w must have the same number of as as bs. y cannot consist only of as or of bs, for then w_2 would not have the same number of as and bs, a contradiction. But if y contains both as and bs then w_2 will contain bs followed by as, which is also a contradiction. ●

Alan Matheson Turing (1912–1954) invented a class of machines called *Turing machines*. While finite-state machines represent the simplest models of computation, Turing machines are recognized as representing the most general models of computation. The languages which these machines accept also represent opposite extremes: the regular languages accepted by finite-state machines being the simplest, and the recursively enumerable languages accepted by Turing machines being the most complicated.

> *Finite-state machines, and the regular languages which they accept, are the simplest models for computation. Turing machines, and the recursively enumerable languages which they accept, are the most general models for computation.*

Turing's seminal paper *On computable numbers with an application to the Entscheidungsproblem* (the Decision Problem) was published in 1936. In this paper, Turing defined the machines which are named for him.

DEFINITION 10.2.3 A *Turing machine* consists of

i. A finite tape alphabet S of symbols, which we will take to be $\{0, 1\}$, and possibly other characters to signal the left and right ends of the string.
ii. A finite control alphabet Q of states, one of which is the start state.
iii. A linear tape extending infinitely in both directions.
iv. The control logic, and a *read-write head*.

In Figure 10.5, we show a typical Turing machine.

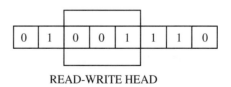

READ-WRITE HEAD

Figure 10.5 A Turing Machine

The machine is given a finite input in S^* on the tape, say starting in position 0. The read-write head is also positioned at location 0, in the start state. The machine executes the following control steps until the machine halts. The final state of the tape (the output) is then an element of the language for the machine.

(i) Read the current symbol and notice the current state.

(ii) Write a symbol in this location.

(iii) Move left, move right, or halt.

The relation between the input and output of a Turing machine corresponds to an algorithm: in fact to *any* algorithm.

> *Turing machines are universal models for algorithms: any algorithm can be computed as the input-output relation for a Turing machine.*

For example, there is a Turing machine which will add two integers, and we show it in Figure 10.6.

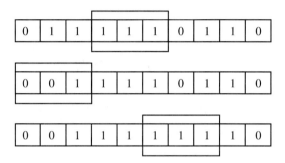

Figure 10.6   A Turing Machine Which Adds

To add the positive integers M and N, this machine proceeds as follows. First it represents M and N as strings of M 1s and N 1s, separated by a single 0. Then the machine scans to the left until it arrives at the last 1 on the left of the string of M 1s. It erases this 1, and then scans to the right until it comes to the 0 separating the string of $M - 1$ 1s from the string of N 1s, and replaces this 0 by a 1. The result is a string of $M + N$ 1s, representing the sum $M + N$, and at this point the machine halts.

We have illustrated these steps in Figure 10.6 with the sum $5 + 2 = 7$.

At the time that Turing was developing his ideas about machines, he was also thinking about the problem of describing *computable numbers*.

> **DEFINITION 10.2.4**   A real number x is *computable* if, for any positive integer n, there is an algorithm which gives the nth digit of the decimal expansion of x in finitely many steps. The input of this algorithm must not mention x.

The condition that x must not be mentioned in the input is important. The *continued fraction algorithm* can be used to represent any real number x. However, x is mentioned in the input of the general continued fraction algorithm, and this is not allowed in the definition of a computable number.

$\sqrt{2}$ is a computable number, because it is represented by the continued fraction expansion

$$\sqrt{2} = 1 + \cfrac{1}{2 + \cfrac{1}{2 + \cfrac{1}{2 + \cfrac{1}{2 + \cdots}}}}$$

Because this continued fraction does not mention $\sqrt{2}$ in its input, it represents $\sqrt{2}$ as a computable number. The expression above means that if we consider the nth digit of the numbers in the sequence

$$1, 1 + \frac{1}{2}, 1 + \frac{1}{2 + \frac{1}{2}}, \ldots$$

it will eventually become the nth digit of the decimal expansion of $\sqrt{2}$. Similar continued fraction expansions show that $\sqrt{n}$ is a computable number for any positive integer n.

Let us represent the denominators in the continued fraction expansion of a real number by $[a, b, c, \ldots]$. Euler found a beautiful continued fraction expansion for e

$$e = 2 + [1, 1, 2, 1, 1, 4, 1, 1, 6, 1, 1, 8, 1, 1, \ldots]$$

which shows that e is a computable number.

The *Wallis product*

$$\pi = 2 \cdot \frac{2 \cdot 2 \cdot 4 \cdot 4 \cdot 6 \cdot 6 \cdot 8 \cdot 8}{1 \cdot 3 \cdot 3 \cdot 5 \cdot 5 \cdot 7 \cdot 7 \cdot 9} \cdots$$

shows that π is a computable number. Notice that the product can be expressed as

$$\pi = 2 \cdot \prod_{n=1}^{\infty} \frac{(2n)^2}{(2n-1)^2}$$

Turing developed Turing machines as models for computation while thinking about how a machine would compute the digits of a computable number.

Turing machines correspond to finite algorithms. A finite algorithm consists of a finite list of sentences in a language whose alphabet we can assume to be finite. Thus a finite algorithm consists of a finite string of symbols on a finite alphabet. Finite strings on a finite alphabet can be listed (we shall outline the proof in the exercises in the next section). For this reason, we can make a list of *all* Turing machines:

$$T_1, T_2, T_3, \ldots, T_n, \ldots$$

The fact that all Turing machines can be listed can be exploited in many ways. For example, the Turing machines on the list can be combined into a single machine U, which will carry out any computation.

> *There is a* universal Turing machine U *which will carry out any computation.*

We shall state and prove *Turing's Theorem*, the basic result in Turing's paper, in the next section. The proof makes essential use of the fact that Turing machines can be listed.

Remember that a Turing machine may halt on a given input. Determining when this happens is a basic question in computer science.

The Halting Problem   Given a Turing machine M, will M halt on a given input?

It can be proved, using Turing's Theorem, that there is no algorithm for solving the Halting Problem.

The Halting Problem is undecidable.

David Hilbert asked in his 1900 address whether there is a finite algorithm to determine whether a sentence in the predicate calculus is a true sentence. This problem is called the Decision Problem. In the propositional calculus, for example, there is such an algorithm: write down the truth table of the sentence, and see whether all the entries in the last column are Ts. Turing proved, using his Theorem, that

The Decision Problem is undecidable.

How can we express the ideal that Turing machines give the most general model for computation? The logician Alonzo Church stated the following conjecture (in a slightly different but equivalent form):

Church's Thesis   Computable functions are exactly those functions which can be computed by a Turing machine.

Church's thesis cannot be proved, because it does not give a precise definition of "computable function." However, there is much evidence for it, and it has been widely accepted. Kurt Gödel, whose work we shall discuss in the next section, said that at first he doubted whether a statement equivalent to Church's Thesis could be true, but that later he came to accept it.

Here is the definition of an important class of machines which is intermediate between the class of finite-state machines and Turing machines.

DEFINITION 10.2.5   A *push-down automaton* is a Turing machine which accepts a context-free language as input.

EXAMPLE 4   Give an example of a push-down automaton which is not a finite-state machine.

Solution   We have seen that the language $L = \{a^n b^n, n > 0\}$ is context-free but not regular. It is clearly recursively enumerable, so there is a Turing machine M which accepts it. M is a push-down automaton but not a finite-state machine.   ●

Recall that a palindrome is a word which is the same in either order. For example, 0010001000100 is a palindrome, but 100100100 is not. The problem of recognizing a palindrome is a standard problem about languages. This problem can be solved by using a suitable push-down automaton, but not by using a finite-state machine. The kind of push-down automaton involved is called a *nondeterministic* push-down automaton.

Remember the Chomsky Heirarchy for languages. Here is the corresponding heirarchy for machines.

The Chomsky Heirarchy for Machines

    I.   Finite-state machines

   II.   Push-down automata

  III.   Machines which accept context-sensitive languages

   IV.   Turing machines

What is the role of the languages that we have studied in computer programs?

The Role of Languages in Computer Programs

    I.   Regular languages describe most of the lexical content of computer languages (the way numbers, names, and other tokens of the language are written).

   II.   Context-free grammars are, in essence, mutually recursive structural inductive definitions.

  III.   Context-sensitivity enters naturally into many computer programs. For example, a variable cannot be used in a program statement until it has been declared. The variable is used in the *context* of being declared.

   IV.   Recursively enumerable languages, because they are the languages accepted by Turing machines, form a kind of limiting case for computer languages, and as such are a continuing object of study.

APPLICATION   *DNA Computing*

The idea that DNA could be used to carry out computations is due to Leonard A. Adleman. Adleman proved that DNA can be used for computing by showing that DNA can be used to solve the Hamiltonian Path Problem. We will give a simplified example which shows how his technique can be applied.

Remember that a Hamiltonian path in a graph is a path which visits each node exactly once. Suppose, for example, that we want to find a route which visits each of the cities in Figure 10.7 exactly once—that is, we want to find a Hamiltonian path in this graph.

You will see easily that the only such path is Atlanta-Boston-Chicago-Detroit. How could we find this path using a computation with DNA?

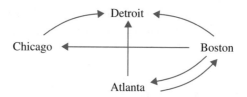

Figure 10.7   A Hamiltonian Path Problem

First, give the cities and the paths between them DNA codes as in Figure 10.8. The codes for the cities have four first letters and four last letters, and the complements of their codes are given by the T-A, G-C pairings. The codes for the paths between cities consist of the last four letters of the code for the first city, then the first four letters of the code for the second city.

City	DNA Name	Complement
Atlanta	ACTTGCAG	TGAACGTC
Boston	TCGGACTG	AGCCTGAC
Chicago	GGCTATGT	CCGATACA
Detroit	CCGAGCAA	GGCTCGTT

Flight	Flight Number
Atlanta − Boston	GCAGTCGG
Atlanta − Detroit	GCAGCCGA
Boston − Chicago	ACTGGGCT
Boston − Detroit	ACTGCCGA
Boston − Atlanta	ACTGACTT
Chicago − Detroit	ATGTCCGA

Figure 10.8   A Code for this Problem

To do the computation, the complements of the cities' codes and the "flight numbers" are synthesised, and put in a water solution. Because complementary DNA sequences bind together, the strings representing the codes and flight numbers bind together, and form longer strings. Among them is the string representing the Hamiltonian path, *GCAGTCGGACTGGGCTATGTCCGA* (Atlanta-Boston-Chicago-Detroit).

The string representing the Hamiltonian path is mixed with many other strings: some of the wrong length, and many of the right length which do not represent Hamiltonian paths. The technique of *gel electrophoresis* is used to select the strings of the right length. Then, the *polymerase chain reaction* technique is used to replicate the string representing the Hamiltonian path (which is picked out by having the right initial and terminal codes) many times, so that the solution consists mostly of copies of this string. Once this is done, the string $GCAGTCGGACTGGGCTATGTCCGA$ can be identified.

The Hamiltonian Path Problem is an NP-complete problem, so any method for solving it applies, in principle, to many other problems.

DNA computing involves the use of many chemical reactions at once, which we can regard as a kind of parallel computing.

EXERCISES

For each of the following regular expressions R, construct a finite-state machine which will accept the regular language determined by R.

1. $(a)^*$
2. $(a^2)^*$
3. $(a^3)^*$
4. $(a+b)^*$
5. $(a)^*(b)^*$
6. $(a+b)^*ab$
7. $(ab)^* + (cd)^*$
8. $a(b)^*c(d)^*$
9. $(a^2)^* + (b^3)^*$
10. $(a^2)^* + (b^3)^* + (c^4)^*$

For each of the following finite-state machines M, find a regular language which M accepts.

11.

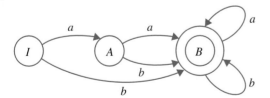

12.

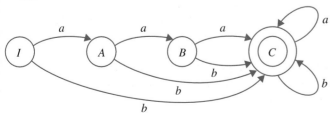

13.

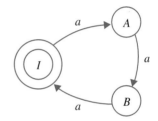

14.

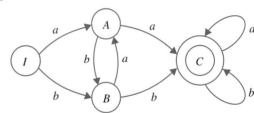

15.

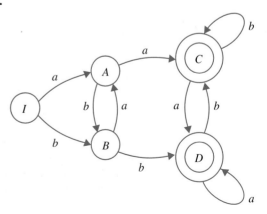

COMPUTER EXERCISE

1. Design a machine which checks whether a finite string on the alphabet {0, 1} is a palindrome.

10.3   GÖDEL AND TURING

> *In the midst of the word he was trying to say,*
> *In the midst of his laughter and glee,*
> *He had softly and silently vanished away—*
> *For the Snark* was *a Boojum, you see.*

In a remarkable passage in the introduction to *The Laws of Thought*, Boole made a statement which can be interpreted as a suggestion that a computer is a logical device, or that we should examine the logical structure of a computer program:

> *Now the actual investigations of the following pages exhibit Logic, in its practical aspect, as a system of processes carried on by the aid of symbols having a definite interpretation, and subject to the laws founded on that interpretation alone.*

The idea that logic is of crucial importance in computer science is due about equally to two men, Alan Turing and John von Neumann (1904–1957).

The emphasis which Turing and von Neumann placed on logic in computer science focused attention on the very original work of the Austrian logician Kurt Gödel (1906–1978). At the end of Chapter 6, we described the Gödel Completeness Theorem (which asserts that every true sentence in the predicate calculus has a proof) and the Gödel Incompleteness Theorem (which asserts that there are true sentences in the augmented predicate calculus which cannot be proved).

Gödel published the Incompleteness Theorem in 1931 in the paper *On formally undecidable propositions of Principia Mathematica and related systems I*. In 1936, Turing published *On computable numbers with an application to the Entscheidungsproblem* (the Decision Problem), which we discussed in the previous section.

Gödel and Turing each saw what is essential, from a mathematical point of view, in the paradoxes of self-reference which we have described throughout this book. Each of their theories uses a form of the *Cantor Diagonal Procedure*.

The paradoxes of self-reference have been a concern of logicians since classical times. The epitaph of one early logician is as follows:

> *Philetas of Cos am I*
> *It was The Liar that caused me to die*
> *And the lost sleep caused thereby*

The English logician William of Sherwood (about A.D. 1200–about A.D. 1270) wrote *De insolubilibus* (On Insoluble Propositions) about the paradoxes of self-reference. Among other paradoxes of this kind, we have discussed the Liar Paradox, the Barber's Paradox, and Russell's Paradox.

Gödel's Theorem uses the technique of *Gödel numbering*, which we shall describe in this section.

The close connection between Turing's results and Gödel's can be understood better if we realize that the Gödel Incompleteness Theorem can be formulated in terms of Turing machines. When it is expressed in this way, the Incompleteness Theorem sets limits to computation in ways which are still not well understood.

Our discussion of logic, languages, and machines has prepared us for a general discussion of recursiveness and recursive enumerability, of computability and unsolvability. We shall discuss *Church's Thesis*, perhaps the most general statement which can be made about computability.

The generality of the Turing machine model is illustrated by the fact that there are *universal Turing machines*.

It might seem that the paradoxes of self-reference lead to purely negative results. This is not true, and to illustrate this we shall describe still another problem with a negative solution, the Tenth Hilbert Problem. Although the solution to this problem is negative, the techniques which were used to prove this lead to a remarkable explicit description of recursively enumerable sets.

The quote at the beginning of this section is the last quatrain of *The Hunting of the Snark*. At the end of this section, we shall explain why we have used quotes from *The Hunting of the Snark* throughout this book.

Groucho Marx remarked that

> *I would never belong to a club which would have me as a member.*

This joke combines self-reference and negation, as in the paradoxes of self-reference which we have studied throughout this book. Self-reference and negation can be combined to give:

I.   A joke, as in this example.

II.   A paradox, such as the Liar Paradox.

III.   A valid proof technique, such as the technique that Cantor used to prove that there can never be a one-to-one correspondence between a set A and the set of all its subsets $P(A)$.

So far, our examples of the paradoxes of self-reference have been far removed from any practical significance, but now we shall give several examples of their use. In order to do this, we shall describe the *Cantor Diagonal Procedure*.

DEFINITION 10.3.1   A set S is *countable* if the elements of S can be listed.

Any finite set is countable, because its elements can be listed by a list that stops.

Cantor proved that the set of rational numbers is countable by observing that the positive rational numbers can be written in a square array as in Figure 10.9 and then listed by using the line in the figure.

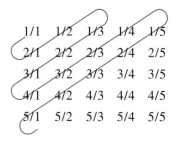

Figure 10.9   The Rational Numbers Are Countable

Each positive rational number occurs infinitely often in the list constructed in this way, but by only listing each rational number the first time it occurs, we can obtain a list of the positive rational numbers in which each number occurs exactly once. Then we can list all the rational numbers by starting the list with 0, and then listing rational numbers a and $-a$ as in Figure 10.9.

The Cantor Diagonal Procedure is used to prove the following theorem. We shall prove the first part of the theorem, and outline the proof of the second part in the exercises.

THEOREM 10.3.1   Let A be any alphabet with at least two elements. Then

    (i)   The set of all strings on A is not countable.

    (ii)   If A is finite, the set of all finite strings on A is countable.

Proof of (i):   Suppose that we can make a list of all strings on A as in Figure 10.10.

$$a_{11}a_{12}a_{13}a_{14}a_{15} \ldots$$
$$a_{21}a_{22}a_{23}a_{24}a_{25} \ldots$$
$$a_{31}a_{32}a_{33}a_{34}a_{35} \ldots$$
$$a_{41}a_{42}a_{43}a_{44}a_{45} \ldots$$
$$a_{51}a_{52}a_{53}a_{54}a_{55} \ldots$$

Figure 10.10   A List of Strings

Now construct a new string $b = b_1b_2b_3 \ldots$ as follows: Let $b_1 \neq a_{11}, b_2 \neq a_{22}, \ldots b_n \neq a_{nn} \ldots$. The string b differs from each string on the list in at least one entry, and therefore does not occur on the list. Thus there cannot be any list of all the strings on A. ●

This theorem can be used to prove that irrational numbers exist.

COROLLARY 10.3.1   Irrational numbers exist.

Proof Since the set of rational numbers is countable, it is enough to prove that the set of real numbers is not countable. We shall prove that even the set of real numbers between 0 and 1 is not countable. If this set is countable, we can make a list of its elements, where each real number on the list is represented by its decimal expansion. Here there is a technical problem, because the decimal expansion of a real number may not be unique. For example,

$$.0020000\ldots = .00199999\ldots$$

We will deal with this difficulty by always choosing the decimal expansion which is smallest in the lexicographical order. For example, of the two decimal expansions above, we choose the one on the right-hand side. This choice is respected by the argument in the next paragraph.

Using the diagonal procedure, we can find a real number whose decimal expansion is different from all the decimal expansions on the list, so this real number cannot be on the list. Therefore the set of real numbers is not countable. ●

In a similar way, we can prove that the set of computable numbers which we discussed in the last section is countable.

THEOREM 10.3.2 The set of computable numbers is countable.

Proof Each computable number is computed by a finite algorithm, which computes only that number. By part (ii) of the previous theorem, the set of finite algorithms is countable. Therefore the set of computable numbers is countable. ●

COROLLARY 10.3.2 There exist numbers which are not computable.

Proof The set of computable numbers is countable, but the set of all real numbers is not. ●

Now we shall explain an idea of Gödel's, which is to represent each sentence or finite set of sentences in a logical theory by a number called a *Gödel number*. We shall give a simplified form of the procedure, which applies only to the propositional calculus, but in a more elaborate form it can be applied to the augmented predicate calculus.

In Figure 10.11, we assign numbers to each of the signs in the propositional calculus (the variables are assigned prime numbers, starting with 7).

Now given a sentence in the propositional calculus, we make a list of the prime numbers in increasing order, one for each sign in the sentence, and multiply powers of these primes together, where the power is the Gödel number of the corresponding sign. This gives the Gödel number of the sentence.

For example, the Gödel number of the sentence

$$(p \rightarrow q)$$

Sign	Gödel Number
~	1
∨	2
∧	3
→	4
(	5
)	6
p	7
q	11
r	13
...	...

Figure 10.11   Gödel Numbers for the Propositional Calculus

is

$$2^5 3^7 5^4 7^{11} 11^6 = 153219249169895698020000$$

and

$$2^5 3^1 5^7 7^6 = 882,367,500,000$$

is the Gödel number of the sentence

$$(\sim p)$$

Finite sets of sentences can be given Gödel numbers in a similar way.

The unique factorization property of the integers implies that each sentence in the propositional calculus is assigned a different Gödel number.

THEOREM 10.3.3   The mapping which assigns a Gödel number to a sentence in the propositional calculus is a one-to-one correspondence of the set of sentences in the propositional calculus with a certain set of positive integers.   ●

Gödel numbers tend to be quite large. On the other hand, the primes that they factor into are relatively small. This means that given a natural number which is a Gödel number, a computer algebra system will factor it quickly so that the sentence that it represents can be identified. That is,

> *Computer algebra systems have made it practical to compute with Gödel numbers.*

Suppose that we have a logical theory S, like the augmented predicate calculus, in which the properties of the integers can be described. Then Gödel numbering can be used to describe the sentences of S, and thus the sentences of S can be referred to in S.

This self-referential property was used by Gödel to prove his famous Incompleteness Theorem, which we will state in a special case. Negation is also involved, so once again we have a logical result which involves self-reference and negation.

The Gödel Incompleteness Theorem   There is a sentence P in the augmented predicate calculus such that neither P nor $\sim P$ can be proved in the augmented predicate calculus.   ●

Why is Gödel's Theorem important?

I.   It seems very plausible that in a reasonable logical system every true sentence should have a proof. Distinguished mathematicians, such as David Hilbert, believed this to be true. Gödel's Theorem showed once and for all that this idea is not correct.

II.   Gödel's Theorem can be stated using Turing machines. When this is done, it imposes limits on computation in ways which are still not understood.

III.   In a broader sense, Gödel's Theorem shows that a system which can be described in finite terms may have behavior which cannot be predicted. This observation applies both to computers and to the brain.

Turing's Theorem   There are well-defined sets of integers for which there is no effective means of testing membership in that set.

Proof   We will use a diagonalization argument. Assume, as we may, that the Turing machines which we consider print a final 1 when they halt. Consider a two-way infinite table, like our table in Figure 10.9 of the rational numbers, in which row i corresponds to Turing machine T_i, and column j corresponds to the input of the integer j. In the table, we will enter a 1 in position (i, j) if Turing machine T_i halts on input j, and a 0 if it does not.

Now consider the diagonal entries of the table, (i, i), and define a new machine T which halts on input i (and so should have a table entry of 1), exactly if the (i, i)-th entry in the table is 0.

Can this be one of the Turing machines on the list? No, because if it is machine T_k, and T halts (does not halt) on input k, then T_k would instead not have halted (have halted) on that input. Thus

$$X = \{i : T_i(i) \neq 1\}$$

is a noncomputable set (not recursively enumerable).   ●

Why is Turing's Theorem important?

I.   Turing's Theorem is a computational counterpart of Gödel's Theorem.

II. Turing's Theorem makes a statement about the fundamental notion of recursive enumerability, or possible types of computation.

III. Like Gödel's Theorem, Turing's Theorem raises numerous questions which still have not been answered. For example, for what special classes of machines is it impossible to define sets which are not recursively enumerable, in terms of machines in the class?

Now we shall describe recursive enumerability from a different point of view.

A *Diophantine equation* is a polynomial equation, perhaps in several variables, with integer coefficients, which is to be solved in the integers. An example is the Fermat equation

$$X^n + Y^n = Z^n$$

which has no solutions in non-zero integers if $n > 2$, as Andrew Wiles proved in 1994.

Some Diophantine equations do have solutions. For example, the equation

$$6Y^2 = X(X + 1)(2X + 1)$$

is important in the theory of supersymmetry in quantum mechanics, and has the solution $X = 24, Y = 70$.

In his famous address in 1900, David Hilbert gave a list of mathematical problems for the new century. Some of these problems have been solved and some have not. One such problem follows.

The Tenth Hilbert Problem   Is there a finite algorithm to determine whether a Diophantine equation has a solution?

It was proved by Martin Davis, Yuri Matijasevic, and Julia Robinson that no such algorithm exists.

THEOREM 10.3.4   There is no finite algorithm to determine whether a Diophantine equation has a solution.   ●

The proof of this theorem is too long to be given here, but let us point out that the proof makes essential use of the fact that the Fibonacci numbers grow exponentially, that is of the formula

$$F_n \approx \phi^n = ((1 + \sqrt{5})/2)^n$$

Diophantine equations seem at first to have little to do with Turing machines or questions about recursive enumerability. For this reason, the following theorem of Yuri Matijasevic is very remarkable.

THEOREM 10.3.5   The following conditions are equivalent:

    (i)   A set S of natural numbers is recursively enumerable.

    (ii)   There is a polynomial $P(t, x_1, \ldots, x_r)$ with integer coefficients such that the equation $P(a, x_1, \ldots, x_r) = 0$ has a solution in natural numbers if and only if $a \in S$. ●

This kind of description of a recursively enumerable set S is called a *Diophantine representation*.

Hilary Putnam gave a simple but useful translation of the condition that a set of natural numbers has a Diophantine representation.

Lemma   The following conditions are equivalent:

    (i)   A set S of natural numbers has a Diophantine representation.

    (ii)   There is a polynomial $Q(t, x_1, \ldots, x_r)$ such that S is precisely the set of positive values of Q as the variables assume values which are natural numbers.

Proof   Set

$$Q(t, x_1, \ldots, x_r) = (t + 1)(1 - P(t, x_1, \ldots, x_r)^2) - 1$$

If $a \in S$, so that $P(a, n_1, \ldots, n_r) = 0$ for some r-tuple of natural numbers, then $Q(a, n_1, \ldots, n_r) = a$. Conversely, if $Q(a, n_1, \ldots, n_r) > 0$, then $P(a, n_1, \ldots, n_r) = 0$, and $Q(a, n_1, \ldots, n_r) = a$. ●

The set of primes is recursively enumerable because there is a finite algorithm to test whether any given natural number is prime. Therefore the set of prime numbers has a Diophantine representation. The simplest such representation which is known is the following, which is due to J. P. Jones.

THEOREM 10.3.6   The set of primes is exactly the set of positive values of the following polynomial, when the values of the variables are natural numbers.

$$(k + 2)\{1 - ([wz + h + j - q]^2 + [(gk + 2g + k + 1)(h + j) + h - z]^2$$
$$+ [16(k + 1)^3(k + 2)(n + 1)^2 + 1 - f^2]^2$$
$$+ [e^3(e + 2)(a + 1)^2 + 1 - o^2)^2]^2 + [(a^2 - 1)y^2 + 1 - x^2]^2$$
$$+ [16x^2y^4(a^2 - 1) + 1 - u^2]^2$$
$$+ [((a + u)^2(u^2 - a)^2 - 1)(n + 4dy)^2 + 1 - (x + cu)^2]^2$$
$$+ [(a^2 - 1)L^2 + 1 - m^2]^2 + [ai + k + 1 - L - i]^2$$
$$+ [n + L + v - y]^2 + [p + L(a - n - 1) + b(2an + 2a - n^2 - 2n - 2) - m]^2$$
$$+ [q + y(a - p - 1) + s(2ap + 2a - p^2 - 2p - 2) - x]^2$$
$$[z + pL(a - p) + t(2ap - p^2 - 1) - pm]^2)\}$$

●

It is believed that this Diophantine representation of the primes can be simplified.

Throughout this book, we have discussed the relationship between computer science and discrete mathematics. How will computer science develop in the future? And how will discrete mathematics be involved in this development?

Themes for the Future of Computer Science

   I.   Architecture: development of better and faster devices

   II. Parallelism, distribution, and networking

   III. Better systems software

   IV. Better codified approaches to software engineering

   V.   More interaction with the natural sciences

   VI. Better interfaces between humans and computers

Consider point I, for example. Computer chips have been improved for some time, but there are limits to how many circuits can be placed on a chip. This may be a fundamental limitation, or an opportunity. *Quantum computing*—the use of chips or other devices which exploit quantum mechanical effects—may make it possible to construct more powerful computers without increasing their size. This possibility was suggested in the 1980s by Richard Feynman and others. Peter Shor has devised a quantum algorithm for factoring integers.

The techniques of discrete mathematics are already being used in quantum mechanics. The graphs called *Feynman diagrams* are used to analyze particle interactions. The lattices which we studied in Chapter 7 are used as models for quantum logic, and perhaps can be used in the study of parallelism in computer science. Let us also mention the *Information Paradox* in quantum mechanics.

It is a basic principle of quantum mechanics that the information in a system can never be destroyed. The information contained in material falling into a black hole seems to contradict this principle, since nothing can escape from a black hole. The same problem exists on a larger scale: all the matter and energy in the early universe are invisible to us now.

Suppose that we compare *observability* in a physical theory with *provability* in a logical theory. Suppose that we compare the information contained in matter which has fallen into a black hole with true but unprovable sentences in a logical theory. Perhaps this gives a hint about how the Information Paradox could be resolved: the true but unprovable sentences are "invisible" to the axioms of the theory. Perhaps we could say, more generally, that a paradox in any theory should be compared with the paradoxes of self-reference.

What about point II? It seems clear that parallel architectures for computers will become increasingly important. But which architecture is the best? If we have a number of chips which are to be connected, we can describe the connection by a graph. Which graph is best? Should the graph of the n-cube be used?

If so, there are problems about the n-cube which should be solved, and are not now properly understood. Here is one: give a simple formula for the number of spanning trees of the n-cube.

Now consider point V. A theme in this book has been the relationship between discrete mathematics and the genetic code. We have seen that there are simple and natural questions about the genetic code which have never been considered, and which could be approached with the techniques of discrete mathematics. From another point of view, we have seen that DNA can be used to carry out computations.

Most DNA is called "junk DNA" and seems to have no function. The genes which really carry genetic information occur in relatively short sequences in their chromosomes. Could we compare "junk DNA" with a random code, such as shift register output, which must then be modified for a specific purpose?

More work is needed on the relationship between the computer and the brain. The brain can store and process enormous amounts of data in ways that are not understood. We mentioned the mysterious vertical columns of cells in the cortex; perhaps they are involved in the storage and processing of information. It is quite possible that the brain is using structures and techniques which are superior to any now being used by computers. When these matters are better understood, it may be possible to design better interfaces between humans and computers.

Whatever the future development of computer science may be, it seems clear that discrete mathematics will contribute to it in an essential way.

APPLICATION   *The Vanishing*

On July 17, 1874, Lewis Carroll (Charles Lutwige Dodgson, (1832—1898)) arrived in Guildford, in Southern England, to help take care of his godson Charles Wilcox. The next day he was walking on a nearby hillside when a single line of a poem occurred to him:

> *For the Snark* was *a Boojum, you see.*

Thirteen years later Carroll wrote

> *I know not what it meant, then: I know not what it means, now; but I wrote it down: and, some time afterwards, the rest of the stanza occurred to me, that being its last line: and so by degrees, at odd moments during the next year or two, the rest of the poem pieced itself together, that being the last stanza.*

Carroll was a logician who taught at Christ Church, Oxford, who besides his literary work is known as the author of *Symbolic Logic*. Despite its title, this book is for the most part a description of Aristotelian logic. One basic goal of symbolic logic, according to Hilbert and Russell, was to prove the completeness of the predicate calculus (augmented or not). It was noticed at the time that Carroll seemed not to want to involve himself in studying questions of this kind.

One aspect of the symbolism of *The Hunting of the Snark* is that the names of all the characters begin with the letter "B." They are hunting a Snark, which turns out to be a Boojum, and when they find it, the Baker vanishes. Here we have self-reference (via the letter "B") and negation (the vanishing of the Baker), the two elements of a paradox of self-reference.

Suppose that Carroll could see, perhaps without realizing it, that efforts to prove the completeness of some logical systems would fail? Could we view this as the symbolic meaning of *The Hunting of the Snark*?

Is the poem a prediction of Gödel's Theorem?   ●

EXERCISES

Here are some exercises about countable sets. Although the set of all strings on any alphabet with two or more elements is not countable, the set of finite strings on any finite alphabet *is* countable. This fact is fundamental in many arguments about computability, so we outline a proof of it here, and ask you to fill in the details.

1. If the set S is countable, prove that any subset of S is countable.

2. If the sets S and T are countable, prove that $S \times T$ is countable.

3. If the sets $S_1, S_2, \ldots, S_r$ are all countable, prove that $S_1 \times S_2 \times \ldots \times S_r$ is countable.

4. If the set S is countable and if $f : S \to T$ is a one-to-one correspondence, prove that T is countable.

5. If we have an increasing chain of countable sets

$$S_1 \subset S_2 \subset \ldots \subset S_k \subset \ldots$$

prove that

$$\cup_{k=1}^{\infty} S_k$$

is countable.

6. If A is a finite set and B is the set of finite strings on A, prove that B is countable.

COMPUTER EXERCISES

These exercises are about Gödel numbers, using the scheme for Gödel numbering given in this section.

Find the Gödel number of each of the following sentences, first in factored form and then multiplied out.

1. $(\sim q)$ **2.** $(q \to r)$

3. $(p \to (\sim q))$ **4.** $((\sim q) \to (\sim p))$

5. $(p \lor q)$ **6.** $(p \land q)$

7. $(p \to (p \lor q))$ **8.** $(p \to (q \to p))$

9. $((p \to (q \to r)) \to ((p \to q) \to (p \to r)))$

10. $(((\sim p) \to (\sim q)) \to (q \to p))$

Each of the following numbers is the Gödel number of a sentence in the propositional calculus. Factor each number, and find the sentence which it represents.

11. 551479687500000

12. 6081271999553160254441380000

13. 16457113838723190132677966689366606140000

14. 6128769966795827920800

15. 30643849833979139604000

16. 18861899519076559895347238623444403078760 2057564533154996640

17. 17119575680491266704969173240203982903512 686444460000

18. 63814764335650020000

19. 12762952867130004000

20. 23048372030003417433112239332469600

Guide to the Literature

A.1 ALGEBRA

1. J. B. Fraleigh, *A First Course in Abstract Algebra, 5th ed.*, Addison-Wesley (1994).

   An excellent general introduction to abstact algebra, contains the facts about groups and finite fields, which are needed in discrete mathematics.

2. R. Lidl and H. Niederreiter, *Introduction to Finite Fields and Their Applications*, Cambridge University Press (1986).

   A complete work on finite fields, gives a thorough mathematical foundation for the use of finite fields in computer science.

A.2 ALGORITHMS

1. A. V. Aho, J. E. Hopcraft, and J. D. Ullman, *The Design and Analysis of Computer Algorithms*, Addison-Wesley (1974).

   A clear treatment of algorithm design and complexity.

2. T. H. Corman, C. E. Leiserson, and R. L. Rivest, *Introduction to Algorithms, 2nd Edition*, McGraw-Hill (1994).

   A complete introduction to the theory of algorithms, includes recent developments.

3. D. E. Knuth, *The Art of Computer Programming, Second Edition: Volume 1, Fundamental Algorithms; Volume 2, Seminumerical Algorithms; Volume 3, Sorting and Searching*, Addison-Wesley (1973).

   A detailed analysis of computer algorithms, with many interesting mathematical details.

4. M. C. Paull, *Algorithm Design: A Recursion-Transformation Framework*, Wiley-Interscience (1988).

    Algorithms from the point of view of structural induction and proof.

A.3 CODES

1. F. J. MacWilliams and N. J. A. Sloane, *The Theory of Error-Correcting Codes*, North-Holland (1981).

    The definitive reference on codes, the bibliography has almost 1500 references.

2. V. Pless, *The Theory of Error-Correcting Codes*, Wiley-Interscience (1989).

    Perhaps the best brief introduction to the theory of codes. The point of view is rather algebraic.

A.4 COMBINATORICS

1. I. Anderson, *Combinatorics of Finite Sets*, Oxford Science Publications (1987).

    An excellent treatment of the theory of set systems, that is, of finite sets of subsets of a finite set.

2. W. W. Rouse Ball and H. S. M. Coxeter, *Mathematical Recreations and Essays, Thirteenth Edition*, Dover (1987).

    A delightful and original collection of problems. This book has gone through many editions, and some editions contain material that is not in later ones. For example, the eleventh edition contains interesting material on the Kirkman Schoolgirl Problem, which was later removed.

3. G. Boole, *Calculus of Finite Differences*, Chelsea (1970).

    Boole's excellent text on the difference calculus.

4. L. Comtet, *Advanced Combinatorics*, D. Reidel Publishing Company (1974).

    The title is a little deceptive: This book is very easy to read, and gives a marvelous survey of the subject.

5. H. S. M. Coxeter, *Regular Polytopes*, Dover (1973).

    These figures, which include the regular solids, have interesting graphs.

6. I. Gessel and G.-C. Rota, *Classic Papers in Combinatorics*, Birkhäuser (1987).

    Gives a clear view of the development of the subject.

7. G. Polya and R. C. Read, *Combinatorial Enumeration of Groups, Graphs, and Chemical Compounds*, Springer-Verlag (1987).

    A classic monograph, translated from the German, the source of Polya's theorem.

8. J. Riordan, *An Introduction to Combinatorial Analysis*, Princeton University Press (1980).

    An exposition of the subject by a master.

9. G.-C. Rota, ed., *Studies in Combinatorics*, M.A.A. Studies in Mathematics, Volume 17 (1978).

   A first-rate collection of papers on combinatorics. A good place to begin studying combinatorics.

10. H. J. Ryser, *Combinatorial Mathematics*, The Carus Mathematical Monographs, Number 14 (1963).

   Short, elegant, and easy to read.

11. N. J. A. Sloane and S. Plouffe, *The Encyclopedia of Integer Sequences*, Academic Press (1995).

   A remarkable study of the recursiveness of sequences of integers, describes 5487 integer sequences.

12. R. P. Stanley, *Enumerative Combinatorics*, Wadsworth and Brooks/Cole (1986).

   An excellent introduction to combinatorics, at a somewhat more advanced level than Ryser's book.

13. H. S. Wilf, *generatingfunctionology*, Academic Press (1990).

   The theory of recursive sequences.

A.5 COMPUTER SCIENCE

1. J. van Leeuwen, ed., *Handbook of Theoretical Computer Science, Volume A, Algorithms and Complexity*, and *Volume B, Formal Models and Semantics*, Elsevier (1992).

   A valuable survey of current issues in theoretical computer science with extensive references.

2. S. V. Pollack, ed., *Studies in Computer Science*, M.A.A. Studies in Mathematics Volume 22 (1982).

   Expository papers about the history and structure of computer science.

3. C. J. Date, *An Introduction to Database Systems, 6th Edition*, Addison-Wesley (1995).

4. J. Ullman, *Principles of Database Systems, 2nd Edition*, Computer Science Press (1982).

   Two standard texts on database systems, largely based on the calculus of relations and logic.

A.6 GRAPHS

1. N. L. Biggs, E. K. Lloyd, and R. J. Wilson, *Graph Theory 1736–1936*, Oxford University Press (1986).

   An exposition of the subject using excerpts from original papers.

2. P. C. Fishburn, *Interval Orders and Interval Graphs*, Wiley-Interscience (1985).

   A useful study of a special class of graphs.

3. D. R. Fulkerson, ed., *Studies in Graph Theory Part I and Part II*, M.A.A. Studies in Mathematics Volume 11 and Volume 12 (1975).

   Anyone who wants to learn graph theory should read this collection of papers.

4. F. Harari, *Graph Theory*, Addison-Wesley (1972).

   An indispensible reference.

5. O. Ore, *Theory of Graphs*, American Mathematical Society (1962).

   Another basic reference, very easy to read.

6. E. M. Palmer, *Graphical Evolution*, Wiley-Interscience (1985).

   Describes random graphs and the mysterious Giant Component of such graphs.

7. R. J. Wilson, *Introduction to Graph Theory, Third Edition*, Longman (1985).

   A very readable introduction to graph theory.

A.7   Biography, Early Works, History

1. A. W. F. Edwards, *Pascal's Arithmetic Triangle*, Oxford University Press (1987).

   Excellent treatment of the topics in Pascal's *Treatise on the Arithmetic Triangle*. Contains a chapter on Jakob Bernoulli's *Ars conjectandi*.

2. L. Euler, *Introduction to Analysis of the Infinite, Book I*, translated by J. D. Blanton, Springer-Verlag (1988).

   A classic of mathematical exposition, unimproved in the areas which it covers. Chapter XIII, "On Recurrent Series," gives a fascinating glimpse at the theory of linear recursive sequences in the middle of the eighteenth century.

3. A. Hald, *A History of Probability and Statistics*, Wiley-Interscience (1990).

   Excellent treatment of the subject, covering the period to 1750, with comments about nineteenth-century developments.

4. A. Hodges, *Alan Turing: The Enigma*, Simon and Schuster (1983).

   An account of Turing's life and scientific work.

5. P. A. MacMahon, *Combinatory Analysis*, Chelsea (1984).

   A very original book that is still influential. Chapter II of Section III describes the *MacMahon Master Theorem*.

6. A. DeMoivre, *The Doctrine of Chances*, Chelsea (1967).

   A reproduction of the last edition of DeMoivre's book, contains many interesting problems.

7. P. R. de Montmort, *Essay d'Analyse sur les Jeux de Hazard*, Chelsea (1980).

   A reproduction of the second edition of Montmort's *On Games of Chance*, in French. Montmort sent a copy to Newton, which Newton seems to have read. Contains many interesting examples.

8. I. Todhunter, *A History of the Mathematical Theory of Probability from the Time of Pascal to That of Laplace*, Chelsea (1965).

   A gold mine of interesting problems.

9. M. N. Cohen, *Lewis Carroll*, Knopf (1995).
   A biography of Lewis Carroll.

10. J. Tanis and J. Dooley, ed., *Lewis Carroll's The Hunting of the Snark*, William Kaufmann (1981).
    A detailed study of the poem that inspired this book, contains Martin Gardner's *The Annotated Snark*.

A.8 LOGIC

1. G. Boole, *An Investigation of the Laws of Thought, on Which Are Founded the Mathematical Theories of Logic and Probabilities*, Dover (1958).
   Boole's timeless book.

2. M. Gardner, *Logic Machines and Diagrams*, Chicago (1982).
   Elegant essays on these topics.

3. S. C. Kleene, *Introduction to Metamathematics*, Van Nostrand (1952).
   A comprehensive treatment of logic, recursiveness, and computability, by a master.

4. E. Mendelson, *Introduction to Mathematical Logic*, Van Nostrand (1964).
   An excellent treatment of mathematical logic.

5. A. N. Whitehead and B. Russell, *Principia Mathematica*, three volumes, Cambridge University Press (1957).
   The authors' concern in writing this book was to develop a logical system in which mathematics could be developed without paradoxes such as Russell's Paradox. Although the system has been simplified, it remains important. The long introduction is worth reading.

A.9 PROBABILITIES

1. W. Feller, *An Introduction to Probability Theory and Its Applications, Volumes I and II, Third Edition*, John Wiley and Sons (1968).
   An excellent introduction to the theory of probabilities. The third chapter of the first volume, about random walks, is a classic of mathematical exposition.

2. S. M. Ross, *Introduction to Probability Models*, Academic Press (1985).
   A good general introduction to probability theory.

A.10 LANGUAGES, AUTOMATA, AND COMPUTABILITY

1. A. Aho, R. Sethi, and J. Ullman, *Compilers: Principles, Techniques, and Tools*, Addison Wesley (1987).
   The standard reference on compilers.

2. A. Aho and J. Ullman, *The Theory of Parsing, Translation, and Compiling*, two volumes, Prentice-Hall (1972, 1973).
   A compendium of language theory, as of the early 1970s.

3. G. Chaitin, *The Limits of Mathematics*, Springer (1998).

The limits of formal reasoning, in the context of algorithmic information theory.

4. M. Davis, *Computability and Unsolvability*, Dover (1973).

Turing machines and undecidable problems.

5. J. Hopcraft and J. Ullman, *Introduction to Automata Theory, Languages, and Computation*, Addison-Wesley (1979).

A standard, comprehensive, but slightly outdated reference on languages, including regular languages and context-free languages.

6. Z. Kohavi, *Switching and Finite Automata Theory*, McGraw-Hill (1978).

Regular expressions, finite automata, and their use in computer architecture.

7. H. Lewis and C. Papademetriou, *Elements of the Theory of Computation*, Prentice-Hall (1998).

A modern reference on formal languages, regular languages, context-free languages, Turing machines, and computability.

B

Notes

In these notes we refer to the Guide to the Literature in Appendix A. The reference IV(3), for example, means the third reference in Section IV of the Guide. References to sections of the book will be II.2 for the second section of Chapter 2, and so on.

Each section of this book begins with a quote from *The Hunting of the Snark*, by Lewis Carroll, VII(9) and VII(10).

In this book we frequently use short quotes from the biographical sketches in *The Dictionary of Scientific Biography*, Charles Scribner's Sons (1970).

Chapter 1

1. The quote at the beginning of this chapter is our own translation of part of a royal edict granting Pascal a monopoly for the manufacture of his calculating machine. The edict was written by a court official for Louis XIV's mother, who was regent at the time.

2. Several anecdotes in Section I.1 are taken from the book by A. W. F. Edwards VII(1).

3. There is a discussion of the general Tower of Hanoi puzzle in the MAA Monthly for March 1941, pp. 216–219. The formula given there for the minimal number of moves still seems not to have been proved completely.

4. The quote at the beginning of Section I.3 and others by Boole are from *The Laws of Thought* VIII(1).

5. Frank Gray was granted U.S. Patent 2632058 on March 17, 1953, for the code that is named for him.

6. We have used the following edition of Venn's book: John Venn, *Symbolic Logic*, Lenox Hill Publishers and Distributors (Burt Franklin), originally published in 1894, reprinted in 1971.

7. In Section I.4 we mention a few notions from set theory. For details, see the book by E. Mendelson VIII(4).

8. The "Towns with Clubs" problem in Section I.4 has far-reaching generalizations in the theory of set systems.

9. The original reference for Hamming codes is R. W. Hamming, "Error Detecting and Error Correcting Codes," *Bell Syst. Tech. J.*, 29 (1950): 147–160.

10. The example of a $5X + 1$ sequence with a cycle in Section I.5 is due to T. J. Marlowe.

11. There is a detailed treatment of the difference calculus in Boole's book IV(3).

12. The lengths of the Collatz sequences in the exercises for Section I.5 are taken from a long table compiled by J. DiVito.

Chapter 2

1. For Euclid's *Elements* we refer to Sir Thomas L. Heath, *The Thirteen Books of Euclid's Elements*, Dover (1956).

2. Several remarks about Fermat and about Mersenne numbers are taken from A. Weil, *Number Theory*, Birkhäuser (1984).

3. For a recent improvement on the Fermat Factorization Algorithm, see V. Lucarelli, *On Factoring n with the b-algorithm*, The College Mathematics Journal 29, no. 4, September 1998, 289–295.

4. In section II.2 and elsewhere there are references to Euler's *Introductio in analysin infinitorum*. We have used the edition VII(2).

5. The treatment of Lamé's Theorem is our own.

6. The material on the Strict Egyptian Fraction Algorithm is from L. Beekmans, "The Splitting Algorithm for Egyptian Fractions," *Journal of Number Theory* 43 (1993): 173–185.

7. The identities for the Fibonacci and Lucas numbers in the Advanced Exercises of Section II.2 are from S. L. Basin and V. E. Hoggatt, Jr., "A Primer on the Fibonacci Sequence, Part I," *Fibonacci Quarterly* 1962: 65–72.

8. The quote from Gauss at the beginning of Section II.3 is from C. F. Gauss, *Disquisitiones Arithmeticae*, English Edition, translated by A. A. Clarke, Springer (1986).

9. In Section II.3 and in several other places we give examples of codes. Our treatment generally follows that in the book by V. Pless III(2).

10. We are grateful to Peter Guidon for explaining a biochemist's view of the genetic code, and for giving us several interesting examples of DNA sequences.

11. In Section II.3 we raise the question, which we return to several times, about the recursiveness of sequences in the genetic code.

Chapter 3

1. The Prüfer correspondence was originally described in E. P. H. Prüfer, "Neuer Beweis eines Satzes über Permutationen (New Proof of a Theorem about Permutations)," *Archiv der Mathematik und Physik* (3) 27 (1918): 142–144.

2. Multinomial coefficients, and of course binomial coefficients, were known to Indian mathematicians well before A.D. 1000.

3. In the exercises for Section III.2 we use several examples from biochemistry, which were suggested to us by Peter Guidon.

4. One of Catalan's original notes about what are now called Catalan numbers is E. Catalan, "Note sur une Équation aux différences finis (Note about a finite-difference equation)," *Journal de Mathématiques Pures et Appliqués* Tome III (Octobre 1838): 508–516.

5. There is a pleasant introduction to Catalan numbers in M. Gardiner, "Catalan Numbers," in *Time Travel and Other Mathematical Bewilderments*, W. H. Freeman (1988): 253–266.

6. The Principle of Inclusion-Exclusion seems to have been stated first by DeMoivre.

7. Our treatment of Jakob Bernoulli's formula for counting sums of kth powers at the end of Section III.3 follows that in D. Struik, *A Source Book in Mathematics, 1200–1800*, Harvard University Press (1969).

Chapter 4

1. The Water, Gas, and Electricity Problem in Section IV.1 was first stated in H. E. Dudeney, *Amusements in Mathematics*, Dover (1958). The point of this problem is that $K_{3,3}$ is not planar. It does not seem to be known when it was first observed that K_5 is not planar.

2. The application at the end of Section IV.1 is our own treatment of the problem of simplifying the graphs of the icosahedron and the dodecahedron for inexperienced students.

3. The problem of the Monk and the Bridges and the Mine Inspector's Problem in the Advanced Exercises of Section IV.2 are from Dudeney's book.

4. The Matrix-Tree Theorem can be used to find the number of spanning trees of the graph of the n-cube for small values of n. It is a very curious fact that there is no simple formula for the number of spanning trees of the n-cube,

although there are such formulas for K_n and for the graph of the n-dimensional octahedron. Can we find a recursion for it?

5. Structural induction is an important topic in computer science. It is widely used in the analysis of algorithms.

Chapter 5

1. The prime-generating polynomial $X^2 - X + 41$ was studied by Euler.

2. We wish to thank Will Washburn for his analysis of the Liar Problem.

3. Aristotle's logical works, which include his analysis of the syllogism, have been been published by Oxford University Press. For example, the Greek text of the Analytics can be found in W. D. Ross, *Aristotelis, Analytica Priora et Posteriora*, Oxford (1964).

4. For all matters relating to mathematical logic, including the propositional calculus, we recommend the book of Mendelson VIII(4).

Chapter 6

1. Huntington's important paper on Boolean algebra is E. V. Huntington, "Postulates for the algebra of logic," *Trans. Amer. Math. Soc.* 5 (1904).

2. There is a very clear treatment of the applications of Boolean algebra to circuits in J. E. Whitesitt, *Boolean Algebra and Its Applications*, Addison-Wesley (1961).

3. For details about the predicate calculus we recommend the book of Mendelson VIII(4).

Chapter 7

1. Our proof of Sperner's Lemma is from D. Lubell, "A short proof of Sperner's lemma," *Journal of Combinatorial Theory* I, 299 (1966).

2. Tverberg's proof of Dilworth's Theorem is from H. Tverberg, "On Dilworth's decomposition theorem for partially ordered sets," *J. Combinatorial Theory* 3 (1967): 305–306.

3. The solution of Dedekind's problem about the number of antichains in the n-cube can be found in A. Kisielewicz, "A solution of Dedekind's problem on the number of isotone Boolean functions," *Journal für die reine und angewandte Mathematik* 386 (1988): 139–144.

4. The paper containing Dilworth's Theorem, and many other interesting results due to Dilworth, together with a biographical sketch, can be found in *The Dilworth Theorems, Selected Papers of Robert P. Dilworth*, edited by K. Bogart, R. Freese, and J. Kung, Birkhäuser (1990).

5. Anyone who is interested in lattices should read G. Birkhoff, *Lattice Theory*, A.M.S. Colloquium Publications Vol. XXV (1967).

Chapter 8

1. The listing algorithms for permutations and subsets in Section VIII.1 are described in the instruction manual for the Macaulay program.

2. Anyone who is interested in algorithms should consult the important text of Knuth II(3).

3. For an up-to-date treatment of all matters relating to algorithms we recommend the text of Corman, Leiserson, and Rivest II(2).

4. The precise form of Stirling's Formula is on page 253 of E. T. Whittaker and G. N. Watson, *A Course of Modern Analysis*, Cambridge (1958).

Chapter 9

1. It is remarkable how little the elementary theory of linear recursive sequences has changed since Euler's treatment of it in the *Introductio in analysin infinitorum*.

2. De Méré's Paradox was discussed by Pascal and Fermat in their correspondence in the summer of 1654.

3. Questions about the dice game of Newton and Pepys originated in a letter from Pepys to Newton, in November of 1693, from which we quote. The idea of varying the number of dice in the game is our own, as is the idea of varying the number of dice in the game associated with De Méré's Paradox. Contrasting the two games gives an excellent lecture about the binomial distribution.

4. The discussion of reliability at the end of Section IX.2 is from C. J. Colbourn, *The Reliability Polynomial*, Computer Communications Networks Group Report E-138, University of Waterloo, Canada, September 1985.

5. André Weil's remarkable work on marriage laws is described in *Sur l'étude de certains types de lois de mariage, (On certain types of marriage laws)*, Complete Works, Vol. I, (390–398), Springer (1979).

Chapter 10

1. Chomsky's original publication about the classification of languages was N. Chomsky, "Three models for the description of language," *IRE Trans. Info. Theory* IT2 (1956): 113–124.

2. Our sketch of the history of computer science is based on S. V. Pollack, "The Development of Computer Science," in *Studies in Computer Science*, MAA Studies in Mathematics (1982).

3. The description of DNA computing at the end of Section X.2 is based on L. M. Adelman, "Computing with DNA," *Scientific American*, (August 1998): 54–61. Adelman ends the article by saying "But biology and computer science life—and computation—are related. I am confident that at their interface great discoveries await those who seek them."

4. There is an excellent edition of Gödel's complete works: Kurt Gödel, *Collected Works*, Oxford University Press (1986).

5. The prime-generating polynomial of J. P. Jones, which we give in Section X.3, is described in M. Davis, Y. Matijasevic, and J. Robinson, *Hilbert's Tenth Problem. Diophantine equations: positive aspects of a negative solution*, in *Mathematical Developments Arising from Hilbert Problems*, American Mathematical Society (1976).

6. We hope that our idea about *The Hunting of the Snark* will serve as a unifying theme for this book.

Answers to Selected Exercises

Section 1.1 Exercises

1. 1   **3.** 10   **5.** 252   **13.** 252   **15.** 924

7. 924   **9.** 1   **11.** 10

17.

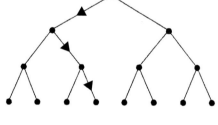

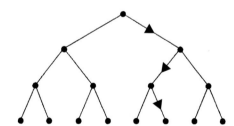

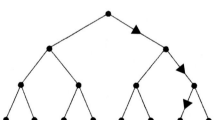

19.

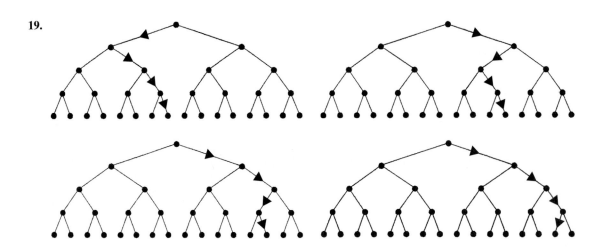

Section 1.1 Advanced Exercises

1.

```
                              1
                          1       1
                      1       2       1
                  1       3       3       1
              1       4       6       4       1
          1       5      10      10       5       1
      1       6      15      20      15       6       1
  1       7      21      35      35      21       7       1
1       8      28      56      70      56      28       8       1
  1   9      36      84     126     126      84      36       9       1
1   10      45     120     210     252     210     120      45      10       1
  11     55     165     330     462     462     330     165      55      11       1
  12     66     220     495     792     924     792     495     220      66      12       1
  13     78     286     715    1287    1716    1716    1287     715     286      78      13       1
1   14     91     364    1001    2002    3003    3432    3003    2002    1001     364      91      14       1
```

3.

```
                              1
                          1       1
                      1       2       1
                  1       3       3       1
              1       4       6       4       1
          1       5      10      10       5       1
      1       6      15      20      15       6       1
  1       7      21      35      35      21       7       1
1       8      28      56      70      56      28       8       1
  1   9      36      84     126     126      84      36       9       1
1   10      45     120     210     252     210     120      45      10       1
  11     55     165     330     462     462     330     165      55      11       1
  12     66     220     495     792     924     792     495     220      66      12       1
  13     78     286     715    1287    1716    1716    1287     715     286      78      13       1
1   14     91     364    1001    2002    3003    3432    3003    2002    1001     364      91      14       1
```

5.

```
                              1
                          1       1
                      1       2       1
                  1       3       3       1
              1       4       6       4       1
          1       5      10      10       5       1
      1       6      15      20      15       6       1
  1       7      21      35      35      21       7       1
1       8      28      56      70      56      28       8       1
    1       9      36      84     126     126      84      36       9       1
        1      10      45     120     210     252     210     120      45      10       1
            1      11      55     165     330     462     462     330     165      55      11       1
        1      12      66     220     495     792     924     792     495     220      66      12       1
    1      13      78     286     715    1287    1716    1716    1287     715     286      78      13       1
1      14      91     364    1001    2002    3003    3432    3003    2002    1001     364      91      14       1
```

Section 1.1 Computer Exercises

1. The values of $n!$ for $0 \le n \le 20$ are given in Figure 1.3.

$21! = 51,090,942,171,709,440,000$

$22! = 1,124,000,727,777,607,680,000$

$23! = 25,852,016,738,884,976,640,000$

$24! = 620,448,401,733,239,439,360,000$

$25! = 15,511,210,043,330,985,984,000,000$

$26! = 403,291,461,126,605,635,584,000,000$

$27! = 10,888,869,450,418,352,160,768,000,000$

$28! = 304,888,344,611,713,860,501,504,000,000$

$29! = 8,841,761,993,739,701,954,543,616,000,000$

$30! = 265,252,859,812,191,058,636,308,480,000,000$

$31! = 8,222,838,654,177,922,817,725,562,880,000,000$

$32! = 263,130,836,933,693,530,167,218,012,160,000,000$

$33! = 8,683,317,618,811,886,495,518,194,401,280,000,000$

$34! = 295,232,799,039,604,140,847,618,609,643,520,000,000$

$35! = 10,333,147,966,386,144,929,666,651,337,523,200,000,000$

$36! = 371,993,326,789,901,217,467,999,448,150,835,200,000,000$

$37! = 13,763,753,091,226,345,046,315,979,581,580,902,400,000,000$

$38! = 523,022,617,466,601,111,760,007,224,100,074,291,200,000,000$

$39! = 20,397,882,081,197,443,358,640,281,739,902,897,356,800,000,000$

$40! = 815,915,283,247,897,734,345,611,269,596,115,894,272,000,000,000$

3. Let $M_n = \binom{2n}{n}$

$M_0 = 1$	$M_9 = 48,620$	$M_{18} = 9,075,135,300$
$M_1 = 2$	$M_{10} = 184,756$	$M_{19} = 35,345,263,800$
$M_2 = 6$	$M_{11} = 705,432$	$M_{20} = 137,846,528,820$
$M_3 = 20$	$M_{12} = 2,704,156$	$M_{21} = 538,257,874,440$
$M_4 = 70$	$M_{13} = 10,400,600$	$M_{22} = 2,104,098,963,720$
$M_5 = 252$	$M_{14} = 40,116,600$	$M_{23} = 8,233,430,727,600$
$M_6 = 924$	$M_{15} = 155,117,520$	$M_{24} = 32,247,603,683,100$
$M_7 = 3,432$	$M_{16} = 601,080,390$	$M_{25} = 126,410,606,437,752$
$M_8 = 12,870$	$M_{17} = 2,333,606,220$	

5. 1   25   300   2,300   12,650   53,130   177,100
480,700 1,081,575 2,042,975 3,268,760 4,457,400
5,200,300   5,200,300   4,457,400   3,268,760
2,042,975 1,081,575 480,700 177,100 53,130
12,650 2,300 300 25 1

7. 1 35 595 6,545 52,360 324,632
1,623,160 6,724,520 23,535,820
70,607,460 183,579,396 417,225,900
834,451,800 1,476,337,800 2,319,959,400
3,247,943,160 4,059,928,950 4,537,567,650
4,537,567,650 4,059,928,950 3,247,943,160
2,319,959,400 1,476,337,800 834,451,800
417,225,900 183,579,396 70,607,460
23,535,820 6,724,520 1,623,160
324,632 52,360 6,545 595 35 1

9. 1 45 990 14,190 148,995 1,221,759
8,145,060 45,379,620 215,553,195 886,163,135
3,190,187,286 10,150,595,910 28,760,021,745
73,006,209,045 166,871,334,960 344,867,425,584
646,626,422,970 1,103,068,603,890
1,715,884,494,940 2,438,362,177,020
3,169,870,830,126 3,773,655,750,150
4,116,715,363,800 4,116,715,363,800
3,773,655,750,150 3,169,870,830,126
2,438,362,177,020 1,715,884,494,940
1,103,068,603,890 646,626,422,970
344,867,425,584 166,871,334,960 73,006,209,045
28,760,021,745 10,150,595,910 3,190,187,286
886,163,135 215,553,195 45,379,620 8,145,060
1,221,759 148,995 14,190 990 45 1

Section 1.2 Exercises

11. 0, 1, 7, 18, 34, 55, 81, 112, 148, 189, 235, 286

Section 1.2 Advanced Exercises

1. 9   **3.** 13

Section 1.2 Computer Exercises

1. $1^2 + 2^2 + 3^2 + \ldots + 24^2 = 4900 = 70^2$

Section 1.3 Exercises

1. True   **3.** False

5. True   **7.** False

9. Possible; we don't know whether it's true or false
without knowing more about the sets.

11. $A = \{a : a \in N$ and $a = 2b$
for $b \in \{1, 2, 3, 4, 5, 6\}\}$

13. $A = \{a : a \in N$ and $a = 4b$
for $b \in \{1, 2, 3, 4, 5, 6\}\}$

15. $A = \{a : a \in N$ and $a = 2^b - 1$
for $b \in \{0, 1, 2, 3, 4, 5\}\}$

17. $A = \{a : a \in N$ and $a = 4b + 1$ for some $b \in N\}$

19. $A = \{a : a \in N$ and $a > a + 1\}$

21. $A \times B = \{\{1, 1\}, \{1, 2\}, \{1, 8\}, \{1, 9\},$
$\{3, 1\}, \{3, 2\}, \{3, 8\}, \{3, 9\},$
$\{5, 1\}, \{5, 2\}, \{5, 8\}, \{5, 9\},$
$\{7, 1\}, \{7, 2\}, \{7, 8\}, \{7, 9\},$
$\{9, 1\}, \{9, 2\}, \{9, 8\}, \{9, 9\}\}$
$A \cup B = \{1, 2, 3, 5, 7, 8, 9\}$
$A \cap B = \{1, 9\}$
$A - B = \{3, 5, 7\}$
$A^c = \{2, 4, 6, 8, 10\}$
$B^c = \{3, 4, 5, 6, 7, 10\}$

23. $A \times B = \{\{1, 4\}, \{1, 8\}, \{1, 12\}, \{1, 16\},$
$\{4, 4\}, \{4, 8\}, \{4, 12\}, \{4, 16\},$
$\{7, 4\}, \{7, 8\}, \{7, 12\}, \{7, 16\},$
$\{10, 4\}, \{10, 8\}, \{10, 12\}, \{10, 16\},$
$\{13, 4\}, \{13, 8\}, \{13, 12\}, \{13, 16\},$
$\{16, 4\}, \{16, 8\}, \{16, 12\}, \{16, 16\}\}$
$A \cup B = \{1, 4, 7, 8, 10, 12, 13, 16\}$
$A \cap B = \{4, 16\}$
$A - B = \{1, 7, 10, 13\}$
$A^c = \{2, 3, 5, 6, 8, 9, 11, 12, 14, 15\}$
$B^c = \{1, 2, 3, 5, 6, 7, 9, 10, 11, 13, 14, 15\}$

25. $(0, 1, 0, 1, 0, 1)$ **27.** $(0, 0, 0, 1, 1, 1, 0)$

29.

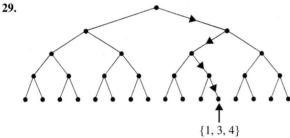

$\{1, 3, 4\}$

31.

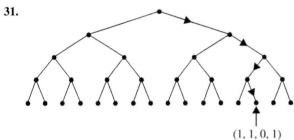

$(1, 1, 0, 1)$

33.

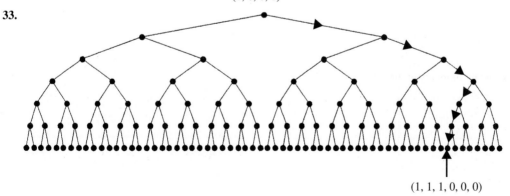

$(1, 1, 1, 0, 0, 0)$

35. Ø {1} {1, 2} {2, 3, 4} {1, 2, 3, 4} **37.** (0, 0, 0) (1, 0, 0) (1, 1, 0) (1, 1, 1)
{2} {1, 3} {1, 3, 4} (0, 1, 0) (1, 0, 1)
{3} {1, 4} {1, 2, 4} (0, 0, 1) (0, 1, 1)
{4} {2, 3} {1, 2, 3}
{2, 4}
{3, 4}

39. (0, 0, 0, 0, 0) (1, 0, 0, 0, 0) (1, 1, 0, 0, 0) (1, 1, 1, 0, 0) (1, 1, 1, 1, 0) (1, 1, 1, 1, 1)
(0, 1, 0, 0, 0) (1, 0, 1, 0, 0) (1, 1, 0, 1, 0) (1, 1, 1, 0, 1)
(0, 0, 1, 0, 0) (0, 1, 1, 0, 0) (1, 0, 1, 1, 0) (1, 1, 0, 1, 1)
(0, 0, 0, 1, 0) (1, 0, 0, 1, 0) (0, 1, 1, 1, 0) (1, 0, 1, 1, 1)
(0, 0, 0, 0, 1) (0, 1, 0, 1, 0) (1, 1, 0, 0, 1) (0, 1, 1, 1, 1)
(0, 0, 1, 1, 0) (1, 0, 1, 0, 1)
(1, 0, 0, 0, 1) (0, 1, 1, 0, 1)
(0, 1, 0, 0, 1) (1, 0, 0, 1, 1)
(0, 0, 1, 0, 1) (0, 1, 0, 1, 1)
(0, 0, 0, 1, 1) (0, 0, 1, 1, 1)

41. (0, 0, 0, 0, 0, 0) (1, 1, 0, 0, 0, 0)
(0, 0, 0, 0, 0, 1) (1, 1, 0, 0, 0, 1)
(0, 0, 0, 0, 1, 1) (1, 1, 0, 0, 1, 1)
(0, 0, 0, 0, 1, 0) (1, 1, 0, 0, 1, 0)
(0, 0, 0, 1, 1, 0) (1, 1, 0, 1, 1, 0)
(0, 0, 0, 1, 1, 1) (1, 1, 0, 1, 1, 1)
(0, 0, 0, 1, 0, 1) (1, 1, 0, 1, 0, 1)
(0, 0, 0, 1, 0, 0) (1, 1, 0, 1, 0, 0)
(0, 0, 1, 1, 0, 0) (1, 1, 1, 1, 0, 0)
(0, 0, 1, 1, 0, 1) (1, 1, 1, 1, 0, 1)
(0, 0, 1, 1, 1, 1) (1, 1, 1, 1, 1, 1)
(0, 0, 1, 1, 1, 0) (1, 1, 1, 1, 1, 0)
(0, 0, 1, 0, 1, 0) (1, 1, 1, 0, 1, 0)
(0, 0, 1, 0, 1, 1) (1, 1, 1, 0, 1, 1)
(0, 0, 1, 0, 0, 1) (1, 1, 1, 0, 0, 1)
(0, 0, 1, 0, 0, 0) (1, 1, 1, 0, 0, 0)
(0, 1, 1, 0, 0, 0) (1, 0, 1, 0, 0, 0)
(0, 1, 1, 0, 0, 1) (1, 0, 1, 0, 0, 1)
(0, 1, 1, 0, 1, 1) (1, 0, 1, 0, 1, 1)
(0, 1, 1, 0, 1, 0) (1, 0, 1, 0, 1, 0)
(0, 1, 1, 1, 1, 0) (1, 0, 1, 1, 1, 0)
(0, 1, 1, 1, 1, 1) (1, 0, 1, 1, 1, 1)
(0, 1, 1, 1, 0, 1) (1, 0, 1, 1, 0, 1)
(0, 1, 1, 1, 0, 0) (1, 0, 1, 1, 0, 0)
(0, 1, 0, 1, 0, 0) (1, 0, 0, 1, 0, 0)
(0, 1, 0, 1, 0, 1) (1, 0, 0, 1, 0, 1)
(0, 1, 0, 1, 1, 1) (1, 0, 0, 1, 1, 1)
(0, 1, 0, 1, 1, 0) (1, 0, 0, 1, 1, 0)
(0, 1, 0, 0, 1, 0) (1, 0, 0, 0, 1, 0)
(0, 1, 0, 0, 1, 1) (1, 0, 0, 0, 1, 1)
(0, 1, 0, 0, 0, 1) (1, 0, 0, 0, 0, 1)
(0, 1, 0, 0, 0, 0) (1, 0, 0, 0, 0, 0)

Section 1.3 Advanced Exercises

1. Use the weights 6, 7, and 18.

3. Use the weights 12, 21, and 28.

5. Use the weights 7, 19, and 25.

Section 1.3 Computer Exercises

1. The first 20 vectors are shown below.

$V_1 = 0000000000$	$V_6 = 0000000111$	$V_{11} = 0000001111$	$V_{16} = 0000001000$
$V_2 = 0000000001$	$V_7 = 0000000101$	$V_{12} = 0000001110$	$V_{17} = 0000011000$
$V_3 = 0000000011$	$V_8 = 0000000100$	$V_{13} = 0000001010$	$V_{18} = 0000011001$
$V_4 = 0000000010$	$V_9 = 0000001100$	$V_{14} = 0000001011$	$V_{19} = 0000011011$
$V_5 = 0000000110$	$V_{10} = 0000001101$	$V_{15} = 0000001001$	$V_{20} = 0000011010$

5. $15 + 21 + 28 + 36 + 45 + 55 = 200$

Section 1.4 Exercises

1.

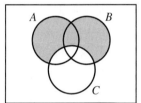

3.

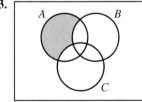

5.

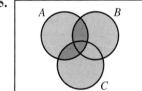

7.

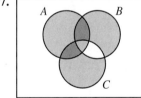

21. 18   **23.** 11   **25.** 12

27. $(0, 0, 1, 1, 1, 0, 0, 1)$

Section 1.4 Advanced Exercises

1.

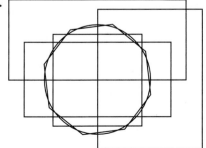

Section 1.4 Computer Exercises

1. There are 5712 such Gray codes.

Section 1.5 Exercises

1. 0, 1, 1, 2, 3, 5, 8, 13, 21, 34, 55, 89, 144, 233, 377, 610, 987, 1597, 2584, 4181, 6765

3. 0, 0, 1, 1, 2, 4, 7, 13, 24, 44, 81, 149, 274, 504, 927, 1705, 3136, 5768, 10609, 19513, 35890

5. $B_6 = \dfrac{1}{42}$ **7.** $B_8 = -\dfrac{1}{30}$

9. 19, 58, 29, 88, 44, 22, 11, 34, 17, 52, 26, 13, 40, 20, 10, 5, 16, 8, 4, 2, 1

11. 123, 370, 185, 556, 278, 139, 418, 209, 628, 314, 157, 472, 236, 118, 59, 178, 89, 268, 134, 67, 202, 101, 304, 152, 76, 38, 19, 58, 29, 88, 44, 22, 11, 34, 17, 52, 26, 13, 40, 20, 10, 5, 16, 8, 4, 2, 1

13. 251, 754, 377, 1132, 566, 283, 850, 425, 1276, 638, 319, 958, 479, 1438, 719, 2158, 1079, 3238, 1619, 4858, 2429, 7288, 3644, 1822, 911, 2734, 1367, 4102, 2051, 6154, 3077, 9232, 4616, 2308, 1154, 577, 1732, 866, 433, 1300, 650, 325, 976, 488, 244, 122, 61, 184, 92, 46, 23, 70, 35, 106, 53, 160, 80, 40, 20, 10, 5, 16, 8, 4, 2, 1

15. 103, 310, 155, 466, 233, 700, 350, 175, 526, 263, 790, 395, 1186, 593, 1780, 890, 445, 1336, 668, 334, 167, 502, 251, 754, 377, 1132, 566, 283, 850, 425, 1276, 638, 319, 958, 479, 1438, 719, 2158, 1079, 3238, 1619, 4858, 2429, 7288, 3644, 1822, 911, 2734, 1367, 4102, 2051, 6154, 3077, 9232, 4616, 2308, 1154, 577, 1732, 866, 433, 1300, 650, 325, 976, 488, 244, 122, 61, 184, 92, 46, 23, 70, 35, 106, 53, 160, 80, 40, 20, 10, 5, 16, 8, 4, 2, 1

17. 27, 82, 41, 124, 62, 31, 94, 47, 142, 71, 214, 107, 322, 161, 484, 242, 121, 364, 182, 91, 274, 137, 412, 206, 103, 310, 155, 466, 233, 700, 350, 175, 526, 263, 790, 395, 1186, 593, 1780, 890, 445, 1336, 668, 334, 167, 502, 251, 754, 377, 1132, 566, 283, 850, 425, 1276, 638, 319, 958, 479, 1438, 719, 2158, 1079, 3238, 1619, 4858, 2429, 7288, 3644, 1822, 911, 2734, 1367, 4102, 2051, 6154, 3077, 9232, 4616, 2308, 1154, 577, 1732, 866, 433, 1300, 650, 325, 976, 488, 244, 122, 61, 184, 92, 46, 23, 70, 35, 106, 53, 160, 80, 40, 20, 10, 5, 16, 8, 4, 2, 1

19.

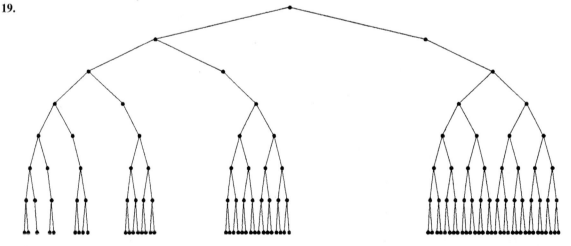

21. T_{n+2} where T_n are the Tribonacci numbers.

23. 65   **25.** 106

27. 505

29. 0, 0, 1, 1, 1, 0, 1

Section 1.5 Advanced Exercises

1.

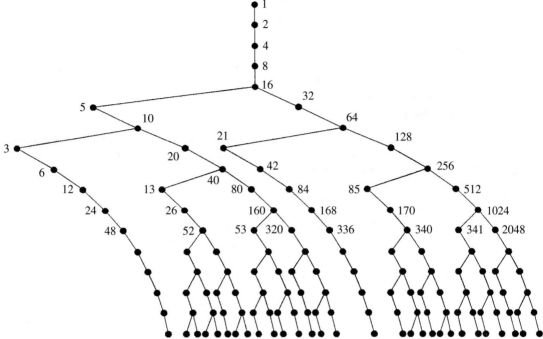

The level sequence is 1, 1, 1, 1, 1, 2, 2, 4, 4, 6, 6, 8, 10, 14, 18, 24, 29

Section 1.5 Computer Exercises

1. The first 10 numbers from each column are shown below.

$$F_0 = 0 \qquad F_{60} = 1,548,008,755,920$$
$$F_1 = 1 \qquad F_{61} = 2,504,730,781,961$$
$$F_2 = 1 \qquad F_{62} = 4,052,739,537,881$$
$$F_3 = 2 \qquad F_{63} = 6,557,470,319,842$$
$$F_4 = 3 \qquad F_{64} = 10,610,209,857,723$$
$$F_5 = 5 \qquad F_{65} = 17,167,680,177,565$$
$$F_6 = 8 \qquad F_{66} = 27,777,890,035,288$$
$$F_7 = 13 \qquad F_{67} = 44,945,570,212,853$$
$$F_8 = 21 \qquad F_{68} = 72,723,460,248,141$$
$$F_9 = 34 \qquad F_{69} = 117,669,030,460,994$$

3. The period is 124.

Section 1.6 Computer Exercises

1. $F_{100} = 354, 224, 848, 179, 261, 915, 075$
$F_{200} = 280, 571, 172, 992, 510, 140, 037,$
$\quad 611, 932, 413, 038, 677, 189, 525$
$F_{300} = 222, 232, 244, 629, 420, 445, 529,$
$\quad 739, 893, 461, 909, 967, 206, 666,$
$\quad 939, 096, 499, 764, 990, 979, 600$
$F_{400} = 176, 023, 680, 645, 013, 966, 468,$
$\quad 226, 945, 392, 411, 250, 770, 384, 383,$
$\quad 304, 492, 191, 886, 725, 992, 896, 575,$
$\quad 345, 044, 216, 019, 675$
$F_{500} = 139, 423, 224, 561, 697, 880, 139,$
$\quad 724, 382, 870, 407, 283, 950, 070, 256,$
$\quad 587, 697, 307, 264, 108, 962, 948, 325,$
$\quad 571, 622, 863, 290, 691, 557, 658, 876,$
$\quad 222, 521, 294, 125$

3. $T_{100} = 53, 324, 762, 928, 098, 149, 064, 722, 658$
$T_{200} = 15, 555, 116, 989, 073, 938, 986, 569, 525,$
$\quad 465, 884, 451, 018, 665, 640, 926, 743, 832$
$T_{300} = 4, 537, 510, 365, 869, 456, 920, 742, 452,$
$\quad 229, 301, 957, 385, 335, 193, 230, 814, 796,$
$\quad 219, 118, 584, 076, 403, 940, 718, 845, 682$
$T_{400} = 1, 323, 615, 909, 467, 905, 705, 275, 587,$
$\quad 535, 093, 751, 741, 849, 628, 450, 560, 688,$
$\quad 425, 852, 266, 218, 524, 071, 810, 108, 710,$
$\quad 154, 572, 661, 993, 481, 263, 106, 493, 760$
$T_{500} = 386, 105, 801, 316, 632, 097, 865, 113, 202,$
$\quad 679, 576, 506, 967, 026, 131, 179, 942, 811,$
$\quad 441, 817, 679, 546, 734, 345, 038, 740, 194,$
$\quad 971, 272, 369, 839, 926, 567, 485, 285, 547,$
$\quad 047, 084, 642, 036, 657, 107, 024, 402$

5.

$C_0 = 1$	$C_9 = 4,862$	$C_{18} = 477, 638, 700$
$C_1 = 1$	$C_{10} = 16,796$	$C_{19} = 1, 767, 263, 190$
$C_2 = 2$	$C_{11} = 58,786$	$C_{20} = 6, 564, 120, 420$
$C_3 = 5$	$C_{12} = 208,012$	$C_{21} = 24, 466, 267, 020$
$C_4 = 14$	$C_{13} = 742,900$	$C_{22} = 91, 482, 563, 640$
$C_5 = 42$	$C_{14} = 2, 674, 440$	$C_{23} = 343, 059, 613, 650$
$C_6 = 132$	$C_{15} = 9, 694, 845$	$C_{24} = 1, 289, 904, 147, 324$
$C_7 = 429$	$C_{16} = 35, 357, 670$	$C_{25} = 4, 861, 946, 401, 452$
$C_8 = 1, 430$	$C_{17} = 129, 644, 790$	

Section 2.1 Exercises

1. g.c.d.$(64, 28) = 4$
l.c.m.$(64, 28) = 448$
$4 = 4 \cdot 64 - 9 \cdot 28$

3. g.c.d.$(130, 23) = 1$
l.c.m.$(130, 23) = 2990$
$1 = 20 \cdot 130 - 113 \cdot 23$

5. g.c.d.$(6765, 4181) = 1$
l.c.m.$(6765, 4181) = 28284465$
$1 = 1597 \cdot 6765 - 2584 \cdot 4181$

7. g.c.d.$(1597, 987) = 1$ (14 steps)
g.c.d.$(1590, 997) = 1$ (8 steps)

Note that $\dfrac{1597}{987}$ is very close to the Golden Ratio.

9. g.c.d.$(93, 47) = 1$ (2 steps)

11. g.c.d.$(605, 322) = 1$ (5 steps)

13. g.c.d.$(70000, 38502) = 2$ (5 steps)

15. $2^6 \cdot 3^3 \cdot 5 \cdot 11$ **17.** $2^7 \cdot 3^2 \cdot 5 \cdot 7 \cdot 11 \cdot 23$

19. $37 \cdot 59$ **21.** $109 \cdot 151$ **23.** $281 \cdot 397$

25.

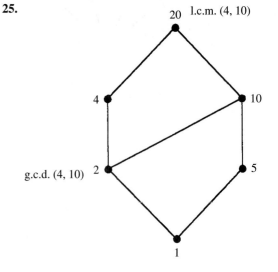

27.

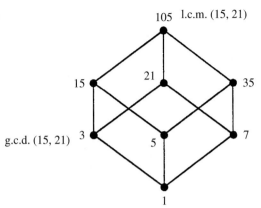

29.

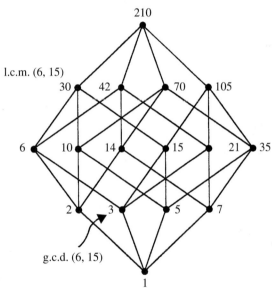

31.

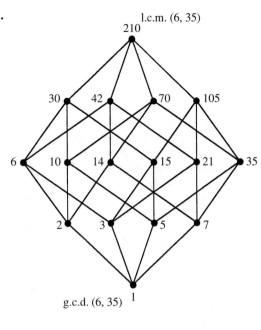

Section 2.1 Advanced Exercises

3. $2^{37} - 1 = 223 \cdot 616318177$

5. $2^{29} - 1 = 233 \cdot 1103 \cdot 2089$

Section 2.1 Computer Exercises

1. The first 20 Mersenne numbers are shown below.

$M_2 = 3$ is prime

$M_3 = 7$ is prime

$M_5 = 31$ is prime

$M_7 = 127$ is prime

$M_{11} = 2047 = (23)(89)$

$M_{13} = 8191$ is prime

$M_{17} = 131,071$ is prime

$M_{19} = 524,287$ is prime

$M_{23} = 8,388,607 = (47)(178,481)$

$M_{29} = 536,870,911 = (233)(1103)(2089)$

$M_{31} = 2,147,483,647$ is prime

$M_{37} = 137,438,953,471$
$= (223)(616,318,177)$

$M_{41} = 2,199,023,255,551$
$= (13367)(164,511,353)$

$M_{43} = 8,796,093,022,207$
$= (431)(9719)(2,099,863)$

$M_{47} = 140,737,488,355,327$
$= (2351)(4513)(13,264,529)$

$M_{53} = 9,007,199,254,740,991$
$= (6361)(69431)(20,394,401)$

$M_{59} = 576,460,752,303,423,487$
$= (179,951)(3,203,431,780,337)$

$M_{61} = 2,305,843,009,213,693,951$ is prime

$M_{67} = 147,573,952,589,676,412,927$
$= (193,707,721)(761,838,257,287)$

$M_{71} = 2,361,183,241,434,822,606,847$
$= (228,479)(48,544,121)(212,885,833)$

7. $N_{19} = (54,730,729,297)(143,581,524,529,603)$. The number of steps for the Fermat Factorization Algorithm is $69,014,857,833,109$.

Section 2.2 Exercises

1. $F_{10} = 55$

$L_{10} = 123$

3. $F_{40} = 102334155$

$L_{40} = 228826127$

5. g.c.d.$(988, 602) = 2 \left(8 \text{ steps; } \dfrac{\ln(602)}{\ln(\phi)} = 13.3 \right)$

7. g.c.d.$(3524580, 2178300) = 60$

$\left(15 \text{ steps; } \dfrac{\ln(2178300)}{\ln(\phi)} = 30.328 \right)$

15. $\dfrac{2}{3} = \dfrac{1}{2} + \dfrac{1}{6}$

17. $\dfrac{5}{7} = \dfrac{1}{2} + \dfrac{1}{5} + \dfrac{1}{70}$

19. $\dfrac{8}{11} = \dfrac{1}{2} + \dfrac{1}{5} + \dfrac{1}{37} + \dfrac{1}{4070}$

Section 2.2 Computer Exercises

11. $5/2 = 1/2 + 1/3 + 1/4 + 1/5 + 1/6 + 1/7 + 1/8 + 1/9 + 1/10 + 1/12$
$+1/13 + 1/14 + 1/20 + 1/21 + 1/30 + 1/42 + 1/43 + 1/44 + 1/45 + 1/56$
$+1/57 + 1/58 + 1/72 + 1/73 + 1/90 + 1/156 + 1/157 + 1/182 + 1/420 + 1/1806$
$+1/1807 + 1/1808 + 1/1892 + 1/1893 + 1/1980 + 1/3192 + 1/3193 + 1/3306$
$+1/5256 + 1/24492 + 1/3263442 + 1/3263443 + 1/3267056 + 1/3581556$
$+1/10192056 + 1/10650056950806$

Section 2.3 Exercises

1. Addition table for Z_2:   Multiplication table for Z_2:

	0	1
0	0	1
1	1	0

	0	1
0	0	0
1	0	1

3. Addition table for Z_4:

	0	1	2	3
0	0	1	2	3
1	1	2	3	0
2	2	3	0	1
3	3	0	1	2

Multiplication table for Z_4:

	0	1	2	3
0	0	0	0	0
1	0	1	2	3
2	0	2	0	2
3	0	3	2	1

5. Addition table for Z_8:

	0	1	2	3	4	5	6	7
0	0	1	2	3	4	5	6	7
1	1	2	3	4	5	6	7	0
2	2	3	4	5	6	7	0	1
3	3	4	5	6	7	0	1	2
4	4	5	6	7	0	1	2	3
5	5	6	7	0	1	2	3	4
6	6	7	0	1	2	3	4	5
7	7	0	1	2	3	4	5	6

Multiplication table for Z_8:

	0	1	2	3	4	5	6	7
0	0	0	0	0	0	0	0	0
1	0	1	2	3	4	5	6	7
2	0	2	4	6	0	2	4	6
3	0	3	6	1	4	7	2	5
4	0	4	0	4	0	4	0	4
5	0	5	2	7	4	1	6	3
6	0	6	4	2	0	6	4	2
7	0	7	6	5	4	3	2	1

7. Addition table for Z_{12}:

	0	1	2	3	4	5	6	7	8	9	10	11
0	0	1	2	3	4	5	6	7	8	9	10	11
1	1	2	3	4	5	6	7	8	9	10	11	0
2	2	3	4	5	6	7	8	9	10	11	0	1
3	3	4	5	6	7	8	9	10	11	0	1	2
4	4	5	6	7	8	9	10	11	0	1	2	3
5	5	6	7	8	9	10	11	0	1	2	3	4
6	6	7	8	9	10	11	0	1	2	3	4	5
7	7	8	9	10	11	0	1	2	3	4	5	6
8	8	9	10	11	0	1	2	3	4	5	6	7
9	9	10	11	0	1	2	3	4	5	6	7	8
10	10	11	0	1	2	3	4	5	6	7	8	9
11	11	0	1	2	3	4	5	6	7	8	9	10

Multiplication table for Z_{12}:

	0	1	2	3	4	5	6	7	8	9	10	11
0	0	0	0	0	0	0	0	0	0	0	0	0
1	0	1	2	3	4	5	6	7	8	9	10	11
2	0	2	4	6	8	10	0	2	4	6	8	10
3	0	3	6	9	0	3	6	9	0	3	6	9
4	0	4	8	0	4	8	0	4	8	0	4	8
5	0	5	10	3	8	1	6	11	4	9	2	7
6	0	6	0	6	0	6	0	6	0	6	0	6
7	0	7	2	9	4	11	6	1	8	3	10	5
8	0	8	4	0	8	4	0	8	4	0	8	4
9	0	9	6	3	0	9	6	3	0	9	6	3
10	0	10	8	6	4	2	0	10	8	6	4	2
11	0	11	10	9	8	7	6	5	4	3	2	1

9. $(1, 2, 2, 1, 2, 2, 1, 2, 1, 1, 1, 2)$

13. $(0, 1, 0, 0, 1, 1, 2, 0, 2, 0, 2, 0)$

11. $(0, 1, 2, 1, 1, 2, 0, 1, 0, 1, 1, 2)$

15. $T\,ACT\,CCT\,GGG\,AG\,AG\,AG\,ACG\,AGT\,GAG\,ACG\,ACC\,GGG\,AC$
20121123330303030130323030130113301
0230330111212121231210121231233311123

17. The period of F_n is 16; the period of L_n is 16.

19. The period of F_n is 24; the period of L_n is 24.

21. The units in Z_{24} are 1, 5, 7, 11, 13, 17, 19, and 23.
Each of these is its own multiplicative inverse.

23. The period is 3.

25. The period is 12.

1	3	4
7	1	8
9	7	6
3	9	2

27. The period is 60.

1	7	8	5	3	8	1	9	0	9	9	8	7	5	2
7	9	6	5	1	6	7	3	0	3	3	6	9	5	4
9	3	2	5	7	2	9	1	0	1	1	2	3	5	8
3	1	4	5	9	4	3	7	0	7	7	4	1	5	6

31. The period of the last two digits is 60.

Section 2.3 Advanced Exercises

1. $X = 4$ **3.** $X = 368$

Section 2.3 Computer Exercises

5. The period is 124.

9. $D = 5091$; encrypted message $T' = 5161$

11. $D = 3, 169, 667$, encrypted message $T' = 1, 532, 762$

Section 2.4 Computer Exercises

1. $D = 237$; encrypted message $T' = 386$

3. $D = 9073$; encrypted message $T' = 4562$

5. $D = 109, 927$; encrypted message $T' = 122, 846$

Section 3.1 Exercises

7. $f(g(x)) = (x + 3)^2$
  $g(f(x)) = x^2 + 3$

9. $f(g(x)) = (x + 1)^3$
  $g(f(x)) = x^3 + 1$

11.

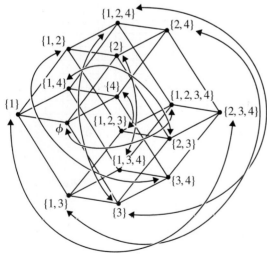

13.

$\emptyset$	$(0, 0, 0, 0)$
$\{1\}$	$(1, 0, 0, 0)$
$\{2\}$	$(0, 1, 0, 0)$
$\{3\}$	$(0, 0, 1, 0)$
$\{4\}$	$(0, 0, 0, 1)$
$\{1, 2\}$	$(1, 1, 0, 0)$
$\{1, 3\}$	$(1, 0, 1, 0)$
$\{1, 4\}$	$(1, 0, 0, 1)$
$\{2, 3\}$	$(0, 1, 1, 0)$
$\{2, 4\}$	$(0, 1, 0, 1)$
$\{3, 4\}$	$(0, 0, 1, 1)$
$\{2, 3, 4\}$	$(0, 1, 1, 1)$
$\{1, 3, 4\}$	$(1, 0, 1, 1)$
$\{1, 2, 4\}$	$(1, 1, 0, 1)$
$\{1, 2, 3\}$	$(1, 1, 1, 0)$
$\{1, 2, 3, 4\}$	$(1, 1, 1, 1)$

15. not reflexive
not symmetric
antisymmetric
not transitive
not a partial ordering
not an equivalence relation

17. reflexive
symmetric
not antisymmetric
transitive
not a partial ordering
an equivalence relation

19. {(1, 1), (2, 2), (3, 3), (4, 4),
(1, 2), (2, 1), (3, 4), (4, 3)}

21. {(1, 1), (2, 2), (3, 3), (4, 4), (5, 5),
(6, 6), (1, 2), (1, 3), (2, 3), (2, 1),
(3, 1), (3, 2), (4, 5), (5, 4)}

23. (2, 3, 3)   **25.** (3, 3, 4, 4)

27.
●————●————●————●————●
1   4   3   2   5

Section 3.1 Advanced Exercises

1.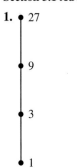

3.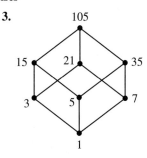

7.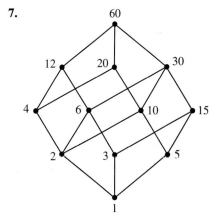

5.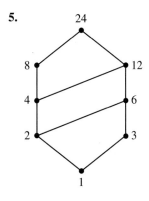

Section 3.2 Exercises

3. 1372879200
5. 64
9. 120
11. 840
13. 720

15. 2162160
17. 670442572800
19. 635013559600
21. 46558512
23. 70

Section 3.2 Advanced Exercises

1. {1, 2, 3}   {3, 3, 1}
    {1, 1, 2}   {3, 3, 2}
    {1, 1, 3}   {1, 1, 1}
    {2, 2, 1}   {2, 2, 2}
    {2, 2, 3}   {3, 3, 3}

3. {1, 2, 3, 4}   {3, 3, 2, 4}   {1, 1, 3, 3}   {1, 1, 1, 4}   {4, 4, 4, 1}
    {1, 1, 3, 4}   {3, 3, 1, 4}   {1, 1, 4, 4}   {2, 2, 2, 1}   {4, 4, 4, 2}
    {1, 1, 2, 4}   {3, 3, 1, 2}   {2, 2, 3, 3}   {2, 2, 2, 3}   {4, 4, 4, 3}
    {1, 1, 2, 3}   {4, 4, 2, 3}   {2, 2, 4, 4}   {2, 2, 2, 4}   {1, 1, 1, 1}
    {2, 2, 3, 4}   {4, 4, 1, 3}   {3, 3, 4, 4}   {3, 3, 3, 1}   {2, 2, 2, 2}
    {2, 2, 1, 4}   {4, 4, 1, 2}   {1, 1, 1, 2}   {3, 3, 3, 2}   {3, 3, 3, 3}
    {2, 2, 1, 3}   {1, 1, 2, 2}   {1, 1, 1, 3}   {3, 3, 3, 4}   {4, 4, 4, 4}

Section 3.2 Computer Exercises

1. 340, 282, 366, 920, 938, 463, 463, 374, 607, 431, 768, 211, 456

3. 13, 407, 807, 929, 942, 597, 099, 574, 024, 998, 205, 846, 127, 479, 365, 820, 592, 393, 377, 723, 561, 443, 721, 764, 030, 073, 546, 976, 801, 874, 298, 166, 903, 427, 690, 031, 858, 186, 486, 050, 853, 753, 882, 811, 946, 569, 946, 433, 649, 006, 084, 096

Section 3.3 Exercises

9.

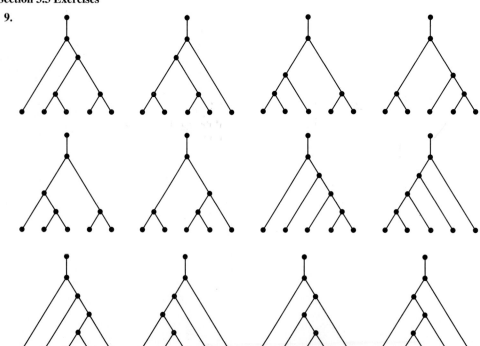

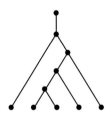 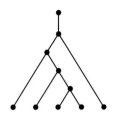

11. $\dfrac{1}{8} \cdot \dfrac{14!}{7! \cdot 7!} = 429$ $\quad$ $\dfrac{14!}{7! \cdot 7!} - \dfrac{14!}{6! \cdot 8!} = 429$

13. $2C_0 \cdot C_5 + 2C_1 \cdot C_4 + 2C_2 \cdot C_3$

15. $2C_0 \cdot C_7 + 2C_1 \cdot C_6 + 2C_2 \cdot C_5 + 2C_3 \cdot C_4$

17. There are 15 primes less than or equal to 47.

19. There are 22 primes less than or equal to 82.

21. $\dfrac{N^2 \cdot (N+1)^2}{4}$ $\qquad$ **23.** 171708332500

Section 4.1 Exercises

1.

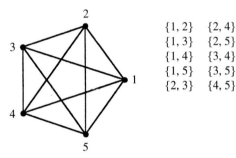

$\{1, 2\}$	$\{2, 4\}$
$\{1, 3\}$	$\{2, 5\}$
$\{1, 4\}$	$\{3, 4\}$
$\{1, 5\}$	$\{3, 5\}$
$\{2, 3\}$	$\{4, 5\}$

For the directed graph, the arrows have been added in such a way that the 1ex order edge list is the same as for the undirected graph.

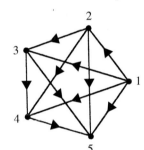

$(1, 2)$	$(2, 4)$
$(1, 3)$	$(2, 5)$
$(1, 4)$	$(3, 4)$
$(1, 5)$	$(3, 5)$
$(2, 3)$	$(4, 5)$

3.

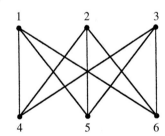

$\{1, 4\}$	$\{2, 4\}$	$\{3, 4\}$
$\{1, 5\}$	$\{2, 5\}$	$\{3, 5\}$
$\{1, 6\}$	$\{2, 6\}$	$\{3, 6\}$

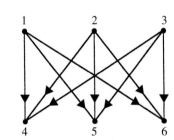

$(1, 4)$	$(2, 4)$	$(3, 4)$
$(1, 5)$	$(2, 5)$	$(3, 5)$
$(1, 6)$	$(2, 6)$	$(3, 6)$

5.

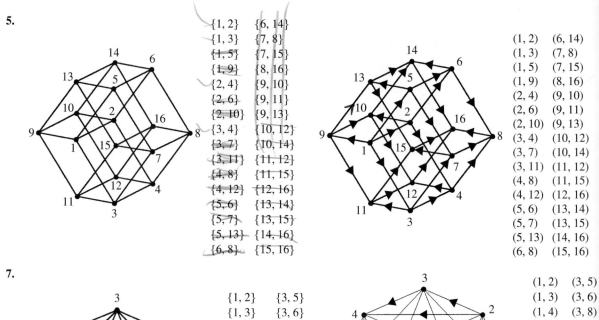

{1, 2}	{6, 14}
{1, 3}	{7, 8}
{1, 5}	{7, 15}
{1, 9}	{8, 16}
{2, 4}	{9, 10}
{2, 6}	{9, 11}
{2, 10}	{9, 13}
{3, 4}	{10, 12}
{3, 7}	{10, 14}
{3, 11}	{11, 12}
{4, 8}	{11, 15}
{4, 12}	{12, 16}
{5, 6}	{13, 14}
{5, 7}	{13, 15}
{5, 13}	{14, 16}
{6, 8}	{15, 16}

(1, 2)	(6, 14)
(1, 3)	(7, 8)
(1, 5)	(7, 15)
(1, 9)	(8, 16)
(2, 4)	(9, 10)
(2, 6)	(9, 11)
(2, 10)	(9, 13)
(3, 4)	(10, 12)
(3, 7)	(10, 14)
(3, 11)	(11, 12)
(4, 8)	(11, 15)
(4, 12)	(12, 16)
(5, 6)	(13, 14)
(5, 7)	(13, 15)
(5, 13)	(14, 16)
(6, 8)	(15, 16)

7.

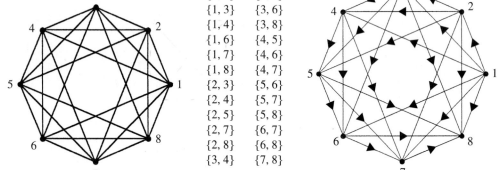

{1, 2}	{3, 5}
{1, 3}	{3, 6}
{1, 4}	{3, 8}
{1, 6}	{4, 5}
{1, 7}	{4, 6}
{1, 8}	{4, 7}
{2, 3}	{5, 6}
{2, 4}	{5, 7}
{2, 5}	{5, 8}
{2, 7}	{6, 7}
{2, 8}	{6, 8}
{3, 4}	{7, 8}

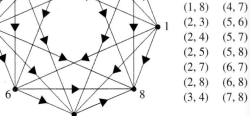

(1, 2)	(3, 5)
(1, 3)	(3, 6)
(1, 4)	(3, 8)
(1, 6)	(4, 5)
(1, 7)	(4, 6)
(1, 8)	(4, 7)
(2, 3)	(5, 6)
(2, 4)	(5, 7)
(2, 5)	(5, 8)
(2, 7)	(6, 7)
(2, 8)	(6, 8)
(3, 4)	(7, 8)

9.   matrix for graph   matrix for digraph

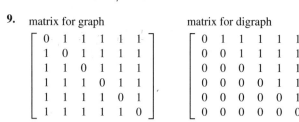

$$\begin{bmatrix} 0 & 1 & 1 & 1 & 1 & 1 \\ 1 & 0 & 1 & 1 & 1 & 1 \\ 1 & 1 & 0 & 1 & 1 & 1 \\ 1 & 1 & 1 & 0 & 1 & 1 \\ 1 & 1 & 1 & 1 & 0 & 1 \\ 1 & 1 & 1 & 1 & 1 & 0 \end{bmatrix} \qquad \begin{bmatrix} 0 & 1 & 1 & 1 & 1 & 1 \\ 0 & 0 & 1 & 1 & 1 & 1 \\ 0 & 0 & 0 & 1 & 1 & 1 \\ 0 & 0 & 0 & 0 & 1 & 1 \\ 0 & 0 & 0 & 0 & 0 & 1 \\ 0 & 0 & 0 & 0 & 0 & 0 \end{bmatrix}$$

11. matrix for graph   matrix for digraph

$$\begin{bmatrix} 0 & 1 & 1 & 0 & 1 & 0 & 0 & 0 \\ 1 & 0 & 0 & 1 & 0 & 1 & 0 & 0 \\ 1 & 0 & 0 & 1 & 0 & 0 & 1 & 0 \\ 0 & 1 & 1 & 0 & 0 & 0 & 0 & 1 \\ 1 & 0 & 0 & 0 & 0 & 1 & 1 & 0 \\ 0 & 1 & 0 & 0 & 1 & 0 & 0 & 1 \\ 0 & 0 & 1 & 0 & 1 & 0 & 0 & 1 \\ 0 & 0 & 0 & 1 & 0 & 1 & 1 & 0 \end{bmatrix} \qquad \begin{bmatrix} 0 & 1 & 1 & 0 & 1 & 0 & 0 & 0 \\ 0 & 0 & 0 & 1 & 0 & 1 & 0 & 0 \\ 0 & 0 & 0 & 1 & 0 & 0 & 1 & 0 \\ 0 & 0 & 0 & 0 & 0 & 0 & 0 & 1 \\ 0 & 0 & 0 & 0 & 0 & 1 & 1 & 0 \\ 0 & 0 & 0 & 0 & 0 & 0 & 0 & 1 \\ 0 & 0 & 0 & 0 & 0 & 0 & 0 & 1 \\ 0 & 0 & 0 & 0 & 0 & 0 & 0 & 0 \end{bmatrix}$$

13. matrix for graph   matrix for digraph

$$
\begin{bmatrix}
0 & 1 & 1 & 0 & 1 & 1 \\
1 & 0 & 1 & 1 & 0 & 1 \\
1 & 1 & 0 & 1 & 1 & 0 \\
0 & 1 & 1 & 0 & 1 & 1 \\
1 & 0 & 1 & 1 & 0 & 1 \\
1 & 1 & 0 & 1 & 1 & 0
\end{bmatrix}
\quad
\begin{bmatrix}
0 & 1 & 1 & 0 & 1 & 1 \\
0 & 0 & 1 & 1 & 0 & 1 \\
0 & 0 & 0 & 1 & 1 & 0 \\
0 & 0 & 0 & 0 & 1 & 1 \\
0 & 0 & 0 & 0 & 0 & 1 \\
0 & 0 & 0 & 0 & 0 & 0
\end{bmatrix}
$$

15.

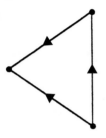

Section 4.1 Advanced Exercises

1. $\dfrac{n \cdot (n-1)}{2}$ **3.** $2n \cdot (n-1)$ **5.**

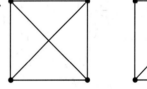

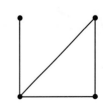

Section 4.1 Computer Exercises

1. The first 20 edges in 1ex order are: {1, 2} {1, 3}
{1.4} {1, 5} {1, 6} {1, 7} {1, 8} {1, 9} {1, 10} {1, 11}
{1, 12} {1, 13} {1, 14} {1, 15} {1, 16} {1, 17} {1, 18}
{1, 19} {1, 20} {2, 3}

7. One possible way to order the vertices of the n-cube
is to use the binary vector at a node as a binary number. For example, node with vector $(0, 0, 0, 0, 0, 0)$
is node 0, node with vector $(0, 0, 1, 0, 0, 1)$ is node
9, and node with vector $(1, 1, 1, 1, 1, 1)$ is node
63. The first 20 edges in this 1ex order are {0, 1}
{0, 2} {0, 4} {0, 8} {0, 16} {0, 32} {1, 3} {1, 5} {1, 9}
{1, 17} {1, 33} {2, 3} {2, 6} {2, 10} {2, 18} {2, 34}
{3, 7} {3, 11} {3, 19} {3, 35}

11. $SG(100, 200)$ is connected.

13. $SG(500, 800)$ is connected.

15. $SG(1000, 1600)$ is connected.

Section 4.2 Exercises

1.

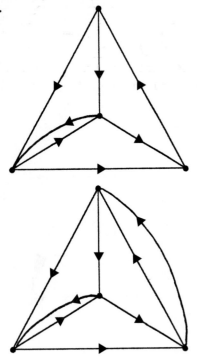

3.

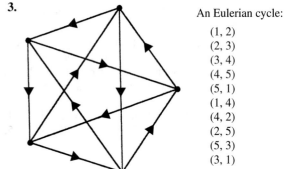

An Eulerian cycle:

(1, 2)
(2, 3)
(3, 4)
(4, 5)
(5, 1)
(1, 4)
(4, 2)
(2, 5)
(5, 3)
(3, 1)

5.

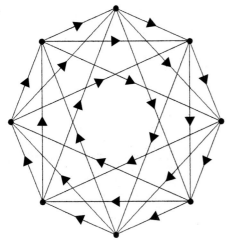

One Eulerian cycle visits the vertices in this order:
1, 7, 5, 3, 1, 8, 6, 4, 2, 8, 7, 6, 3, 2, 7, 4, 3, 8, 5, 4, 1, 6, 5, 2, 1

7. Label the vertices of the 5-cube with the binary vectors of length 5. Connect vertices whose labels differ in a single coordinate. Convert the binary labels to their decimal equivalents and visit the vertices in the following order:
0, 16, 17, 1, 2, 18, 19, 3, 7, 23, 22, 6, 5, 21, 20, 4, 12, 28, 29, 13, 14, 30, 31, 15, 11, 27, 26, 10, 9, 25, 24, 8, 0

9.

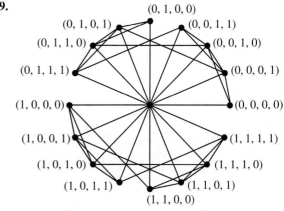

11.

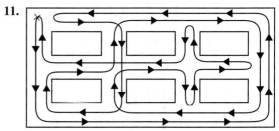

13. A minimal tour uses the edges of length 15, 10, 14, and 18.

Section 4.2 Advanced Exercises

3.

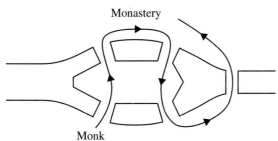

5. 3800 yards

7.

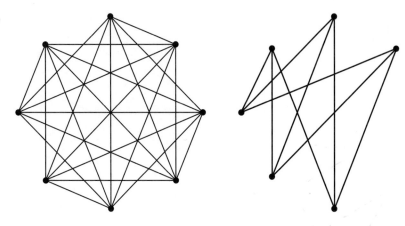

Section 4.3 Exercises

1.   **3.**

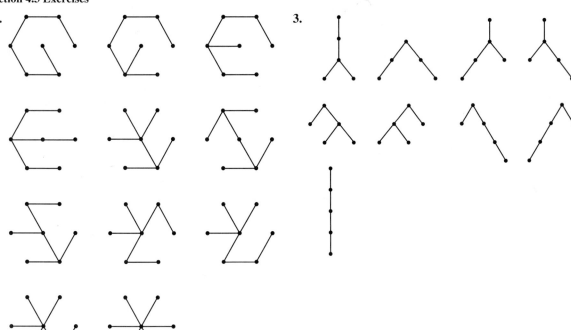

5.

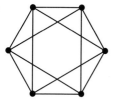

Five spanning trees for the octahedron:

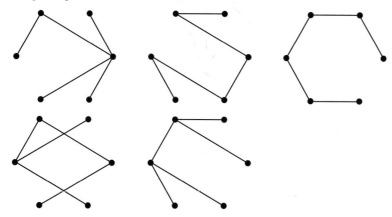

7.

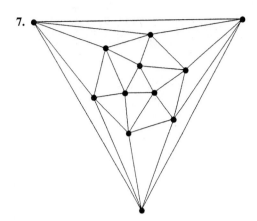

Five spanning trees for the graph of the icosahedron:

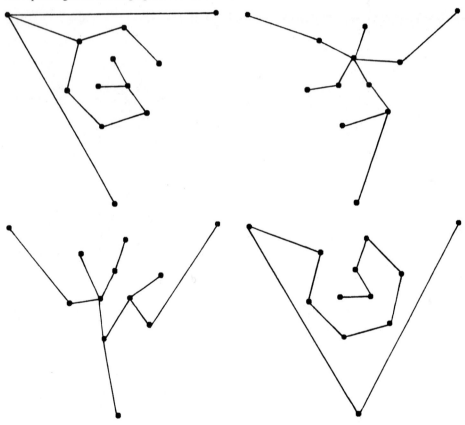

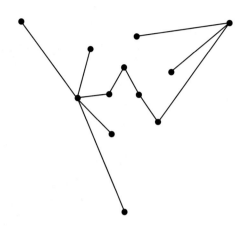

9. 15   **11.** 121   **13.** 384

19. Inorder:

15.

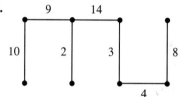

17. Breadth first:

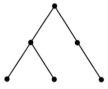

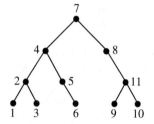

Preorder:

Depth first:

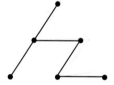

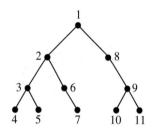

Postorder:

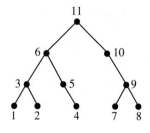

Section 4.3 Advanced Exercises

1. The trees are listed by their Prüfer codes.

(1, 1, 1, 1, 1, 1)	(2, 2, 1, 1, 1, 1)	(2, 2, 2, 1, 1, 1)
(2, 1, 1, 1, 1, 1)	(1, 2, 1, 2, 1, 1)	(2, 2, 1, 2, 1, 1)
	(2, 1, 1, 2, 1, 1)	(2, 1, 2, 1, 2, 1)

(2, 3, 1, 1, 1, 1)	(1, 1, 2, 2, 3, 3)	(1, 1, 1, 2, 2, 3)
(1, 2, 1, 3, 1, 1)	(1, 2, 1, 3, 2, 3)	(1, 3, 1, 2, 2, 1)
(2, 1, 1, 3, 1, 1)	(1, 2, 1, 2, 3, 3)	

(1, 2, 3, 4, 1, 1)	(1, 1, 2, 3, 4, 5)	(1, 2, 3, 4, 5, 6)
(1, 2, 1, 3, 4, 1)	(1, 2, 1, 3, 4, 5)	
(1, 2, 1, 3, 1, 4)	(1, 2, 3, 1, 4, 5)	

3. Label the vertices 1 through 7, starting in the lowest level at the leftmost node in that level. Go across each level and then to the leftmost node of the level above. The root will always be labeled 7. Here are the Prüfer codes.

(5, 5, 6, 6, 7)	(2, 3, 5, 5, 7)	(4, 4, 5, 6, 6)	(3, 3, 5, 5, 6)
(2, 5, 5, 6, 7)	(2, 4, 5, 5, 7)	(4, 5, 5, 6, 6)	(4, 4, 5, 5, 6)
(3, 5, 5, 6, 7)	(2, 3, 6, 6, 7)	(2, 4, 4, 6, 6)	(3, 4, 5, 5, 6)
(4, 5, 5, 6, 7)	(2, 4, 6, 6, 7)	(3, 4, 4, 6, 6)	(2, 3, 5, 5, 6)
(2, 5, 6, 6, 7)	(2, 3, 5, 6, 7)	(2, 5, 5, 6, 6)	(2, 4, 5, 5, 6)
(3, 5, 6, 6, 7)	(2, 4, 5, 6, 7)	(3, 5, 5, 6, 6)	(2, 4, 4, 5, 6)
(4, 5, 6, 6, 7)	(2, 4, 4, 5, 7)	(2, 4, 5, 6, 6)	(3, 4, 4, 5, 6)
(3, 4, 5, 5, 7)	(3, 4, 4, 5, 7)	(3, 4, 5, 6, 6)	(3, 3, 4, 5, 6)
(3, 3, 5, 5, 7)	(3, 4, 4, 6, 7)	(3, 3, 4, 6, 6)	(2, 3, 4, 5, 6)
(4, 4, 5, 5, 7)	(2, 4, 4, 6, 7)	(3, 3, 5, 6, 6)	
(3, 4, 6, 6, 7)	(3, 3, 4, 5, 7)	(2, 3, 4, 6, 6)	
(4, 4, 6, 6, 7)	(3, 3, 4, 6, 7)	(2, 3, 5, 6, 6)	
(3, 3, 6, 6, 7)	(2, 3, 4, 5, 7)		
(3, 3, 5, 6, 7)	(2, 3, 4, 6, 7)		
(4, 4, 5, 6, 7)			
(3, 4, 5, 6, 7)			

5. 841   **7.** 209   **9.** 82944

Section 4.3 Computer Exercises

1. $K_{5,5}$ has 390, 625 spanning trees.

3. The 8-wheel has 2205 spanning trees.

5. The ladder graph has 780 spanning trees.

7. The icosahedron has 5,184,000 spanning trees.

11. The Prüfer codes for the first 20 spanning trees are 112248 112488 113488 113834 115856 115857 115868 115878 121268 121334 121348 121556 121557 121568 121578 122157 122434 122437 122656 122657

Section 5.1 Exercises

1. No.

3. Since there are eighteen syllables in the phrase, it does not name a number.

5. $(p \wedge q) \rightarrow r$ **7.** $(p \vee r) \rightarrow q$

9. $(p \vee \sim p) \rightarrow \sim r$ **11.** Joan is not at the office.

13. If Joan is at the office or Laura is at the office, then Joan is at the office.

15. If whenever Joan is not at the office Laura is at the office, then John is at the office.

17. John is not at the office and either Joan or Laura is at the office.

Section 5.1 Advanced Exercises

1. The traveler should ask two questions similar to the solution in the Application on page 167. First he asks "If you belonged to the other tribe would you say that the road on the right leads to the city?" If the answer is "no," he takes the road on the right. If the answer is "yes," he asks "If you belonged to the other tribe would you say that the road on the left leads to the city?" If the answer is "no," he takes that road, if it's "yes" he takes the road in the middle.

Section 5.1 Computer Exercises

3. The smallest value of x for which $\pi_3(x) < \pi_1(x)$ is 26,861.

Section 5.2 Exercises

1. Universal affirmative; A

$A \subset B$ where A is the set of Cretans and B the set of liars.

3. Particular affirmative; I.

$A \cap B \neq \emptyset$, where A is the set of members of the Committee on Bylaws and B is the set of members of the Committee on Committees.

5. Universal negative; E.

$A \cap B = \emptyset$, where A is the set of members of the Rank and Tenure Committee and B is the set of members of the Committee on Bylaws.

7. Particular negative; O.

$A \cap B^c \neq \emptyset$, where A is the set of members of the Committee on Committees and B is the set of members of the Committee on Bylaws.

9. The type is AAA.   **11.** The type is AAA.

13. The type is AAA. (Reversing the order of the first two statements: "All stories that make me yawn are uninteresting. Your story makes me yawn. Therefore your story is uninteresting.")

15. Some faculty are wealthy.
All wealthy faculty are on the CC.
Some wealthy faculty are on the CC.
    (This is IAI)
All wealthy faculty on the CC are on the SS.
All members of the SS are senior.
Some senior faculty are wealthy.
    (This is AAI, with the first syllogism establishing the necessary fact that there are some wealthy faculty on the CC.)

Section 5.3 Exercises

1. Well-formed **3.** Not well-formed **9.** Not a tautology **11.** A tautology

5. Not well-formed **7.** Not a tautology **21.** No **23.** No

Section 5.3 Advanced Exercises

7. $a(b(c(d(ef))))$ $((ab)c)(d(ef))$
$a(b(c((de)f)))$ $((ab)c)((de)f)$
$a(b((cd)(ef)))$ $(a(b(cd)))(ef)$
$a(b((c(de))f))$ $(a((bc)d))(ef)$
$a(b(((cd)e)f))$ $((ab)(cd))(ef)$
$a((bc)(d(ef)))$ $((a(bc))d)(ef)$
$a((bc)((de)f))$ $(((ab)c)d)(ef)$
$a((b(cd))(ef))$ $(a(b(c(de))))f$
$a(((bc)d)(ef))$ $(a(b((cd)e)))f$
$a((b(c(de)))f)$ $(a((bc)(de)))f$
$a((b((cd)e))f)$ $(a((b(cd))e))f$
$a(((bc)(de))f)$ $(a(((bc)d)e))f$
$a(((b(cd))e)f)$ $((ab)(c(de)))f$
$a((((bc)d)e)f)$ $((ab)((cd)e))f$
$(ab)(c(d(ef)))$ $((a(bc))(de))f$
$(ab)(c((de)f))$ $(((ab)c)(de))f$
$(ab)((cd)(ef))$ $((a(b(cd)))e)f$
$(ab)((c(de))f)$ $((a((bc)d))e)f$
$(ab)(((cd)e)f)$ $(((ab)(cd))e)f$
$(a(bc))(d(ef))$ $(((a(bc))d)e)f$
$(a(bc))((de)f)$ $((((ab)c)d)e)f$

Section 6.1 Exercises

7. Not a Boolean algebra

9. Not a Boolean algebra

11. Not a Boolean algebra

13.
x	y	$x+y$
0	0	0
1	0	1
0	1	1
1	1	1

15.
x	y	xy
0	0	0
1	0	0
0	1	0
1	1	1

19.
x	y	z	$\overline{x}+yz$
0	0	0	1
1	0	0	0
0	1	0	1
1	1	0	0
0	0	1	1
1	0	1	0
0	1	1	1
1	1	1	1

17. $x+y+z$ is equal to 1 unless x, y, and z are all 0.

21. $x\overline{y}+xy+\overline{x}y$
$x+y$

23. $\overline{x}yz+\overline{x}y\overline{z}+xyz+x\overline{y}z+\overline{x}\,\overline{y}z$
$(x+y+z)(\overline{x}+y+z)(\overline{x}+\overline{y}+z)$

25. $xy\overline{z}+xyz+x\overline{y}z+\overline{x}yz$
$(x+y+\overline{z})(x+y+z)(x+\overline{y}+z)(\overline{x}+\overline{y}+x)$

27. $(x\vee y)$ **29.** $(x\vee(y\wedge z)\vee(x\wedge\sim z))$

Section 6.1 Computer Exercises

1. $x+y+z = xyz+xy\overline{z}+x\overline{y}z+x\overline{y}\overline{z}+\overline{x}yz+\overline{x}y\overline{z}+\overline{x}\overline{y}z$

7. $xyz + x\overline{y}z + xy\overline{z} + x\overline{y}\overline{z} + \overline{x}yz + \overline{x}y\overline{z} = x + y$

3. $x + y + z + w = xyzw + xyz\overline{w} + xy\overline{z}w + xy\overline{z}\overline{w} + x\overline{y}zw + x\overline{y}z\overline{w} + x\overline{y}\overline{z}w + x\overline{y}\overline{z}\overline{w} + \overline{x}yzw + \overline{x}yz\overline{w} + \overline{x}y\overline{z}w + \overline{x}y\overline{z}\overline{w} + \overline{x}\overline{y}zw + \overline{x}\overline{y}z\overline{w} + \overline{x}\overline{y}\overline{z}w$

Section 6.2 Exercises

1. Table:

x	y	$x + y$
0	0	0
1	0	1
0	1	1
1	1	1

Gray code:

(x, y)	$x + y$
(0, 0)	0
(0, 1)	1
(1, 1)	1
(1, 0)	1

Binary tree:

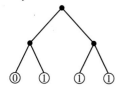

n-cube:

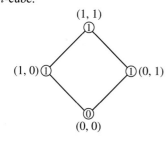

3. Table:

x	y	$x\overline{y}$
0	0	0
1	0	1
0	1	0
1	1	0

Gray code:

(x, y)	$x\overline{y}$
(0, 0)	0
(0, 1)	0
(1, 1)	0
(1, 0)	1

Binary tree:

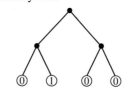

n-cube:

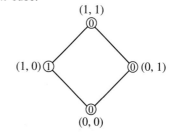

5. Table:

x	y	z	$x + y + \overline{z}$
0	0	0	1
1	0	0	1
0	1	0	1
1	1	0	1
0	0	1	0
1	0	1	1
0	1	1	1
1	1	1	1

Gray code:

(x, y, z)	$x + y + \overline{z}$
(0, 0, 0)	1
(0, 0, 1)	0
(0, 1, 1)	1
(0, 1, 0)	1
(1, 1, 0)	1
(1, 1, 1)	1
(1, 0, 1)	1
(1, 0, 0)	1

Binary tree:

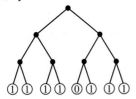

n-cube:

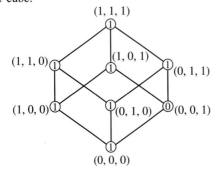

7. Table:

x	y	z	$\overline{x} + yz + xyz$
0	0	0	1
1	0	0	0
0	1	0	1
1	1	0	0
0	0	1	1
1	0	1	0
0	1	1	1
1	1	1	1

Gray code:

(x, y, z)	$\overline{x} + yz + xyz$
(0, 0, 0)	1
(0, 0, 1)	1
(0, 1, 1)	1
(0, 1, 0)	1
(1, 1, 0)	0
(1, 1, 1)	1
(1, 0, 1)	0
(1, 0, 0)	0

Binary tree:

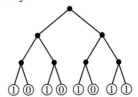

n-cube:

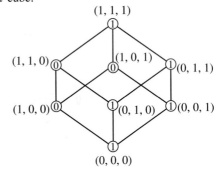

9. Table:

x	y	z	w	$xy + xz + xw + yz + yw + zw$
0	0	0	0	0
1	0	0	0	0
0	1	0	0	0
1	1	0	0	1
0	0	1	0	0
1	0	1	0	1
0	1	1	0	1
1	1	1	0	1
0	0	0	1	0
1	0	0	1	1
0	1	0	1	1
1	1	0	1	1
0	0	1	1	1
1	0	1	1	1
0	1	1	1	1
1	1	1	1	1

Gray code:

(x, y, z, w)	$xy + xz + xw + yz + yw + zw$
$(0, 0, 0, 0)$	0
$(0, 0, 0, 1)$	0
$(0, 0, 1, 1)$	1
$(0, 0, 1, 0)$	0
$(0, 1, 1, 0)$	1
$(0, 1, 1, 1)$	1
$(0, 1, 0, 1)$	1
$(0, 1, 0, 0)$	0
$(1, 1, 0, 0)$	1
$(1, 1, 0, 1)$	1
$(1, 1, 1, 1)$	1
$(1, 1, 1, 0)$	1
$(1, 0, 1, 0)$	1
$(1, 0, 1, 1)$	1
$(1, 0, 0, 1)$	1
$(1, 0, 0, 0)$	0

Binary tree:

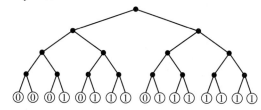

n-cube:

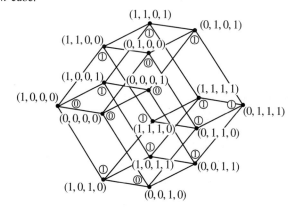

11. $x \cdot y + x\overline{y} + \overline{x}y$ **13.** $x\overline{y}$

15. $x \cdot y \cdot z + x \cdot y \cdot \overline{z} + x \cdot \overline{y} \cdot z + x \cdot \overline{y} \cdot \overline{z} + \overline{x} \cdot y \cdot z + \overline{x} \cdot y \cdot \overline{z} + \overline{x} \cdot \overline{y} \cdot z$

17. $\overline{x} \cdot y \cdot z + \overline{x} \cdot y \cdot \overline{z} + \overline{x} \cdot \overline{y} \cdot z + \overline{x} \cdot \overline{y} \cdot \overline{z} + x \cdot y \cdot z$

19. $x \cdot y \cdot z \cdot w + \overline{x} \cdot y \cdot z \cdot w + x \cdot \overline{y} \cdot z \cdot w + x \cdot y \cdot \overline{z} \cdot w + x \cdot y \cdot z \cdot \overline{w} + \overline{x} \cdot \overline{y} \cdot z \cdot w + \overline{x} \cdot y \cdot \overline{z} \cdot w + \overline{x} \cdot y \cdot z \cdot \overline{w} + x \cdot \overline{y} \cdot \overline{z} \cdot w + x \cdot \overline{y} \cdot z \cdot \overline{w} + x \cdot y \cdot \overline{z} \cdot \overline{w}$

21. See Example 5.

23. $\begin{bmatrix} 0 & 1 \\ 0 & 0 \end{bmatrix}$

25. $\begin{bmatrix} 1 & 1 & 1 & 1 \\ 1 & 1 & 1 & 0 \end{bmatrix}$

27. $\begin{bmatrix} 1 & 0 & 0 & 0 \\ 1 & 1 & 1 & 1 \end{bmatrix}$ **29.** $\begin{bmatrix} 1 & 1 & 1 & 1 \\ 1 & 1 & 0 & 1 \\ 1 & 0 & 0 & 0 \\ 1 & 1 & 0 & 1 \end{bmatrix}$

31. This circuit can be implemented as an OR of three inputs consisting of AND(x, y), AND(x, NOT y), AND(NOT(x, y)). In simplified form it is a single AND gate with inputs x and y.

33. This requires only two gates:

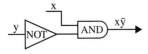

The Karnaugh map has a single entry, so this circuit is simplified.

35. This circuit can be implemented as an OR gate with seven inputs; each of these is a three-input AND gate with inputs defined by the following sum-of-products expression, where the barred letters go through a NOT gate first.

$$x \cdot y \cdot z + x \cdot y \cdot \overline{z} + x \cdot \overline{y} \cdot z + x \cdot \overline{y} \cdot \overline{z} + \overline{x} \cdot y \cdot z + \overline{x} \cdot y \cdot \overline{z} + \overline{x} \cdot \overline{y} \cdot \overline{z}$$

The simplified form terminates with a single three-input OR gate with inputs x, y, and NOT z.

37. This circuit can be implemented as an OR gate with five inputs; each of these is a three-input AND gate with inputs defined by the following sum-of-products expression, where the barred letters go through a NOT gate first.

$$\overline{x} \cdot y \cdot z + \overline{x} \cdot y \cdot \overline{z} + \overline{x} \cdot \overline{y} \cdot z + \overline{x} \cdot \overline{y} \cdot \overline{z} + x \cdot y \cdot z$$

The original Boolean function in Exercise 7 shows one way of simplifying the circuit: a three-input OR gate with inputs NOT x, AND(y, z), and AND(x, y, z).

39. This circuit can be implemented as an OR gate with eleven inputs; each of these is a four-input AND gate with inputs defined by the following sum-of-products expression, where the barred letters go through a NOT gate first.

$$x \cdot y \cdot z \cdot w + \overline{x} \cdot y \cdot z \cdot w + x \cdot \overline{y} \cdot z \cdot w + x \cdot y \cdot \overline{z} \cdot w + x \cdot y \cdot z \cdot \overline{w} + \overline{x} \cdot \overline{y} \cdot z \cdot w \ldots + \overline{x} \cdot y \cdot \overline{z} \cdot w + \overline{x} \cdot y \cdot z \cdot \overline{w} + x \cdot \overline{y} \cdot \overline{z} \cdot w + x \cdot \overline{y} \cdot z \cdot \overline{w} + x \cdot y \cdot \overline{z} \cdot \overline{w}$$

This requires one OR, 11 ANDs, and 16 NOTs. The original Boolean function in Exercise 9 can be implemented directly as six two-input AND gates feeding into a six-input OR gate.

Section 6.2 Advanced Exercises

1. NAND(NAND(NAND(NAND(NAND(x, x), y),NAND(NAND(x, x), y)),
NAND(NAND(NAND(x, x), y),NAND(NAND(x, x), y))),
NAND(NAND(NAND(x, y),NAND(x, y), NAND(NAND(x, y),NAND(x, y))))

NOR(NOR(NOR(NOR(x, x),NOR(y, y)), NOR(NOR(NOR(x, x),
NOR(x, x)),NOR(y, y))), NOR(NOR(NOR(x, x), NOR(y, y)),
NOR(NOR(NOR(x, x), NOR(x, x)), NOR(y, y))))

The circuit also simplifies to y.

3. NAND(NAND(NAND(NAND(NAND(NAND(NAND(x, x),
   NAND(x, x)), NAND(NAND(y, y), NAND(y, y))),
   NAND(NAND(NAND(x, x), NAND(x, x)), NAND(NAND(y, y),
   NAND(y, y)))), NAND(z, z)), NAND(NAND(NAND(NAND(x, x),
   NAND(NAND(y, y), NAND(y, y))), NAND(NAND(x, x),
   NAND(NAND(y, y), NAND(y, y)))), NAND(NAND(z, z),
   NAND(z, z)))), NAND(NAND(NAND(NAND(NAND(NAND(x, x),
   NAND(x, x)), NAND(NAND(y, y), NAND(y, y))),
   NAND(NAND(NAND(x, x), NAND(x, x)), NAND(NAND(y, y),
   NAND(y, y)))), NAND(z, z)), NAND(NAND(NAND(NAND(x, x),
   NAND(NAND(y, y), NAND(y, y))), NAND(NAND(x, x),
   NAND(NAND(y, y), NAND(y, y)))), NAND(NAND(z, z),
   NAND(z, z)))))

   NOR(NOR(NOR(NOR(NOR(NOR(NOR(x, x), NOR(x, x)),
   NOR(y, y)), z), NOR(NOR(NOR(NOR(x, x), NOR(x, x)), NOR(y, y)), z)),
   NOR(NOR(NOR(NOR(NOR(x, x), NOR(x, x)), NOR(y, y)), z),
   NOR(NOR(NOR(NOR(x, x), NOR(x, x)), NOR(y, y)), z))),
   NOR(NOR(NOR(NOR(NOR(x, x), NOR(NOR(y, y), NOR(y, y))),
   NOR(z, z)), NOR(NOR(NOR(x, x), NOR(NOR(y, y), NOR(y, y))),
   NOR(z, z))), NOR(NOR(NOR(NOR(x, x), NOR(NOR(y, y),
   NOR(y, y))), NOR(z, z)), NOR(NOR(NOR(x, x), NOR(NOR(y, y),
   NOR(y, y))), NOR(z, z)))))

5. For the Boolean function describing the region,
   $f(1, 0, 1) = 1$ and $f(0, 1, 0) = 1$ and f is 0 for all
   other inputs.

Section 6.3 Exercises

1. Not a wff; needs parentheses.
3. A wff.   5. A wff.   7. A wff.
9. Not a wff, needs parentheses.
11. $(\forall x)(A(x) \rightarrow B(x))$, where A is the property of
    being the square of a number and B is the property
    of being nonnegative. This is true if S is the set of
    real numbers but not true if S is the set of complex
    numbers.
13. $(\forall x)(B(x) \rightarrow A(x))$
    $(\exists x)(C(x) \wedge B(x))$
    conclude that
    $(\exists x)(C(x) \wedge A(x))$
    All pets are loved. Some cats are pets. Therefore
    some cats are loved. $A(x)$ means "x is loved," $B(x)$
    means "x is a pet," and $C(x)$ means "x is a cat."

15. $(\forall x)(A(x) \rightarrow B(x))$
    $(\exists x)(C(x) \wedge (\sim (B(x))))$
    conclude that
    $(\exists x)(C(x) \wedge (\sim (A(x))))$
    All dogs are loyal. Some cats are not loyal. There-
    fore, some cats are not dogs. $A(x)$ means "x is a
    dog," $B(x)$ means "x is loyal," and $C(x)$ means "x
    is a cat."

17. $(\exists x)(B(x) \wedge A(x))$
    $(\forall x)(B(x) \rightarrow C(x))$
    conclude that
    $(\exists x)(C(x) \wedge A(x))$
    Some butterflies are beautiful. All butterflies are in-
    sects. Therefore some insects are beautiful. $A(x)$
    means "x is beautiful," $B(x)$ means "x is a butter-
    fly," and $C(x)$ means "x is an insect."

19. $(\forall x)(A(x) \to B(x))$
$(\forall x)(B(x) \to (\sim (C(x))))$
conclude that
$(\forall x)(C(x) \to (\sim (A(x))))$
All flying creatures are birds. No fish are birds. Therefore, no fish fly. $A(x)$ means "x is creature that flies," $B(x)$ means "x is a bird," and $C(x)$ means "x is a fish."

21. $(\forall x)(\forall y)(\forall z)(M(x) \to (F(x, y) \land B(y, z) \to N(x, z)))$ with $M(x)$ meaning "x is a man," $N(x, z)$ meaning "x is the nephew of z," and F and B as in Example 4.

23. $(\forall x)(A(x) \to (B(x) \to C(x)))$ where $A(x)$ means "x is a number," $B(x)$ means "x is larger than 1," and $C(x)$ means "there is an integer between x and $2x$."

25. $(\exists x)(A(x) \land (\sim (B(x))))$

27. $(\forall x)(\exists y)(\exists z)(A(z) \land (\sim (B(x, y))))$

Section 7.1 Exercises

1.

3.

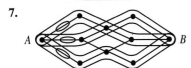

5.

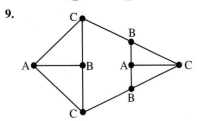

7.

A

B

9.

11.

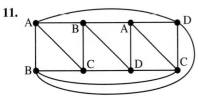

13.

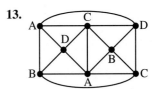

15.

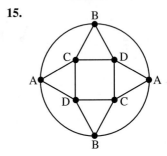

17.

19.

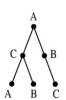   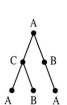   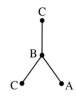

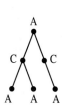

Section 7.1 Advanced Exercises

1. 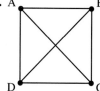   **3.**   **5.**

7. No.   **9.** Yes.

Section 7.2 Exercises

1. The maximum length of an antichain is 3.

3. The maximum length of an antichain is 4.

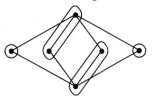

5. The maximum length of an antichain is 3.

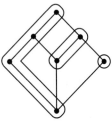

7. This is a lattice, complemented but not distributive.

9. Not a lattice: the two elements at the middle level are covered by two different minimal elements, so they don't have a join.

11. The complementary parts are $(2, 15)$, $(3, 10)$, $(5, 6)$, and $(1, 30)$.

13. The element corresponding to the rightmost vertex has the two opposite elements as complements.

15. There are 16 subspaces.

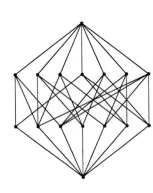

The bottom vertex corresponds to the subspace consisting of the binary vector $(0, 0, 0)$. The first row of seven vertices represents the one-element subspaces, labeled with the binary vectors $(0, 0, 1)$, $(0, 1, 0)$, $(0, 1, 1)$, $(1, 0, 0)$, $(1, 0, 1)$, $(1, 1, 0)$, and $(1, 1, 1)$ in this order from left to right. Each vertex in the next row represents a three-element subspace consisting of the three second-row elements to which it is connected. The top vertex represents the space of all binary vectors of length three.

Section 8.1 Exercises

1. The steps for the Mergesort are
$(1, 4)(3, 7)(2, 5)(8, 9)$
$(1, 3, 4, 7)(2, 5, 8, 9)$
$(1, 2, 3, 4, 5, 7, 8, 9)$

The steps for the Quicksort are
$(2, 1, 3)(7, 4, 5, 9, 8)$
$(1)(2, 3)(7, 4, 5, 9, 8)$
$(1)(2, 3)(5, 4)(7, 9, 8)$
$(1)(2, 3)(4)(5)(7, 9, 8)$
$(1)(2, 3)(4)(5)(7)(9, 8)$
$(1)(2, 3)(4)(5)(7)(8)(9)$

3. The steps for the Mergesort are
$(1, 2)(4, 5)(3, 9)(6, 7)$
$(1, 2, 4, 5)(3, 6, 7, 9)$
$(1, 2, 3, 4, 5, 6, 7, 9)$

The steps for the Quicksort are
$(1)(2, 5, 4, 9, 3, 7, 6)$
$(1)(2)(5, 4, 9, 3, 7, 6)$
$(1)(2)(3, 4)(9, 5, 7, 6)$
$(1)(2)(3, 4)(6, 5, 7)(9)$
$(1)(2)(3, 4)(5)(6, 7)(9)$

5. The steps for Mergesort are

$(7, 11)(13, 19)(1, 18)(5, 12)(4, 16)(3, 17)(2, 15)(10, 14)$
$(7, 11, 13, 19)(1, 5, 12, 18)(3, 4, 16, 17)(2, 10, 14, 15)$
$(1, 5, 7, 11, 12, 13, 19, 18)(2, 3, 4, 10, 14, 15, 16, 17)$
$(1, 2, 3, 4, 5, 7, 10, 11, 12, 13, 14, 15, 16, 17, 18, 19)$

The steps for the Quicksort are

$(2, 3, 4, 5, 1)(18, 19, 12, 13, 16, 17, 11, 15, 7, 14, 10)$
$(1)(3, 4, 5, 2)(18, 19, 12, 13, 16, 17, 11, 15, 7, 14, 10)$
$(1)(2)(4, 5, 3)(18, 19, 12, 13, 16, 17, 11, 15, 7, 14, 10)$
$(1)(2)(3)(5, 4)(18, 19, 12, 13, 16, 17, 11, 15, 7, 14, 10)$
$(1)(2)(3)(4)(5)(18, 19, 12, 13, 16, 17, 11, 15, 7, 14, 10)$
$(1)(2)(3)(4)(5)(10, 14, 12, 13, 16, 17, 11, 15, 7)(19, 18)$
$(1)(2)(3)(4)(5)(7)(14, 12, 13, 16, 17, 11, 15, 10)(19, 18)$
$(1)(2)(3)(4)(5)(7)(10, 12, 13, 11)(17, 16, 15, 14)(19, 18)$
$(1)(2)(3)(4)(5)(7)(10)(12, 13, 11)(17, 16, 15, 14)(19, 18)$
$(1)(2)(3)(4)(5)(7)(10)(11)(13, 12)(17, 16, 15, 14)(19, 18)$
$(1)(2)(3)(4)(5)(7)(10)(11)(12)(13)(17, 16, 15, 14)(19, 18)$
$(1)(2)(3)(4)(5)(7)(10)(11)(12)(13)(14, 16, 15)(17)(19, 18)$
$(1)(2)(3)(4)(5)(7)(10)(11)(12)(13)(14)(16, 15)(17)(19, 18)$
$(1)(2)(3)(4)(5)(7)(10)(11)(12)(13)(14)(15)(16)(17)(19, 18)$
$(1)(2)(3)(4)(5)(7)(10)(11)(12)(13)(14)(15)(16)(17)(18)(19)$

11. $\{1, 2\}, \{1, 3\}, \{1, 4\}, \{2, 3\}, \{2, 4\}, \{3, 4\}$

13. $\{1, 2\}, \{1, 3\}, \{1, 4\}, \{2, 3\}, \{2, 4\}, \{3, 4\}$

15. $\{1, 2, 3\}, \{1, 2, 4\}, \{1, 2, 5\}, \{1, 2, 6\}, \{1, 3, 4\},$
$\{1, 3, 5\}, \{1, 3, 6\}, \{1, 4, 5\}, \{1, 4, 6\}, \{1, 5, 6, \},$
$\{2, 3, 4\}, \{2, 3, 5\}, \{2, 3, 6\}, \{2, 4, 5\}, \{2, 4, 6\},$
$\{2, 5, 6\}, \{3, 4, 5\}, \{3, 4, 6\}, \{3, 5, 6\}, \{4, 5, 6\}$

17. $\{1, 2, 3\}, \{1, 2, 4\}, \{1, 2, 5\}, \{1, 3, 4\}, \{1, 2, 6\},$
$\{1, 3, 5\}, \{2, 3, 4\}, \{1, 3, 6\}, \{1, 4, 5\}, \{2, 3, 5\},$
$\{1, 4, 6\}, \{2, 3, 6\}, \{2, 4, 5\}, \{1, 5, 6\}, \{2, 4, 6\},$
$\{3, 4, 5\}, \{2, 5, 6\}, \{3, 4, 6\}, (3, 5, 6\}, \{4, 5, 6\}$

Section 8.1 Advanced Exercises

1. As you move through the Standard Gray Code in order the coordinates that change are 3, 2, 3, 1, 3, 2, 3. Calling the largest disk 1, the middle-sized disk 2, and the smallest disk 3, this says: move the smallest disk to either empty peg, move the middle disk to the other empty peg, move the small disk on top of the middle disk, move the large disk to the empty peg, move the small disk to the empty peg, move the middle disk on top of the large disk, and, finally, move the small disk on top of the middle disk.

Section 8.1 Computer Exercises

7. The first 20 subsets are shown below.

$S_1 = \{1, 2, 3, 4\}$ $S_{11} = \{1, 3, 5, 6\}$
$S_2 = \{1, 2, 3, 5\}$ $S_{12} = \{2, 3, 5, 6\}$
$S_3 = \{1, 2, 4, 5\}$ $S_{13} = \{1, 4, 5, 6\}$
$S_4 = \{1, 3, 4, 5\}$ $S_{14} = \{2, 4, 5, 6\}$
$S_5 = \{2, 3, 4, 5\}$ $S_{15} = \{3, 4, 5, 6\}$
$S_6 = \{1, 2, 3, 6\}$ $S_{16} = \{1, 2, 3, 7\}$
$S_7 = \{1, 2, 4, 6\}$ $S_{17} = \{1, 2, 4, 7\}$
$S_8 = \{1, 3, 4, 6\}$ $S_{18} = \{1, 3, 4, 7\}$
$S_9 = \{2, 3, 4, 6\}$ $S_{19} = \{2, 3, 4, 7\}$
$S_{10} = \{1, 2, 5, 6\}$ $S_{20} = \{1, 2, 5, 7\}$

9. The first 20 subsets are shown below.

$S_1 = \{1, 2, 3, 4\}$ $S_{11} = \{1, 2, 4, 7\}$
$S_2 = \{1, 2, 3, 5\}$ $S_{12} = \{1, 2, 3, 8\}$
$S_3 = \{1, 2, 4, 5\}$ $S_{13} = \{2, 3, 4, 6\}$
$S_4 = \{1, 2, 3, 6\}$ $S_{14} = \{1, 3, 5, 6\}$
$S_5 = \{1, 3, 4, 5\}$ $S_{15} = \{1, 3, 4, 7\}$
$S_6 = \{1, 2, 4, 6\}$ $S_{16} = \{1, 2, 5, 7\}$
$S_7 = \{1, 2, 3, 7\}$ $S_{17} = \{1, 2, 4, 8\}$
$S_8 = \{2, 3, 4, 5\}$ $S_{18} = \{1, 2, 3, 9\}$
$S_9 = \{1, 3, 4, 6\}$ $S_{19} = \{2, 3, 5, 6\}$
$S_{10} = \{1, 2, 5, 6\}$ $S_{20} = \{1, 4, 5, 6\}$

Section 8.2 Exercises

1. Kruskal and Prim both give

3. Kruskal and Prim both give

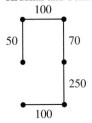

5. Kruskal and Prim both give

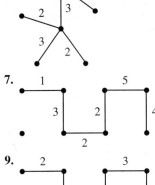

7.

9.

13. The maximal flow from A to B is 11.

15. The maximal flow from A to B is 4.

Section 8.3 Exercises

17. 2.15 seconds

19. 26 days

21. 8.33 minutes

23. 5.56 minutes

25. $2.9 \cdot 10^{15}$ years

27. 39902 by Stirling's formula (error -1%); 40274 with series (error -0.1%)

29. $2.422787 \cdot 10^{18}$ by Stirling's formula (error -0.4%); $2.43242 \cdot 10^{18}$ with series (error -0.02%)

Section 8.3 Advanced Exercises

1. No.   **3.** No.

Section 8.3 Computer Exercises

1. $f(n) \le g(n)$ for $n \ge 1007$; this is the smallest such value.

3. $f(n) \le g(n)$ for $n \ge 10011$; this is the smallest such value.

5. $f(n) \le g(n)$ for $n \ge 35100$; this is the smallest such value.

7. $f(n) \le g(n)$ for $n \ge 439$; this is the smallest such value.

Section 8.4 Computer Exercises

1. For $n = 10$, the absolute difference is approximately 0.951663×10^{-1} and the relative difference is approximately 0.578542×10^{-1}. For $n = 100$, the absolute difference is approximately 0.995017×10^{-2} and the relative difference is approximately 0.604896×10^{-2}. For $n = 1000$, the absolute difference is approximately 0.9995×10^{-3} and the relative difference is approximately 0.607623×10^{-3}.

3. For $n = 10$, the absolute difference is approximately 0.154954×10^{-5} and the relative difference is approximately 0.152313×10^{-5}. For $n = 100$, the absolute difference is approximately 0.19505×10^{-10} and the relative difference is approximately 0.191725×10^{-10}. For $n = 1000$, the absolute difference is approximately 0.1995×10^{-15} and the relative difference is approximately 0.196099×10^{-15}.

Section 9.1 Exercises

1. $a_0 = 1, a_1 = 2, a_n = 2a_{n-1} - a_{n-2}$

3. $a_0 = 1, a_1 = 3, a_n = 0 \cdot a_{n-1} + a_{n-2}$

5. $a_0 = 1, a_1 = 4, a_2 = 9, a_n = 3a_{n-1} - 3a_{n-2} + a_{n-3}$

7. $a_0 = 2, a_1 = 2, a_n = a_{n-1} + a_{n-2}$

9. $a_0 = 1, a_1 = -1, a_2 = 0, a_n = a_{n-1} + a_{n-2} + a_{n-3}$

11. $\displaystyle\sum_{n=0}^{\infty} x^{2n}$ **13.** $\displaystyle\sum_{n=0}^{\infty} x^{3n}$

15. $\displaystyle\sum_{n=0}^{\infty} \frac{(n+1) \cdot (n+2)}{2} \cdot x^n$

17. $1 - x + x^3 - x^4 + x^6 - x^7 + x^9 - x^{10} + x^{12} - x^{13} + \ldots$

or $\displaystyle\sum_{n=0}^{\infty} \left[\frac{u^{n+1} - (\overline{u})^{n+1}}{i \cdot \sqrt{3}} \right] \cdot x^n$

where $u = -\dfrac{1}{2} + i \cdot \dfrac{\sqrt{3}}{2}$.

19. $\displaystyle\sum_{n=0}^{\infty} \frac{x^{2n}}{2^{n+1}}$

21. The generating function is $\dfrac{1}{1 - 2x}$ and the nth term is given by $a_n = 2^n$.

23. The generating function is $\dfrac{3 - 4x}{(1 - x) \cdot (1 - 2x)}$ and the nth term is given by $a_n = 2^{n+1} + 1$.

25. The generating function is $\dfrac{1}{(1 - x) \cdot (1 - x^3)}$ and the nth term is given by $a_n = \dfrac{1}{3} \cdot \left(\dfrac{\overline{u}^n - u^n}{i \cdot \sqrt{3}} + n \right)$ where $u = -\dfrac{1}{2} + i \cdot \dfrac{\sqrt{3}}{2}$.

27. $T(n) = \Theta\left(n^{\frac{\log(2)}{\log(3)}} \right)$ **29.** $T(n) = \Theta(n^2 \lg n)$

Section 9.1 Advanced Exercises

1. 1, 1, 2, 2, 3, 4, 4, 4, 5, 6, 7, 7, 8, 8, 8, 8, 9, 10, 11, 12, 12, 13, 14, 14, 15, 15, 15, 16, 16, 16

3. 1, 1, 2, 3, 3, 4, 5, 5, 6, 6, 6, 8, 8, 8, 10, 9, 10, 11, 11, 12, 12, 12, 12, 16, 14, 14, 16, 16, 16, 16

5. The first six tangent numbers are 1, 2, 16, 272, 7936, and 353792. The first six secant numbers are 1, 1, 5, 61, 1385, and 50521.

Section 9.1 Computer Exercises

1. $\tan(x) = x + 2x^3/3! + 16x^5/5! + 272x^7/7! + 7936x^9/9! + 353792x^{11}/11! + 22368256x^{13}/13! + 1903757312x^{15}/15! + 209865342976x^{17}/17! + 29088885112832x^{19}/19! + \ldots$

$\sec(x) = 1 + x^2/2! + 5x^4/4! + 61x^6/6! + 1385x^8/8! + 50521x^{10}/10! + 2702765x^{12}/12! + 199360981x^{14}/14! + 19391512145x^{16}/16! + 2404879675441x^{18}/18! + \ldots$

Section 9.2 Exercises

1. $\dfrac{5}{36}$ **3.** $\dfrac{1}{6}$ **5.** 0.313

7. 0.164   **9.** 0.5   **11.** 0.711

13. 0.711   **15.** 0.132   **17.** 0.647

19. 0.522   **21.** 0.011

23. Player A's chance is 0.767 and Player B's chance is 0.773.

25. 0.499   **27.** 0.444   **29.** 0.023

31. 0.081   **33.** 0.159   **35.** 0.0047

Section 9.3 Exercises

1. $(1, 2)$ **3.** $(2, 3)$ **5.** $(1, 3, 2)$

7. $(1, 4, 3, 2)$ **9.** $(1, 4, 3)$

11. 1, 2, 3, 6, and 7 are odd; 4, 5, 8, 9, and 10 are even.

13. e $(1, 2, 3)$ $(1, 3, 4)$
   $(1, 2)(3, 4)$ $(1, 3, 2)$ $(1, 4, 3)$
   $(1, 3)(2, 4)$ $(1, 2, 4)$ $(2, 3, 4)$
   $(1, 4)(2, 3)$ $(1, 4, 2)$ $(2, 4, 3)$

15. e $(1, 3)(4, 6)$
   $(1, 2, 3, 4, 5, 6)$ $(2, 6)(3, 5)$
   $(1, 3, 5)(2, 4, 6)$ $(1, 5)(2, 4)$
   $(1, 4)(2, 5)(3, 6)$ $(1, 2)(3, 6)(4, 5)$
   $(1, 5, 3)(2, 6, 4)$ $(1, 4)(2, 3)(5, 6)$
   $(1, 6, 5, 4, 3, 2)$ $(1, 6)(2, 5)(3, 4)$

17. Since $|D_6| = 12$, there are $6!/12$ or 60 inequivalent configurations.

29. 10   **21.** 120   **23.** 15

25. 10   **27.** 6   **29.** 15

Section 9.3 Computer Exercises

1. The first 20 values of $d(n)$ are shown below.

$d(1) = 0$	$d(8) = 14,833$	$d(15) = 481,066,515,734$
$d(2) = 1$	$d(9) = 133,496$	$d(16) = 7,697,064,251,745$
$d(3) = 2$	$d(10) = 1,334,961$	$d(17) = 130,850,092,279,664$
$d(4) = 9$	$d(11) = 14,684,570$	$d(18) = 2,355,301,661,033,953$
$d(5) = 44$	$d(12) = 176,214,841$	$d(19) = 44,750,731,559,645,106$
$d(6) = 265$	$d(13) = 2,290,792,932$	$d(20) = 895,014,631,192,902,121$
$d(7) = 1854$	$d(14) = 32,071,101,049$	

3. The first 20 values of T_n are shown below.

$T_0 = 1$	$T_5 = 9$	$T_{10} = 719$	$T_{15} = 87,811$
$T_1 = 1$	$T_6 = 20$	$T_{11} = 1842$	$T_{16} = 235,381$
$T_2 = 1$	$T_7 = 48$	$T_{12} = 4766$	$T_{17} = 634,847$
$T_3 = 2$	$T_8 = 115$	$T_{13} = 12,486$	$T_{18} = 1,721,159$
$T_4 = 4$	$T_9 = 286$	$T_{14} = 32,973$	$T_{19} = 4,688,676$

Section 10.1 Exercises

1. The set of all words in a and b that end in ab.

3. The set of all words in 00 and 11.

5. The set of all words in a, b, and c in which all the cs come after all the as and bs.

7. The set of all words in 01 and 10.

9. The set of all words in a and bc.

11. $L(G)$ consists of the words described by $\epsilon + a(a)^*b$. The language is regular.

13. $L(G)$ consists of ϵ and the set $\{(a + ab(a)^*)^n(b(a)^*)^n, n > 0\}$. The language is context-free.

15. $L(G) = \{\epsilon, a, b, ab\}$. The language is regular.

17. $L(G) = \{\epsilon, abc, accc, abbc, abccc, accbc, accccc, bcbc, bcccc\}$. The language is regular.

19. $L(G)$ consists of the words described by $\epsilon + b + a(a)^*b(a)^*$. The language is regular.

Section 10.2 Exercises

1.

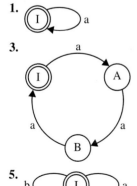

3.

5.

7.

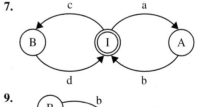

9.

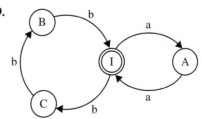

11. $(a^2 + ab + b)((a)^* + (b)^*)$

13. $(a^3)^*$

15. $(a(ba)^*a+b(ab)^*b+ab(ab)^*b+ba(ba)^*a)((a)^*+$
$(b)^*)$

Section 10.3 Computer Exercises

1. $2^5 3^1 5^{11} 7^6 = 551479687500000$

3. $2^5 3^7 5^4 7^5 11^1 13^{11} 17^6 19^6$
$= 164571138387231901326779666893666066140000$

5. $2^5 3^7 5^2 7^{11} 11^6 = 6128769966795827920800$

7. $2^5 3^7 5^4 7^5 11^7 13^2 17^{11} 19^6 23^6 =$
577871447333928142322920807295665554614719014540000

9. $2^5 3^5 5^7 7^4 11^5 13^{11} 17^4 19^{13} 23^6 29^6 31^4 37^5 41^5 43^7 47^4 53^{11} 59^6 61^4 67^5 71^7 73^4 79^{13} 83^6 89^6 97^6$
$= 152827884675337667917811776899443726466954809518473400399212806895501653770217779794357199$
111055892694309307722232200017096822131632876224929458260449359211004339089679963803431961
294146667664256910669113515555664920157116053797500000

11. $(\sim q)$　　　　**13.** $(p \to (\sim q))$　　**15.** $(p \wedge q)$

17. $(p \to (q \to p))$　　　　**19.** $(p \wedge p)$

Index